COATED VESICLES

EDITED BY

C. D. OCKLEFORD AND A. WHYTE

*Department of
Anatomy, University of
Leicester Medical School*

*Department of
Pathology, University
of Cambridge*

T0296978

CAMBRIDGE UNIVERSITY PRESS

CAMBRIDGE

LONDON   NEW YORK   NEW ROCHELLE

MELBOURNE   SYDNEY

CAMBRIDGE UNIVERSITY PRESS
Cambridge, New York, Melbourne, Madrid, Cape Town, Singapore, São Paulo, Delhi

Cambridge University Press
The Edinburgh Building, Cambridge CB2 8RU, UK

Published in the United States of America by Cambridge University Press, New York

www.cambridge.org
Information on this title: www.cambridge.org/9780521105743

First published 1980
This digitally printed version 2009

A catalogue record for this publication is available from the British Library

Library of Congress Cataloguing in Publication data
Main entry under title:
Coated vesicles.

Includes indexes.
1. Coated vesicles. I. Ockleford, C. D., 1948–   II. Whyte, A.
[DNLM: 1. Cells–Physiology. QH604 C652]
QH603.C63C6   574.8'74   79-17280

ISBN 978-0-521-22785-8 hardback
ISBN 978-0-521-10574-3 paperback

CONTENTS

Preface

Coated vesicles are structurally distinct cellular organelles with a variable geometry limited by, amongst other factors, Eulers theorem.* They occur in organisms as diverse as pigs and Protozoa. Just as other organelles, e.g. lysosomes and microtubules, became the object of increased research effort subsequent to their first successful isolation, we believe that coated vesicles, which have recently been isolated successfully and consistently in different laboratories, are poised ready for examination by a range of refined molecular and immunological techniques.

Our aims for this volume include the gathering together of relevant information which will act as a reference for researchers in and entering this field. We will definitely produce the best source book on the subject, for the simple reason that there is no other! The fact that we can make this claim reflects what we think is an extremely serious imbalance. Whereas we would not wish to promote the use of weight of published material as a yardstick for the quality of the science it contains, we cannot help but notice that the published volumes on lysosomes and mitochondria already fill complete shelves in biomedical reference libraries; we wish to go a small way towards redressing this balance by producing one volume on coated vesicles.

A critic might be tempted to suggest that there is as yet no volume on coated vesicles because not enough is known about them to make one worthwhile. We hope that the volume itself will refute this argument, and we anyway believe that it is also worthwhile to indicate areas of ignorance which are ripe for effective research.

What then are the major questions to be addressed? If the topics were to be

*This theorem, named after its discoverer the Swiss mathematician L. Euler (1707–1783), deals with the topology of curves and surfaces. It follows from the theorem that for all simple polyhedrons: $V - E + F = 2$, where V is the number of vertices, E the number of edges, and F the number of faces.

arranged in a spectrum from greatest density of factual information to most speculative we could perhaps begin with the comparative cell biology which suggests the proposed functional roles; continue to the experimental cell biology which demonstrates these; progress to the molecular biology of vesicle components and their supramolecular arrangements; continue on to models of vesiculisation; and ultimately venture tentatively into the discussion of organelle dysfunction and related human pathology.

Apart from the areas covered in these chapters there are a number of exciting subjects related to coated vesicles which have not even been broached academically. These notions, fit only for the Preface, include speculation on the evolutionary origin of coated vesicles. For example, will adherents to Margulis's endosymbiont theory of the origin of organelles be capable of constructing a strong case for the evolution of coated vesicles from viruses similar to C-type or retroviruses? The parallels in structure and formation of coated vesicles and these viruses are quite startling and the similarity has led more than one fine-structuralist into confusion. Equally far removed from the everyday constraints of evidence and experiment are the possibilities of exploiting coated vesicles as a means of guiding biologically active molecules to particular cell compartments in particular cell types. Employment of coated vesicles in this role must have crossed more than one scientific mind. In this respect the development of the field depends greatly on the general development of receptor chemistry.

We are encouraged by the prospect that there is much of value yet to be discovered in the subject area of this volume — much more than all the authors put together will be able to research. Should someone consider joining us as a result of reading our volume we will have accomplished one of the things we have set out to do: point the way to an interesting, potentially very active field of research.

In conclusion we would like to emphasise that in order to speed publication our editorial policy in the production of this volume has been autocratic rather than consultative. To avoid more overlap than is useful for purposes of integrating chapters we have deleted material from individual chapters unilaterally. If this has been damaging to the style of the individual contributions we apologise, accept the blame, and offer by way of excuse our best intention to improve the whole volume.

December, 1978   C. D. O.

                                                                           A. W.

FOREWORD

B.M.F. PEARSE

Coated vesicles occur in most eukaryotic cells. They are characterised by the polygonal coats they have on the cytoplasmic surfaces of their lipid membranes; their diameters vary between 50 and 150 nm. Of their various functions, one of the most striking examples is seen in developing chicken oocytes, where coated vesicles can pinocytose up to 1 g of yolk proteins a day in a single cell. Another example is the recycling of membrane after transmitter release in excised frog neuromuscular junction preparations. These terminals can secrete hundreds of times more quanta of neurotransmitter than the number of vesicles they initially contain; yet they still look full of vesicles. Here the replenishment of these secretory vesicles from the presynaptic membrane is mediated by coated vesicles. In fibroblasts, coated vesicles are responsible for the uptake of several macromolecules, including low-density lipoprotein and epidermal growth factor, which ultimately reach lysosomes and are degraded. In other cells coated vesicles appear to bud from the Golgi apparatus and seem to be involved in secretion. In general, the function of coated vesicles apparently entails the transfer of membrane and selected macromolecules between different membranous sites within the cell and in a defined direction. However, much has still to be learned about their full role.

Biochemically, coated vesicles are characterised by the major structural protein of their coats — clathrin, a polypeptide of molecular weight 180 000. They can be disrupted *in vitro* to give defined subunits containing several molecules of clathrin which, on suitable treatment, reassemble to form coat-like particles. Electron micrographs of tilted specimens of coated vesicles show that their coats are based on polyhedral lattices constructed from 12 pentagons (required by Euler's theorem) plus a variable number of hexagons. Among the smaller particles, three such structures have been identified, two containing 108 molecules of clathrin each and a third containing 84. The coats of larger particles are thought to be constructed on similar principles. Thus, clathrin forms flexible lattices ideally suited to participating in the formation of vesicles during their budding from membranes.

Clathrin from several different animal sources shows a high degree of conservation of amino acid sequence as shown by one-dimensional peptide mapping. Although the precise extent of this conservation must await detailed sequence studies, the picture which has emerged so far is reminiscent of actin, another 'scaffold' protein. The reason that actin sequences are highly conserved is presumably because this molecule forms many associations with other proteins (i.e. with myosin, tropomyosin and so on) and this makes it difficult for its sequence to drift without the protein–protein contacts being disturbed. Perhaps the same constraints apply to clathrin, since it is evident that here, too, several interactions between a clathrin molecule and other proteins must exist. During a cycle of pinocytosis, this would include: (1) the binding of the receptor (protein) for the pinocytosed molecule by clathrin, (2) assembly of clathrin subunits to form a flexible cage, and (3) the association of clathrin with other unknown proteins which may be involved in the transport of the coated vesicle to another membrane, and its specific fusion with that membrane. Since the clathrin of coated vesicles seems to operate in a wide variety of contexts in different cells, where different receptors are involved and the vesicles shunted between different membranes, the case for associations of clathrin with many other proteins would seem to be strong.

Purified coated vesicles (e.g. from chick oocytes) seem to lack specific contents and receptor molecules, and therefore represent only a restricted form of the intact cellular system of coated membranes. The quantity of clathrin isolated from lymphoma cells (E.L.4) suggests that about 1 % of these cells' lipid could be in coated vesicles at any one moment. If coated vesicles have a half-life of minutes only, as estimated in fibroblasts, they could therefore be responsible for a considerable transfer of membrane components within cells. Future biochemical studies on this interesting cell organelle should be rewarding.

CONTRIBUTORS

F. Beck
Department of Anatomy
Medical Sciences Building,
University of Leicester,
University Road,
Leicester LE1 7RH, UK.

A.L. Blitz
Department of Physiology & Biochemistry,
Boston University School of Medicine,
80 East Concord Street,
Boston,
Massachusetts 02118, USA

R. Duncan
Biochemistry Research Laboratory,
University of Keele,
Staffordshire ST5 5BG, UK

R.E. Fine
Department of Physiology & Biochemistry,
Boston University School of Medicine,
80 East Concord Street,
Boston,
Massachusetts 02118, USA

F. Giorgi
Istituto di Istologia e Embriologia,
Università degli studi di Pisa,
Via A. Volta 4,
56100 Pisa, Italy

J. Kartenbeck
Division of Membrane Biology & Biochemistry,
Institute of Experimental Pathology,
German Cancer Research Centre,
Im Neuenheimer Feld 280,
6900 Heidelberg, GFR

Contributors

J.B. Lloyd
Biochemistry Research Laboratory,
University of Keele,
Staffordshire ST5 5BG, UK

E.A. Munn
Agricultural Research Council Institute of Animal Physiology,
Babraham,
Cambridge CB2 4AT, UK

A.J. Nevorotin
Laboratory of Electron Microscopy,
Department of Morphology,
Central Research Laboratory,
1st Medical Institute,
197089 Leningrad, USSR

E.H. Newcomb
Department of Botany,
University of Wisconsin,
Madison,
Wisconsin 53706, USA

C.D. Ockleford
Department of Anatomy,
University of Leicester Medical School,
University Road,
Leicester LE1 7RH, UK

B.M.F. Pearse
Medical Research Council Laboratory of Molecular Biology,
Hills Road,
Cambridge CB2 2QH, UK

M.K. Pratten
Biochemistry Research Laboratory,
University of Keele,
Staffordshire ST5 5BG, UK

A. Rees
Laboratory of Molecular Biophysics,
Department of Zoology
University of Oxford,
South Parks Road,
Oxford, OX1 3PS, UK

R. Rodewald
Department of Biology,
Gilmer Hall,
University of Virginia,
Charlottesville,
Virginia 22901, USA

Contributors

K. Wallace
Laboratory of Molecular Biophysics,
Department of Zoology,
University of Oxford,
South Parks Road,
Oxford OX1 3PS, UK

A. Whyte
Department of Pathology,
University of Cambridge,
Tennis Court Road,
Cambridge CB2 1QP, UK

A.E. Wild
Department of Biology,
Medical and Biological Sciences Building,
University of Southampton,
Southampton SO9 3TU, UK

1

Coated vesicles: a morphologically distinct subclass of
endocytic vesicles

ARTHUR E.WILD

The first workers to use the term 'coated vesicles' in a definitive sense appear
to have been Rosenbluth & Wissig (1963, 1964), in their case to describe
characteristic vesicles in toad spinal ganglia which had the ability to endocytose
exogenous ferritin. Other workers have used different names to describe the
same type of vesicle in a wide variety of tissues and cell types. For example, in
their now classic study of yolk protein uptake by oocytes of the mosquito,
Roth & Porter (1964) described them as bristle-coated. Casley-Smith (1969),
in studies on macrophages, referred to them as rhopheosomes or rhopheocytic
vesicles, and in so doing was following a precedent set by Policard & Bessis
(1958) in what was one of the earliest studies on these vesicles. Gray (1961)
referred to them as complex vesicles, Palay (1963) as alveolate, Wolfe (1965)
as acanthosomes, and Maunsbach (1963) as 'possessing an amorphous coat'.
In most cases these terms refer to a distinct organised layer on the cytoplasmic
aspect of the vesicle membrane which is a characteristic feature of what are
now commonly known as coated vesicles. Much interest has centred on them
in relation to endocytosis and it is true to say that they are ubiquitous in
tissues and cells actively engaged in protein uptake. In making such a statement
I do not wish to imply that protein uptake is necessarily their sole function.
In some cells, described in later chapters, coated vesicles play a secretory role
and, in addition, regenerate membrane from the cell surface in order that
secretory functions may continue. Neither do I wish to imply that coated
vesicles are the only means of entry for proteins into cells. They can, however,
provide what is coming to be recognised as a distinct, selective pathway in
some cases which is best considered in relation to other endocytic processes.
This opening chapter starts therefore, with a brief account of endocytosis in
general, which will lay the foundations for a more detailed consideration of
coated vesicle structure and function in later chapters.

Classification of endocytosis

Endocytosis describes the process whereby extracellular materials are taken up by cells within membrane-bound vesicles. It can be broadly classified (see reviews by Fawcett, 1965; Allison & Davies, 1974) into phagocytosis and pinocytosis. Phagocytosis involves the ingestion of extracellular particles such as bacteria and erythrocytes by extensions of pseudopodia or filopodia to entrap the particle, which as a result becomes enclosed within a vesicle formed from the plasma membrane. The particle more or less fills the vesicle, which is referred to as a 'phagosome' and which is of such dimensions (of the order of 1–2 μm) as to be visible in the light microscope. Phagocytosis occurs in all classes of animals and is a function of macrophages and polymorphonuclear leucocytes in mammals. Pinocytosis, a term first introduced by Lewis (1931), describes the uptake of droplets of fluid containing dissolved substrates and colloids too finely divided to be seen under the microscope. He first observed its occurrence in tissue culture cells but there is now ample evidence that it occurs *in vivo* in cells. Pinocytosis can be divided into macropinocytosis and micropinocytosis, depending upon the size of the vesicle formed following pinocytic activity.

Macropinocytic vesicles can be visualised in the light microscope and have a lower limit therefore of about 0.3 μm in diameter. They are prominent in cells constituting absorptive epithelia and in which protein degradation can be expected to occur to a greater or lesser extent. Endodermal cells of rabbit yolk sac splanchnopleur and gut enterocytes of suckling mammals (see Wild, 1973*a*), renal proximal tubule cells (Maunsbach, 1969) and thyroid epithelial cells (Williams & Wolff, 1971) are good examples. Fluorescent protein tracing (see Wild, 1973*b*) is a particularly useful way of visualising macropinocytosis, and the appearance of macropinocytic vesicles following immunofluorescent localisation of immunoglobulin in rabbit yolk sac endoderm is shown in Plate 1(*f*). Macropinocytosis by macrophages, both *in vivo* (Rhodes & Lind, 1968) and *in vitro* (Dingle *et al.*, 1973) has also been well demonstrated by fluorescent protein tracing. It will be appreciated that prior to the introduction of the term 'macropinocytic vesicle' (Rodewald, 1973; Allison & Davies, 1974), 'endocytic vesicle', 'absorptive vesicle', 'pinosome', 'large apical vesicle' and similar descriptive terms were, and still are, used to describe such vesicles.

Micropinocytosis involves the formation of vesicles which can be seen only in the electron microscope and two distinct types or subclasses can be recognised, namely smooth-walled micropinocytic vesicles and coated micropinocytic vesicles. The former type are commonly found in capillary endothelia (see Plate 1*b*) and were in fact first described by Palade (1953) in early

electron-microscopical investigations of such cells. They are also a striking
feature of peritoneal mesothelial cells (Fedorko & Hirsch, 1971) and are
present in macrophages (Casley-Smith, 1969) and other cell types. They have
a diameter of about 70–100 nm. There is good evidence that smooth-walled
micropinocytic vesicles (in some cell types at least) pinocytose tracer macromolecules and, by implication, plasma proteins, but whether they do this
non-selectively is not at all clear. This is a point I shall return to after considering coated vesicle structure and when I discuss the selective uptake and
transport of proteins by coated vesicles.

Different endocytic events are also separable on the basis of their energy
requirements. The formation of phagosomes and macropinocytic vesicles
appears to be dependent upon an energy-requiring contractile system, whereas
the position with regard to the formation of micropinocytic vesicles is less
clear (see Chapter 10).

Structure of coated micropinocytic vesicles
Nature of the coating
Coated vesicles can be distinguished from smooth-walled micropinocytic
vesicles by the coating on the cytoplasmic aspect of the vesicle membrane to
which I have already alluded. When the plane of section in electron micrographs is more or less median through the vesicle, the coating appears as
radially arranged projections (Plate 1c). Roth & Porter (1964) described these
as bristles, but I shall refer to them as projections. Other terms such as 'spikes',
'spines', 'hair-like projections', 'striae' and 'fuzzy coats' are words which have
frequently been used to describe them (see Appendix 1).

Estimates of the length of the projections on coated vesicles in different
cell types vary between 15 and 25 nm, although projections appear to be
fairly uniform in size on any one coated vesicle. The distance between
projections is of the order of 20–25 nm and the width of each projection
about 5 nm. Small coated vesicles (< 75 nm in diameter) have been reported
to have 8–13 projections and larger coated vesicles (> 100 nm in diameter)
18–20 projections (Friend & Farquhar, 1967). In some cases, especially
where sectioned coated vesicles in nerve endings are concerned, the distal
endings of the projections appear to be connected by bars, thus giving the
appearance of a coronet (Gray, 1961; Kanaseki & Kadota, 1969). When the
plane of the section is tilted (Ockleford, 1976) or when surface views of
coated vesicles have been fortuitously obtained in sections, the projections
have been inferred to represent a polygonal network. Palay (1963) observed
such oblique sections of coated vesicles (alveolate vesicles) in Purkinje cells
of rat cerebellum and suggested that the projections (or 'striae', as he referred

to them) represented the vertical walls of a honeycomb which made up the
limiting membrane of the vesicle. It was, of course, for this reason that he
used the term 'alveolate' to describe the vesicles. Bowers (1964) also observed
the coating on vesicles in surface view in pericardial cells of the aphid and was
the first to suggest that the honeycomb probably consisted of hexagons combined with pentagons to cover the spherical surface. More recent pictures of
the surface of tangentially sectioned coated vesicles have been obtained by
Ockleford (1976) in studies of human syncytiotrophoblast. These show very
clearly the polygons making up the polyhedral structure (Plate 1e). From his
observations and measurements, Ockleford (1976) has constructed a three-
dimensional model which, in keeping with the earlier suggestions of Palay
(1963), incorporates ridges of material along the surface of the vesicle membrane to make up the honeycomb pattern. Such a model must of course
consist of a mixture of regular hexagons and pentagons, since a pattern
composed of regular or irregular hexagons alone, will not enclose space.

Another model of coated vesicle structure, 'the vesicle in a basket', has
been proposed by Kanaseki & Kadota (1969). Using a partial purification
technique to isolate coated vesicles from nerve endings of guinea pig brain
and from liver, these workers were able to make a detailed study of negatively
stained coated vesicles in the electron microscope. They deduced that coated
vesicles consisted of an inner vesicle with a membrane-to-membrane diameter
of 50 nm, which was surrounded by a structure resembling a spherical basket.
The basketwork measured 100 nm in diameter and appeared to be composed
of a network of regular hexagons and pentagons each with sides (subunits)
24 nm in length. Kanaseki & Kadota (1969) could not confirm the existence
of bristles extending from the surface of the vesicle and considered these to be
an artifact due, in median sections, to the superimposed images of parts of
the polygonal network composing the basket. It was supposed by these
workers that some substance existed that separated the basket and the
vesicle, but they could not refute that it was either the honeycomb-like
structure or the projections that filled the space. Stained coated vesicles
isolated from rabbit yolk sac (Plate 1d) present an appearance somewhat
similar to the model proposed by Kanaseki & Kadota (cf. Fig. 28 of their
original paper).

Following on from the partial purification technique described by Kanaseki
& Kadota (1969), improvements have been made by Pearse (1975) in the
isolation of coated vesicles so that very pure fractions can now be obtained
from a variety of tissues (Pearse, 1976). This has enabled not only detailed
analysis of coated vesicle comparative morphology to be carried out, but
also some degree of biochemical characterisation. An analysis by Pearse (1975)

of coated vesicles from pig brain on sodium dodecyl sulphate polyacrylamide gel electrophoresis (SDS-PAGE), revealed a major protein band of about 180 000 mol. wt. This protein was shown to be on the cytoplasmic surface of the vesicle and to be a major constituent of the coating, since digestion with trypsin or pronase produced naked vesicles when viewed by electron microscopy, and an absence of bands on SDS-PAGE. The protein was named 'clathrin' by Pearse, to indicate the lattice-like structure it formed, and was found to constitute the coating of vesicles isolated from bullock adrenal medulla, bullock and pig brain, and a non-secreting mouse lymphoma cell line (Pearse, 1976). Coated vesicles from chicken oocytes (Woods, Woodward & Roth, 1978), rabbit brain (Blitz, Fine & Toselli, 1977) and human placenta (Ockleford & Whyte, 1977) also have a major protein of 180 000 mol. wt. It has thus become fashionable to refer to the coating on these vesicles as the 'clathrin coat', although Matus (1976*a*) has argued in *Nature* that 'cytonexin' would have been a better name and that clathrin is a misnomer. This challenge arose from the findings of Gray (1972) that presynaptic nerve terminals contained a fibrous network which he called the 'cytonet'. This was also claimed to be present on the vesicles, being indistinguishable from the coating. Gray (1975) later repudiated the evidence of the cytonet and the coating on the vesicles on the grounds that they were fixation artifacts, possibly arising from precipitation of cytoplasmic protein. It is now clear that the coating is not an artifact (the isolation involves no use of fixatives), although, as I have tried to indicate, it is still a matter of debate as to how the coating represented by the polygonal basketwork of negatively stained vesicles and the bristles seen in median sections, relate to each other. Matus (1976*b*), in reply to a defence of clathrin by Pearse & Bretscher (1976), has argued that the cytonet should be resurrected, but since we know precisely what we mean now when referring to the polygonal basketwork, it seems better to use clathrin to describe its molecular composition until further information about the real existence of the cytonet is obtained.

This seems an appropriate point at which to mention the coating that can sometimes be seen on the external surface of the plasma membrane of the vesicle (i.e. directed towards the vesicle lumen) and which is presumed to represent the glycocalyx. This is not a feature peculiar to coated vesicles; all endocytic vesicles are likely to possess it to a varying degree. In the micro-pinocytic vesicles (or tubules) of the absorptive cells of *Hydra*, Slauterback (1967) described a highly structured glycocalyx which led him to refer to them as 'coated vesicles'. They are not of course coated vesicles in the currently accepted sense. Coated vesicles in macrophages have been inferred to have a particularly abundant glycocalyx, since they were reported to form at caveolae on the cell surface which in turn showed dense labelling when reacted with ruthenium red (Lagunoff & Curran, 1972). Similarly, glycocalyx stains have been found to react strongly in the caveolae which form coated vesicles in human syncytiotrophoblast (Ockleford & Whyte, 1977). It has also been suggested (Ockleford, 1976) that the glycocalyx, in its distribution on the vesicle membrane, may reflect the distribution of the cytoplasmic projections making up the coat (see Plate 1c). Perhaps as an outcome of this, Bowers (1964) suggested that the inner and outer coatings may be the binding sites needed for protein uptake; others hold that the clathrin coat has more to do with infolding of the plasma membrane although, as is discussed later, it may also form an integral part of specific receptor molecules.

Size variations

Unlike smooth-walled micropinocytic vesicles, coated vesicles appear to have a more variable size range. For example, those apparently involved in the recycling of synaptic vesicle membrane during transmitter release at frog neuromuscular junctions were reported by Heuser & Reese (1973) to have external diameters (i.e. coating included) of 50–55 nm; those involved in immunoglobulin transport across proximal enterocytes of suckling rat gut (Rodewald, 1973) and across rabbit yolk sac endoderm (Moxon, Wild & Slade, 1976) ranged from 100 to 200 nm and from 50 to 200 nm in diameter respectively. Some attention has been focussed on the possibility that coated vesicles within a single cell type may fall into distinct size ranges, since this could indicate possible functional variation. This is not an easy thing to determine from sectioned material since the plane of section will not always be median. Nevertheless, several workers have in the past described coated vesicles as falling into two size ranges. Bowers (1964), in her study of aphid pericardial cells, found coated vesicles of diameter about 75 nm at the outer cell surface membrane and 110 nm in more deeply infolded regions. In bone marrow sinusoidal endothelium, De Bruyn, Michelson & Becker (1975) found coated vesicles with an average diameter of 40 nm in the cytoplasm and larger ones, of average diameter 140 nm, at the luminal and abluminal surfaces; coated vesicles with dimensions clearly intermediate between these two groups were not found. Two populations of coated vesicles were also described by Friend & Farquhar (1967) in a study of the epithelium of rat vas deferens. The larger ones, with diameters greater than 100 nm, were found concentrated in the apical cytoplasm and pinocytosed administered horseradish peroxidase (HRP), but smaller ones (less than 75 nm) were found primarily in the Golgi region and never contained HRP. From this study it has been suggested that larger coated vesicles are primarily pinocytic and smaller ones lysosomal in nature. Similar inferences might be made about coated vesicles in hamster peritoneal macrophages, since Dumont (1969) found that the larger peripheral ones accumulated colloidal gold, but that the smaller Golgi-associated ones did not. In considering the possible lysosomal nature of the smaller Golgi-associated coated vesicles, however, it is perhaps worth noting that De Bruyn *et al.* (1975) could not detect acid phosphatase in coated vesicles of any size range, although this enzyme, which is often taken as a marker of lysosomes, was detected in dense bodies and Golgi cisternae. This does not of course necessarily preclude the smaller vesicles from being lysosomal in nature but, if they are, suggests some rather special function.

Variation in size has also been found when purified fractions of coated vesicles isolated from different tissues have been compared (Pearse, 1976), but as yet no substantiated functional difference can be attributed to it. In terms of coat structure, larger vesicles have an increased number of hexagonal units making up their clathrin coats. Larger coated vesicles also have a larger inner vesicle and therefore more space to enclose with polygons. From considerations of this sort it would appear that size differences arise at the time of vesicle formation and not afterwards as a result of osmotic effects, although as Allison & Davies (1974) have postulated, osmotic effects might account for the slight variation in size seen with smooth-walled micropinocytic vesicles.

If two distinct size populations of coated vesicles coexisted in cells and tissues, as implied earlier, one would expect the distribution curve of measured diameters or volumes of large numbers of isolated, negatively stained vesicles to be bimodal in appearance. Two groups of workers, however – Ockleford, Whyte & Bowyer (1977) working with human placenta as source material, and Woods *et al.* (1978) working with porcine brain and chicken oocytes – have both found only a unimodal distribution. Clearly, a similar analysis is now needed for such tissues and cells as rat vas deferens, bone marrow endothelium and macrophages, although this may be difficult to achieve because of the large quantity of material needed for efficient isolation of pure coated vesicle fractions.

Formation of coated micropinocytic vesicles

So far my remarks have been concerned primarily with fully formed coated micropinocytic vesicles. There is no doubt that these vesicles arise in many cell types as coated pits, i.e. areas of the plasma membrane, probably containing a condensed layer of glycocalyx, which are in the process of invaginating and which possess a coat of projections on the convex surface (Plate 1*a*). These pits appear to 'pinch off' from the plasma membrane, and observation of different stages in the endocytosis of low-density lipoprotein labelled with ferritin (LDL-ferritin) by human fibroblasts (Anderson, Brown & Goldstein, 1977) gives some idea as to the probable sequence of events. Thus the coated pit first appears to become more indented, the two portions at the opening of the pit move closer together, and finally fusion and separation from the plasma membrane take place. Coated vesicles forming in mosquito oocytes (Roth & Porter, 1964; Roth, Cutting & Atlas, 1976) and hen oocytes (Roth *et al.*, 1976) differ slightly in that they have a long neck in some cases (0–150 nm in length) which does not possess a cytoplasmic coating. In jejunal enterocytes of suckling rat gut (Rodewald, 1973) and in rabbit yolk sac endoderm (Wild, 1975; Moxon *et al.*, 1976) coated vesicles form at the base of microvilli, but may also form from elements of the sub-apical canalicular system in gut enterocytes (Rodewald, 1973).

How the plasma membrane invaginates, and what the relevance of the coating is in the process, have been matters for some speculation. An early suggestion by Roth & Porter (1964) was that the free ends of the projections from the cytoplasmic surface might gradually repel each other and that this might generate the process of infolding of the plasma membrane with subsequent formation of a vesicle. Kanaseki & Kadota (1969) considered the coating to be exclusively a mechanism for infolding of the membrane and have suggested that the driving force for invagination arises from the conversion of certain fixed hexagons, which, as I described earlier, make up the coating, into pentagons. Thus, in their opinion, the first stage of transformation occurs in regions of the plasma membrane already possessing clathrin coats composed entirely of hexagons in a plane surface. (It is geometrically impossible for a plane surface to be composed of a mixture of regular hexagons and pentagons.) This transformation will result in a hemisphere forming which is finally converted into a vesicle by the process continuing into the neck of the invagination. What the stimulation is for this conversion to pentagons is not discussed, but in a modification of this hypothesis Ockleford (1976) considers the first step in coated vesicle formation to be specific binding of extracellular molecules to the glycocalyx. Assuming this to be composed of transmembrane proteins, or of proteins combined with other molecules that penetrate the membrane, he suggests the next step is for membrane-associated proteins to aggregate into a membrane complex. It is this complex which spontaneously forms a polygonal pattern, curvature of the membrane being generated simultaneously until finally a vesicle is formed. I mentioned previously that the location of the glycocalyx within the vesicles may reflect the positions of the projections on the cytoplasmic surface; Ockleford (1976) regards this as possible supporting evidence for his interpretation of events. Adding weight to Ockleford's concept of vesicle formation is the recent report by Anderson, Goldstein & Brown (1977) that mislocation of receptors for LDL, and for some protein needed for incorporation into coated pits, may occur in mutant human fibroblasts. In a model to explain how LDL is normally internalised by fibroblasts, these workers have suggested that the receptor may be a transmembrane protein which is inserted randomly in the fibroblast plasmalemma. Specificity for LDL resides in that portion of the molecule which is external, whilst the cytoplasmic portion contains the recognition sequences necessary for incorporation into coated pits through peripheral membrane (possibly clathrin) protein. The peripheral membrane proteins gather the receptor molecules into the forming pit. A mutant form of fibroblast has been discovered by Anderson, Goldstein & Brown (1977) in which binding of LDL to the cell membrane can occur, but not incorporation into coated vesicles. Formation of coated vesicles still takes place in this mutant, presumably because there are other substances present in the culture medium capable of initiating it. In considering coated vesicle formation, however, it seems important not to overlook the fact that smooth-walled micropinocytic vesicles infold from the plasma membrane without the 'inside help' that clathrin is postulated to provide in those hypotheses previously mentioned. This could be interpreted as evidence that the formation of the clathrin coat is a consequence of infolding of a particular part of the plasma membrane, and not a cause. Smooth-walled micropinocytic vesicles perhaps lack the coat because they do not contain, in their particular region of plasmalemma, specific receptor molecules which clathrin is capable of aggregating. To help resolve these points we need to know whether or not coated vesicle formation can take place in the absence of specific proteins. From their studies on uptake of ferritin by macrophages, Lagunoff & Curran (1972) claim that binding takes place to pre-existing coated regions. However, it seems possible from their treatment of cells (isolation and washing were with cold medium) that coated vesicles could have been blocked during their process of formation *in vivo*, especially since Anderson, Brown & Goldstein (1977) found that the effect of chilling fibroblasts was to cause an arrest in 'pinching-off' of coated membrane regions.

It has been suggested that coated vesicles may differ from smooth-walled micropinocytic vesicles in terms of rate of formation, the sequence of events being smooth-running in the former type but exhibiting a delay in separation from the cell surface in the latter (Fawcett, 1965). No quantitative data are available to substantiate this point.

Selective uptake and transport of proteins by coated micropinocytic vesicles

Certain cell types have a particular need to either store, break down or transport *specific* proteins present in the external medium and it is significant that coated vesicles have been implicated either directly or indirectly in all cases as a means of doing this.

The oocyte of a variety of animals represents a cell type with a need to store, in large quantities in some cases, yolk proteins that have been synthesised elsewhere in the female body and secreted into the blood. That these are sequestered in coated vesicles has been well established for oocytes of the mosquito. Roth & Porter (1964) observed a 15-fold increase in the number of coated pits forming from the oolema of vitellogenic mosquito oocytes (i.e. those sequestering yolk proteins whose synthesis had been stimulated by a blood meal) compared to resting, previtellogenic oocytes. Since these pits became filled with electron-dense material, they suggested that they were sites of selective yolk protein uptake. Direct evidence for this has come from studies made by Roth et al. (1976) in which vitellogenic oocytes were cultured *in vitro* in the presence of either yolk proteins conjugated to ferritin, or ferritin alone. Only the conjugate, as evidenced by the ferritin label, became localised in coated pits and vesicles. In birds, immunoglobulins are also transported to the vitellogenic oocyte and, later on in development of the embryo, gain access to the circulation by selective transport across the yolk sac; in this way the chick acquires passive immunity from the mother (Brambell, 1970). When isolated plasma membranes of chicken oocytes were exposed to chicken IgG conjugated to ferritin, or to free ferritin, again, only the conjugate appeared in coated pits and vesicles (Roth et al., 1976). Yolk proteins (phosvitin and lipovitellin) did not compete for binding by ^{125}I-labelled chicken IgG to such membranes and in other studies (Yusoko & Roth, 1976) chicken IgG and bovine serum albumin did not compete for binding of ^{125}I-labelled yolk proteins. These findings indicate the presence of separate specific receptors for yolk proteins and IgG on the plasma membrane and point to their localisation in coated pits. An interesting observation made by Roth and his co-workers was that less than 20% of the coated pits in the chicken oocyte plasma membrane contained conjugated protein; this raised the possibility that receptors for these two proteins may be associated with different coated vesicles. This needs further investigation since despite the fact that the membranes were washed, pre-formed coated pits containing bound endogenous yolk protein and IgG might have been present, thus blocking receptor sites. Conjugation with ferritin might also impair the reactivity of many molecules (see Slade & Wild, 1971; Moxon et al., 1976), perhaps leading to unsaturation of receptors.

The fact that cultured human fibroblasts endocytose LDL via coated pits and vesicles has already been mentioned. LDL is the major cholesterol-carrying protein of human plasma and through such endocytosis and subsequent interaction with lysosomes, free cholesterol is made available for membrane synthesis and other metabolic functions (Goldstein & Brown, 1976). In relation to my previous comments about possible differences between smooth-walled and coated micropinocytic vesicles in terms of specific receptor location, it is interesting that LDL-ferritin could not be detected in the smooth-walled type, despite their abundance in cultured human fibroblasts (Anderson, Brown & Goldstein, 1977). Also, through the use of freeze fracture/deep etching techniques, Orci *et al.* (1978) have shown that the membrane comprising coated pits on human fibroblasts differs from that comprising smooth-walled micropinocytic vesicles and the remainder of the plasma membrane in that it contains a greater number of intramembrane particles which are of a larger size.

Further evidence for the specific nature of the receptors present on human fibroblasts was that human IgG−ferritin conjugate did not bind to the cell surface. However, this did not become localised in smooth-walled vesicles either, and when ferritin (but only in high concentration) or HRP was added to fibroblast cultures, both of these proteins could subsequently be detected in coated vesicles (Anderson, Brown & Goldstein, 1977). These latter findings raise questions concerning the *degree* to which uptake of proteins into coated vesicles is selective and to what extent ferritin and HRP, convenient tracers though they are, truly reflect what happens to endogenous proteins in the environment of the cell.

Non-selective fluid endocytosis (Jacques, 1969) might be expected to occur if all the space within the forming vesicle is not taken up by glycocalyx and bound protein. Roth *et al.* (1976) have calculated an approximate void volume equal to 22 % of the total pit volume for coated vesicles forming in mosquito oocytes, which could explain why, in some cases, uptake of ferritin and HRP into the same coated vesicle occurred when mosquito oocytes were exposed simultaneously to these exogenous proteins in the presence of endogenous yolk protein (Anderson & Spielman, 1971).

There are a number of reports, some already cited, in which free ferritin at least, has not been detected in coated vesicles. We observed no uptake of free ferritin, or ferritin conjugated to IgG for that matter, in coated vesicles formed in rabbit yolk sac endoderm (Moxon *et al.*, 1976); nor did Rodewald (1973) observe free ferritin uptake into coated vesicles in jejunal enterocytes of suckling rat gut. In vitelline vessel endothelial cells of rabbit and rat yolk sac splanchnopleur, ferritin can occasionally be seen localised in smooth-walled, but not coated, micropinocytic vesicles (Moxon, 1976). In macrophages, however, ferritin becomes localised in both types of vesicle (Casley-Smith, 1969; Lagunoff & Curran, 1972). Endogenous ferritin is readily seen in coated pits and vesicles in erythroblasts (Fawcett, 1965). This is not a surprising finding since ferritin is incorporated into haemoglobin and needs to be specifically incorporated from other sites into the cell. From what has been learnt about other cells, one can well believe there are specific receptors for ferritin localised in coated pits and vesicles in erythroblasts. It may be that ferritin can bind (by positively charged groups to the negatively charged glycocalyx) to a varying extent to other receptor molecules present on different cells, which might help to explain these various findings, added to which fluid endocytosis might also occur to a greater or lesser extent. The same considerations would also apply to HRP.

The small size of coated vesicles and the fact that the glycocalyx where they form is particularly dense on the plasma membrane, can be regarded as aids to selective endocytosis. There might also be some contraction of the vesicle as it 'nips off' from the plasma membrane, causing an expulsion of unbound proteins (a possible interpretation which Anderson, Brown & Goldstein (1977) place on one of their electron micrographs when deducing sequences of endocytosis of LDL-ferritin).

In other cell types there is a specific requirement for proteins to be transported intact across the cell. How this is achieved may depend upon the degree of selectivity required. It is well established that selective transport of immunoglobulins takes place in the human chorioallantoic placenta, rabbit yolk sac splanchnopleur, and the gut of the suckling rat and mouse (Brambell, 1970; Wild, 1973a; Wild, 1974). Such selection is independent of molecular size, related to immunoglobulin class, structure and species of origin, and in my view, but contrary to that of Brambell (1970) − is not intimately related with proteolysis (Wild, 1975, 1976). The cellular mechanisms involved in such selective transport were thought by Brambell (1970) to involve non-selective endocytosis and binding to specific receptors for the Fc region of the immunoglobulin molecule. These were postulated to provide protection for immunoglobulins against proteolysis when transporting endocytic vesicles (macropinocytic vesicles) fused with lysosomes in the cells concerned (syncytiotrophoblast, yolk sac endoderm and gut enterocytes). Resulting phagolysosomes were then presumed to exocytose such protected protein in some way, so that it was then free to diffuse into the blood capillaries underlying the cells. I have presented evidence elsewhere (Wild, 1975, 1976) that in rabbit yolk sac endoderm at least, phagolysosomes do not exocytose their contents, but represent 'dead ends' as far as protein transport is concerned.

Macropinocytic vesicles form in yolk sac endoderm (as they do in many other types of absorptive epithelia) from a submicroscopic apical canalicular system (see Plate 2*a*). Pinocytosis of proteins via this system is non-selective in the sense that different proteins, including immunoglobulins from different species, become localised in the same macropinocytic vesicle and resulting phagolysosome (Wild, Stauber & Slade, 1972; Wild, 1974), i.e. binding to specific receptors does not appear to be associated with uptake and transport in the tubules that make up the canalicular system. This is not to say that binding to the glycocalyx that is evident on the tubules comprising the canalicular system cannot take place so that absorptive endocytosis is facilitated, or that such binding cannot have a degree of selectivity. Indeed, Lloyd *et al.* (1975) found differences in endocytic indices for rat yolk sac, exposed *in vitro* to various forms of bovine serum albumin, which are hard to explain in any way but this.

Contrary to the 'Brambell Hypothesis', I have suggested that selective transport of immunoglobulins in rabbit yolk sac endoderm takes place not in phagolysosomes but in coated vesicles, that selection is a cell surface event and not an intracellular one, and that Fc receptors serve not to protect against proteolysis but to segregate, at the cell surface, immunoglobulin destined for transport in coated vesicles from that destined for proteolysis in phagolysosomes (Wild, 1975, 1976). These events are depicted in Fig. 1. Evidence for this comes from the fact that coated vesicles (pits) are the only vesicle type to be seen in confluence with the lateral and basal plasmalemma, suggesting they are exocytosing contained protein. More direct evidence for this has come from experiments in which rabbit yolk sac endoderm was exposed *in utero* to IgG–HRP and IgG–ferritin conjugates, rabbit, rat and bovine anti-HRP antibodies (IgG), free ferritin, HRP, and human IgG, and the localisation of these proteins in endodermal cells examined at the ultrastructural level (Moxon *et al.*, 1976; Moxon, 1976). All proteins except free ferritin and IgG–ferritin conjugates could be detected in coated vesicles, although there were differences with respect to which proteins were seen in coated vesicles in confluence with the basal and lateral plasmalemma compared with those that were free within the intercellular space. The most significant results were those obtained with rabbit, rat and bovine anti-HRP antibodies and with human IgG localisation, since the methods employed to trace these immunoglobulins (reaction of specific antibody-containing tissue with HRP, and human IgG-containing tissue with HRP-labelled Fab fragments of sheep anti-human IgG) involved no possibility of steric hindrance or non-specific binding. This might well occur however, with HRP–protein and ferritin–protein conjugates. One drawback with HRP as an ultrastructural protein label is that the reaction product it forms has a tendency to obscure the coating on vesicles,

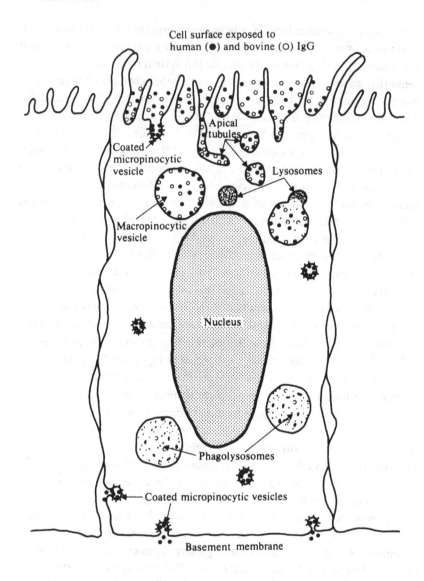

Fig. 1. Diagrammatic representation of the mechanism whereby
selective transport of immunoglobulins is thought to take place across
rabbit yolk sac endoderm. The cell surface is shown exposed to a
mixture of human and bovine IgG. Both of these immunoglobulins
are taken up non-selectively into macropinocytic vesicles via elements
of the apical tubular canalicular system. Complete degradation of
these proteins then occurs when the vesicles fuse with lysosomes —
there is no exocytosis via phagolysosomes. Human IgG (but little, if
any, bovine IgG) may also bind to receptors (specific for the Fc region)

sometimes making conclusive identification of vesicle type difficult. This was certainly a problem we encountered in localising immunoglobulins at the apical cell surface of yolk sac endoderm (Plate 2*a*), and it was also mentioned by Anderson, Brown & Goldstein (1977) as a difficulty when studying uptake of HRP—LDL conjugate by fibroblasts. Proteins do not penetrate the junctional complex between cells and we are certain that an intercellular route cannot account for the subsequent occurrence of immunoglobulin in intercellular spaces (Plate 2*b*). Discharge of immunoglobulin from coated vesicles that have fused with the lateral plasmalemma seems more likely than pinocytosis of immunoglobulin that had reached the intercellular space by some other mechanism. Exocytosis rather than pinocytosis is also our explanation for the occurrence of coated vesicles containing rabbit, and human, HRP conjugates, close to and in confluence with, the basal plasmalemma (Plate 3*a*). A finding similar to that reported by Roth *et al.* (1976) for chick oocyte plasma membrane, was that not all coated vesicles contained detectable immunoglobulin (see Plates 2*b* and 3*a*) and we too have speculated that this could be due to a segregation of different proteins into different coated vesicles, but other explanations are also possible (Moxon *et al.*, 1976). In these experiments we could not detect bovine anti-HRP antibodies either in coated vesicles or in the intercellular space, although 'empty' coated vesicles were readily detected in the cells. As for all proteins studied, however, we could readily find bovine anti-HRP antibody in apical tubules, macropinocytic vesicles and phagolysosomes. It is well established that unlike rabbit and human IgG, bovine IgG is very poorly transported from yolk sac endoderm to the foetal blood. Furthermore, whilst it has been possible to demonstrate the presence of Fc receptors for rabbit IgG (with which human IgG also binds) on rabbit yolk sac endoderm, no such receptor could be detected for bovine IgG (Wild & Dawson, 1977). Thus the link between Fc receptors, coated vesicles, and selective immuno-globulin transport across rabbit yolk sac endoderm is a strong one. There is also a similar strong link for transport across jejunal enterocytes of suckling rat gut. Rodewald (1973) found rat and bovine IgG—ferritin conjugates (but not free ferritin or chicken IgG—ferritin) became localised in, and were subsequently transported by, coated vesicles in these cells. Furthermore Rodewald

---

localised in regions of the apical plasmalemma where coated vesicles will form. These coated vesicles then transport the contained im-munoglobulin across the cell by fusing with the lateral plasmalemma and discharging it into the intercellular space, or by fusing with the basal plasmalemma and discharging it into the basement membrane. Coated vesicles do not fuse with lysosomes during such transport. (After Wild, 1976.)

(1976) found that whilst jejunal cells pinocytosed and transported rabbit IgG—HRP and rabbit IgG Fc—HRP in coated vesicles, they did not pinocytose rabbit IgG Fab—HRP. Clearly, there are differences in the nature of the Fc receptors present on rat jejunal cells and rabbit yolk sac endoderm with respect to binding of ferritin conjugates. Our findings correlated well with a previous study (Slade & Wild, 1971) in which it was shown that human IgG was prevented from being transported to foetal rabbit blood by conjugation to ferritin.

Uptake and transport of guinea pig IgG—HRP via coated vesicles also takes place in endodermal cells of guinea pig yolk sac splanchnopleur when this is cultured *in vitro* (King, 1977a). A low level of ferritin transport also appears to take place across syncytiotrophoblast of guinea pig chorioallantoic placenta in coated vesicles (King & Enders, 1971).

Transport via smooth-walled micropinocytic vesicles

It is well known that smooth-walled micropinocytic vesicles can shuttle large protein tracers such as HRP (see Plate 3b), ferritin and myoglobin, and particles such as colloidal iron, across capillary endothelial cells (Bruns & Palade, 1968; Simionescu, Simionescu & Palade, 1973). These vesicles have been considered to represent the morphological counterparts of the 'large pores' which physiologists have postulated to be present in capillary endothelium. Similarly, smooth-walled micropinocytic vesicles transport ferritin and particles such as thorotrast, but not colloidal gold, across mesothelial cells of the omentum (Fedorko & Hirsch, 1971). One cannot presume that transport in these vesicles is by non-selective fluid endocytosis (bulk transport, to put it another way) since there has been insufficient study of transport of labelled endogenous plasma proteins to enable definite conclusions to be drawn. Indeed, an element of selective adsorptive endocytosis is implied by the findings of Fedorko & Hirsch (1971). The point has also been made (Wissig & Williams, 1978) that bulk transport is inconsistent with physiological data relating molecular size and the degree to which substances are transported from blood across capillary endothelium. These authors have produced evidence for transport of low molecular weight protein (microperoxidase) through endothelial cell junctions (the 'small pores' of the physiologists) and it seems likely that total transport represents a summation of ultrafiltration through junctions and bulk transport, together with absorptive endocytosis, through vesicles. Unlike the situation for coated vesicles, however, there is no evidence that smooth-walled micropinocytic vesicles are associated with *specific* receptors for proteins.

Fate of coated vesicles

Coated vesicles containing yolk proteins lose their clathrin coats when they move into the cytoplasm from the oocyte surface. The naked vesicles then fuse to form larger yolk granules (Roth *et al.*, 1976). Coated vesicles formed from the surface of aphid pericardial cells (Bowers, 1964), proximal tubule kidney cells (Maunsbach, 1966), rat macrophages (Lagunoff & Curran, 1972) and human fibroblasts (Anderson, Brown & Goldstein, 1977) also lose their clathrin coats and in fibroblasts at least, fuse with lysosomes. These are all cells in which one would suppose that little subsequent export of endocytosed protein took place, and in fusing with lysosomes, the naked vesicles are little different from macropinocytic vesicles or smooth-walled micropinocytic vesicles that have aggregated to form larger vesicles. The fate of endocytic vesicles in general, is, however, potentially very variable in different cells (see Jacques, 1975, and Chapter 2). The mechanism controlling loss of coat material is unknown but Pearse (1976) has suggested that the absence of amino-sugars (and therefore possibly carbohydrate) from clathrin is what one might expect of a protein which can associate with, and dissociate from, membranes in a cyclic fashion. Possessing carbohydrate would be a hindrance in that it might lead to more stable bonds forming. Coats appear to be retained in yolk sac endoderm (Moxon *et al.*, 1976) and jejunal enterocytes (Rodewald, 1973), as might be expected if the vesicles are engaged in diacytosis (transepithelial transport). I have suggested that the coating might be crucial in *preventing* fusion with lysosomes (Wild, 1975, 1976). Some loss of coat material would be expected to occur when the vesicles fused with the lateral and basal plasmalemma. Coated vesicles in syncytiotrophoblast of human placenta have been found to have a distribution much restricted to the cell surface and which has been interpreted as possibly indicating loss of coats as vesicles move inwards (Ockleford & Whyte, 1977). When human placental villi are exposed to human IgG–HRP *in vitro*, the conjugate becomes localised in coated vesicles in syncytiotrophoblast (King, 1977*b*), but whether these transport IgG selectively across the cells, or merely provide a means of selecting it for digestion, is still unclear.

Concluding remarks

By classifying pinocytosis into macropinocytosis and micropinocytosis, and in recognising functional as well as morphological differences in these processes, Allison & Davies (1974) made considerable progress in promoting the understanding and facilitating the communication that Fawcett (1965) had earlier called for in his essay on 'Surface specializations of absorbing cells'. In this essay, Fawcett drew attention to the fact that the term

pinocytosis had been applied to a variety of different kinds of surface activity
in which (as he saw it) the only common feature was uptake of fluid in bulk.
He felt that different terms should be applied to these different surface
activities. Even though Fawcett recognised coated vesicles as being different
in structure (and possibly function) to smooth-walled micropinocytic vesicles,
all too often the term micropinocytosis has subsequently been used to encompass both types of vesicle formation as if they were the same. What I have
tried to do in this opening chapter is portray coated vesicles as being sufficiently
different to warrant their promotion to a distinct subclass of endocytic
vesicle. In so doing I have indicated many of the subjects to be developed in
later chapters, and hope that this introduction will serve to awaken the interest
of the reader.

I am grateful to Dr C.D. Ockleford, Department of Anatomy, University of
Leicester, for supplying the photographs for Plates 1(a), (c) and (e); to my
former student Dr Lesley A. Moxon for permission to publish Plate 1(b); and
to Mr F. Al-Khafaji, Department of Human Morphology, University of
Southampton, for supplying the photographs for Plate 3(b). Some of my own
work referred to here was supported by the Medical Research Council.

References

Allison, A.C. & Davies, P. (1974). Mechanisms of endocytosis and exocytosis.
    In *Transport at the Cellular Level,* ed. M.A. Sleigh & D.H. Jennings,
    Symposium of the Society for Experimental Biology 28, pp. 521–
    46. Cambridge University Press.
Anderson, R.G.W., Brown, M.S. & Goldstein, J.L. (1977). Role of the coated
    endocytic vesicle in the uptake of receptor-bound low density
    lipoprotein in human fibroblasts. *Cell,* 10, 351–64.
Anderson, R.G.W., Goldstein, J.L. & Brown, M.S. (1977). A mutation that
    impairs the ability of lipoprotein receptors to localise in coated pits
    on the cell surface of human fibroblasts. *Nature, London,* 270,
    695–9.
Anderson, W.A. & Spielman, A. (1971). Permeability of the ovarian follicle
    of *Aedes aegypti* mosquitoes. *Journal of Cell Biology,* 50, 201–21.
Blitz, A.L., Fine, R.E. & Toselli, P.A. (1977). Evidence that coated vesicles
    isolated from brain are calcium sequestering organelles resembling
    sarcoplasmic reticulum. *Journal of Cell Biology,* 75, 135–47.
Bowers, B. (1964). Coated vesicles in the pericardial cells of the aphid (*Myzus
    persicae* Sulz). *Protoplasma,* 59, 351–67.
Brambell, F.W.R. (1970). The Transmission of Passive Immunity from Mother
    to Young. *Frontiers of Biology* 18. Amsterdam: North-Holland.
Bruns, R.R. & Palade, G.E. (1968). Studies on blood capillaries. Transport of
    ferritin molecules across the wall of muscle capillaries. *Journal of
    Cell Biology,* 37, 277–99.

Casley-Smith, J.R. (1969). Endocytosis: the different energy requirements for the uptake of particles by small and large vesicles into peritoneal macrophages. *Journal of Microscopy*, **90**, 15–30.

Crowther, R.A., Finch, J.T. & Pearse, B.M.F. (1976). On the structure of coated vesicles. *Journal of Molecular Biology*, **103**, 785–98.

De Bruyn, P.P.H., Michelson, S. & Becker, R.P. (1975). Endocytosis, transfer tubules and lysosomal activity in myeloid sinusoidal endothelium. *Journal of Ultrastructural Research*, **53**, 133–51.

Dingle, J.T., Poole, A.R., Lazarus, G.S. & Barrett, A.J. (1973). Immunoinhibition of intracellular protein digestion in macrophages. *Journal of Experimental Medicine*, **137**, 1124–41.

Dumont, A. (1969). Ultrastructural study of the maturation of peritoneal macrophages in hamster. *Journal of Ultrastructural Research*, **29**, 191–209.

Fawcett, D.W. (1965). Surface specialisation of absorbing cells. *Journal of Histochemistry and Cytochemistry*, **13**, 75–91.

Fedorko, M.E. & Hirsch, J.G. (1971). Studies on transport of macromolecules and small particles across mesothelial cells of the mouse omentum. I. Morphological aspects. *Experimental Cell Research*, **69**, 113–27.

Friend, D.S. & Farquhar, M.G. (1967). Functions of coated vesicles during protein absorption in the rat vas deferens. *Journal of Cell Biology*, **35**, 357–76.

Goldstein, J.L. & Brown, M.S. (1976). The LDL pathway in human regulation of cholesterol metabolism. In *Current Topics in Cellular Regulation*, ed. B.L. Horecker & E.R. Stadtman, vol. 2, pp. 147–81. New York & London: Academic Press.

Gray, E.G. (1961). The granule cells, mossy synapses and Purkinje spine synapses of the cerebellum: light and electron microscope observations. *Journal of Anatomy, London*, **95**, 345–56.

Gray, E.G. (1972). Are the coats of coated vesicles artifacts? *Journal of Neurocytology*, **1**, 363–82.

Gray, E.G. (1975). Synaptic fine structure and nuclear, cytoplasmic and extracellular networks. The stereoframework concept. *Journal of Neurocytology*, **4**, 315–39.

Heuser, J.E. & Reese, T.S. (1973). Evidence for recycling of synaptic vesicle membrane during transmitter release at the frog neuromuscular junction. *Journal of Cell Biology*, **57**, 315–44.

Jacques, P.J. (1969). Endocytosis. In *Lysosomes in Biology and Pathology*, ed. J.T. Dingle & H.B. Fell, vol. 2, pp. 395–420. Amsterdam: North-Holland.

Jacques, P.J. (1975). Cell biological processes involved in transport of matter across tissue membranes. In *Maternofoetal Transmission of Immunoglobulins*, ed. W.A. Hemmings, *Clinical and Experimental Immunoreproduction 2*, pp. 201–23, Cambridge University Press.

Kanaseki, T. & Kadota, K. (1969). The vesicle in a basket. A morphological study of the coated vesicle isolated from nerve endings in the guinea pig brain with special reference to membrane movements. *Journal of Cell Biology*, **42**, 202–20.

King, B.F. (1977*a*). An electron microscopic study of absorption of peroxidase-conjugated immunoglobulin G by guinea pig visceral yolk sac *in vitro*. *American Journal of Anatomy*, 148, 447–56.

King, B.F. (1977*b*). In vitro absorption of peroxidase conjugated IgG by human placental villi. *Anatomical Record*, 187, 624–5.

King, B.F. & Enders, A.C. (1971). Protein absorption by the guinea pig chorio-allantoic placenta. *American Journal of Anatomy*, 130, 409–30.

Lagunoff, D. & Curran, D.E. (1972). Role of bristle coated membrane in uptake of ferritin. *Experimental Cell Research*, 75, 337–46.

Lewis, W.H. (1931). Pinocytosis. *Bulletin of the Johns Hopkins Hospital*, 49, 17–27.

Lloyd, J.B., Williams, K.E., Moore, A.T. & Beck, F. (1975). Selective uptake and intracellular digestion of protein by rat yolk sac. In *Maternofoetal Transmission of Immunoglobulins*, ed. W.A. Hemmings, *Clinical and Experimental Immunoreproduction 2*, pp. 169–78. Cambridge University Press.

Matus, A.J. (1976*a*). The cytonet protein. *Nature, London*, 262, 176.

Matus, A.J. (1976*b*). Coated vesicles and clathrin. A.J. Matus replies. *Nature, London*, 263, 95.

Maunsbach, A.B. (1963). Electron microscopic observations on ferritin absorption in microperfused renal proximal tubules. *Journal of Cell Biology*, 19, 48A.

Maunsbach, A.B. (1966). Absorption of ferritin by rat kidney proximal tubule cells: electron microscopic observations of the initial uptake phase in cells of microperfused single proximal tubules. *Journal of Ultrastructural Research*, 16, 1–12.

Maunsbach, A.B. (1969). Functions of lysosomes in kidney cells. In *Lysosomes in Biology and Pathology*, ed. J.T. Dingle & H.B. Fell, vol. 1, pp. 117–54. Amsterdam: North-Holland.

Moxon, L.A. (1976). Cellular mechanisms involved in specific protein transport across yolk sac splanchnopleur. Unpublished Doctoral Thesis, University of Southampton.

Moxon, L.A., Wild, A.E. & Slade, B.S. (1976). Localisation of proteins in coated micropinocytotic vesicles during transport across rabbit yolk sac endoderm. *Cell and Tissue Research*, 171, 175–93.

Ockleford, C.D. (1976). A three dimensional reconstruction of the polygonal pattern on placental coated-vesicle membranes. *Journal of Cell Science*, 21, 83–91.

Ockleford, C.D. & Whyte, A. (1977). Differentiated regions of human placental cell surface associated with exchange of materials between maternal and foetal blood: coated vesicles. *Journal of Cell Science*, 25, 293–312.

Ockleford, C.D., Whyte, A. & Bowyer, D.E. (1977). Variation in volume of coated vesicles isolated from human placenta. *Cell Biology International Reports*, 1, 137–46.

Orci, L., Carpentier, J.-L., Perrelet, A., Anderson, R.G.W., Goldstein, J.L. & Brown, M.S. (1978). Occurrence of low density lipoprotein receptors within large pits on the surface of human fibroblasts as demonstrated by freeze-etching. *Experimental Cell Research*, **113**, 1–13.

Palade, G.E. (1953). The fine structure of blood capillaries. *Journal of Applied Physiology*, **24**, 1–24.

Palay, S.F. (1963). Alveolate vesicles in Purkinje cells of the rat cerebellum. *Journal of Cell Biology*, **19**, 89A.

Pearse, B.M.F. (1975). Coated vesicles from pig brain. Purification and biochemical characterization. *Journal of Molecular Biology*, **97**, 93–8.

Pearse, B.M.F. (1976). Clathrin: a unique protein associated with intracellular transfer of membrane by coated vesicles. *Proceedings of the National Academy of Sciences, USA*, **73**, 1255–9.

Pearse, B.M.F., & Bretscher, M.S. (1976). Coated vesicles and clathrin. *Nature, London*, **263**, 95.

Policard, A. & Bessis, M. (1958). Sur un mode d'incorporation des macromolecules par la cellule, visible au microscope electronique; la rhopheocytose. *Comptes Rendus Hebdomadaire des Séances de l'Academie des Sciences, Paris*, **246**, 3194.

Rodewald, R.B. (1973). Intestinal transport of antibodies in the newborn rat. *Journal of Cell Biology*, **58**, 189–211.

Rodewald, R.B. (1976). Intestinal transport of peroxidase-conjugated IgG fragments in the neonatal rat. In *Maternofoetal Transmission of Immunoglobulins*, ed. W.A. Hemmings, *Clinical and Experimental Immunoreproduction 2*, pp. 137–53. Cambridge University Press.

Rhodes, J.M. & Lind, I. (1968). Antigen uptake *in vivo* by peritoneal macrophages from normal mice, and those undergoing primary or secondary responses. *Immunology*, **14**, 511–25.

Rosenbluth, J. & Wissig, S.L. (1963). The uptake of ferritin by toad spinal ganglion cells. *Journal of Cell Biology*, **19**, 91A.

Rosenbluth, J. & Wissig, S.L. (1964). The distribution of exogenous ferritin in toad spinal ganglia and the mechanism of its uptake by neurons. *Journal of Cell Biology*, **23**, 307–25.

Roth, T.F., Cutting, J.A. & Atlas, S.B. (1976). Protein transport: a selective membrane mechanism. *Journal of Supramolecular Structure*, **4**, 487–508.

Roth, T.F. & Porter, K.R. (1964). Yolk protein uptake in the oocyte of the mosquito *Aedes aegypti* L. *Journal of Cell Biology*, **20**, 313–32.

Simionescu, N., Simionescu, M. & Palade, G.E. (1973). Permeability of muscle capillaries to exogenous myoglobin. *Journal of Cell Biology*, **57**, 424–52.

Slade, B.S. & Wild, A.E. (1971). Transmission of human γ-globulin to rabbit foetus and its inhibition by conjugation with ferritin: *Immunology*, **20**, 217–23.

Slauterback, D.B. (1967). Coated vesicles in absorptive cells of *Hydra*. *Journal of Cell Science*, **2**, 563–72.

Wild, A.E. (1970). Protein transmission across the rabbit foetal membranes. *Journal of Embryology and Experimental Morphology*, **24**, 313–30.

Wild, A.E. (1973a). Transport of immunoglobulins and other proteins from mother to young. In *Lysosomes in Biology and Pathology*, ed. J.T. Dingle, vol. 3, pp. 169–215. Amsterdam: North-Holland.

Wild, A.E. (1973b). Fluorescent protein tracing in the study of endocytosis. In *Lysosomes in Biology and Pathology*, ed. J.T. Dingle, vol. 3, pp. 522–36. Amsterdam: North-Holland.

Wild, A.E. (1974). Protein transport across the placenta. In *Transport at the Cellular Level*, ed. M.A. Sleigh & D.H. Jennings, *Symposium of the Society for Experimental Biology* 28, pp. 521–46. Cambridge University Press.

Wild, A.E. (1975). Role of the cell surface in selection during transport of proteins from mother to foetus and newly born. *Philosophical Transactions of the Royal Society of London*, 271B, 395–410.

Wild, A.E. (1976). Mechanism of protein transport across the rabbit yolk sac endoderm. In *Maternofoetal Transmission of Immunoglobulins*, ed. W.A. Hemmings, *Clinical and Experimental Immunoreproduction* 2, pp. 155–67. Cambridge University Press.

Wild, A.E. & Dawson, P. (1977). Evidence for Fc receptors on rabbit yolk sac endoderm. *Nature, London*, 268, 443–45.

Wild, A.E., Stauber, V.V. & Slade, B.S. (1972). Simultaneous localisation of human γ-globulin, I^{125} and ferritin during transport across the rabbit yolk sac splanchnopleur. *Zeitschrift für Zellforschung*, 123, 168–77.

Williams J.A. & Wolff, J. (1971). Thyroid secretion *in vitro*: multiple actions of agents affecting secretion. *Endocrinology*, 88, 206–17.

Wissig, S.L. & Williams, M.C. (1978). Permeability of muscle capillaries to microperoxidase. *Journal of Cell Biology*, 76, 341–59.

Wolfe, D.E. (1965). The epiphyseal cell: an electron microscopic study of its intercellular relationships and intracellular morphology in the pineal body of the albino rat. *Progress in Brain Research*, 10, 332–86.

Woods, J.W., Woodward, M.P. & Roth, T.F. (1978). Common features of coated vesicles from dissimilar tissues: composition and structure. *Journal of Cell Science*, 30, 87–97.

Yusoko, S.C. & Roth, T.F. (1976). Binding to specific receptors on oocyte plasma membranes by serum phosvitin–lipovitellin. *Journal of Supramolecular Structure*, 4, 89–97.

Plates

Plate 1. (*a*) Surface region of human syncytiotrophoblast showing
coated pits (arrowed) in the convoluted plasma membrane. (From
Ockleford, 1976.) (*b*) Endothelial cell (EC) of a vitelline vessel in
rabbit yolk sac splanchnopleur showing smooth-walled micropinocytic
vesicles. VL, vessel lumen. (Courtesy of Dr L.A. Moxon.) (*c*) Median
section through a fully formed coated vesicle in human syncytiotrophoblast. Note that the dense projections from its circumference
coincide with slight increases in density of the internal glycoprotein
coat. (From Ockleford, 1976.) (*d*) Coated vesicles isolated from
rabbit yolk sac splanchnopleur by the technique of Pearse (1976)
and positively stained with uranyl acetate. Note the presence of an
inner vesicle and outer polygonal basketwork. (*e*) Polygonal pattern
on the surface of a tangentially sectioned coated vesicle in human
syncytiotrophoblast. (From Ockleford, 1976.) (*f*) Macropinocytic
vesicles containing human IgG in the apical region of rabbit yolk sac
endoderm. Detection was by means of FITC-labelled rabbit anti-human
IgG applied to de-waxed sections of ethanol-fixed tissue. (From
Wild, 1970.)

Plate 2. (*a*) Detail of the apical region of an endodermal cell of rabbit
yolk sac splanchnopleur exposed *in vivo* to rabbit anti-HRP (IgG) for
1 hour and subsequently treated with HRP after glutaraldehyde
fixation. What are interpreted as being coated vesicles (CV) containing
reaction product indicative of IgG can be seen forming from, and
lying close to, the apical plasmalemma. Profiles of the apical tubular
canalicular system in transverse and longitudinal section (arrows) are
also evident. (From Moxon *et al.*, 1976.) (*b*) Detail of the lateral
region of endodermal cells of rabbit yolk sac splanchnopleur exposed
in vivo to rabbit anti-HRP (IgG) for 1 hour and subsequently treated
with HRP after glutaraldehyde fixation. Reaction product indicative
of IgG can be seen in the intercellular space (ICS). Coated vesicles
free of (asterisked) and containing, reaction product, are present in
the cytoplasm. (From Moxon *et al.*, 1976.)

Plate 3. (*a*) Detail of the basal region of an endodermal cell from
rabbit yolk sac splanchnopleur exposed *in vivo* to human IgG–HRP
conjugate for 1 hour. Reaction product indicative of the conjugate
is present in two coated vesicles (arrowed) lying close to, and
apparently fusing with, the basal plasmalemma. Coated vesicles
devoid of reaction product (asterisked) are also present. (In part from
Moxon *et al.*, 1976.) (*b*) Blood capillary in an inferior mesenteric
ganglion of a guinea pig that had been injected intravenously with
HRP. Reaction product indicative of HRP can be seen in the collapsed
lumen of the capillary (CL) and in smooth-walled vesicles (see insert)
in the endothelial cells (EC). (Courtesy of Mr F. Al-Khafaji.)

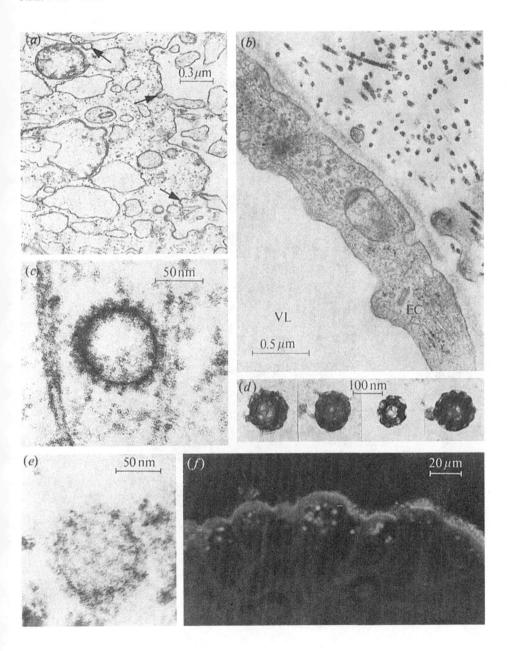

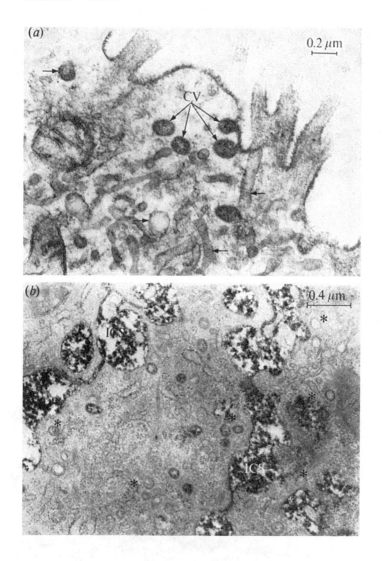

A. E. Wild   Plate 3

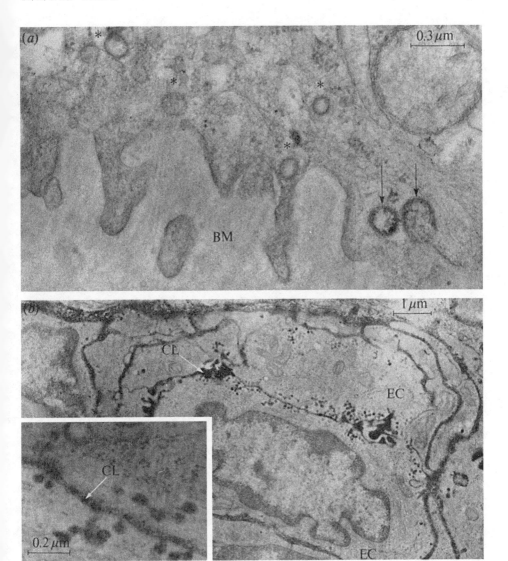

2

Coated vesicles in different cell types: some functional implications

A.J. NEVOROTIN

Ample evidence indicates that coated vesicles are commonly occurring structures in different cell types in a diversity of animal species. Coated vesicles have been demonstrated in cells under physiological and experimental conditions, *in vivo* and *in vitro*, as well as in pathological situations. It is necessary, therefore, to examine the possibility that coated vesicles are ubiquitous structures and to record their occurrence in different animals, different cell types, and under varying cell-environmental conditions.

In a recent article an attempt has been made to review the information pertinent to coated vesicles and their function (Nevorotin, 1977). From an analysis of this and of further original observations it may be concluded that there are several distinct groups of coated vesicles. These all share a similar if not identical structural organisation of an external lattice structure and membranous vesicle wall, but differ in their vesicular contents, the route the coated vesicle apparently follows within the cytoplasm, and their contribution to cellular metabolism.

This chapter presents a comparative analysis of coated vesicles and their function but deals only briefly with the literature in the specific areas dealt with in later chapters.

Morphological aspects: general comments

There is more than one ultrastructurally distinct entity which has been termed a coated vesicle. The structure generally observed and tentatively related here to a principal class is characterised, when viewed in median ultrathin section, as a sphere bounded by unit membrane, with several projections, 15–20 nm in length, radiating from the external surface. In certain situations vesicles have projections which are very short and slim, as in the case of 'fuzz-coated vesicles', a type of vesicle found in a common cytoplasm with 'true' coated vesicles in differentiating rat odontoblasts (Weinstock & Leblond,

1974); similar projections from the cytoplasmic surface of the elongated invaginations in absorptive cells of *Hydra* imply some relationship between these structures and coated vesicles (Slautterback, 1967). In addition a pattern of cup-shaped coated vesicles with internally located projections has been described recently by McKanna (1973) in a peritrich ciliate. These structures, although all similar to coated vesicles, are not identical to them and will therefore not be considered here.

According to the three-dimensional model for the coated vesicles of the principal class, each vesicle consists of two major parts. These are a smooth microvesicle indistinguishable from, for example, an ordinary synaptic vesicle in a nerve ending, and a structure which encloses the microvesicle. The enclosing structure is made up of a continuous polygonal network, the sides of the polygons being the projections (earlier called bristles). Since first demonstrated in guinea pig synaptosomes (Kanaseki & Kadota, 1969), analogous organisation of coated vesicles has been shown in the synaptic apparatus of the central nervous system (CNS) in many representatives of vertebrate classes (Gray & Willis, 1970; Pearse, 1975; Blitz & Fine, 1976), the presynaptic bags of guinea pig and rat retina (Gray & Pearse, 1971), the neurosecretory terminals of rat and hamster posterior pituitary (Douglas, Nagasawa & Schulz 1971; Nagasawa, Douglas & Schulz, 1971), tick midgut epithelial cells (Raikhel, 1974), ovine skeletal muscle cell (Morton & Rowe, 1974), and human placenta (Ockleford, 1976). In this chapter the characteristic polygonal network is demonstrated in grazing sections of the giant coated vesicles containing the secretory granules in rat adenohypophysial mammatrophs (Plate 1*a*). In the median sections of these granule-containing coated vesicles the typical projections can be seen intermingled with the polygons of the shell, so that the identity of both the projections and the polygon sides is apparent (Plate 1*b*). It can be concluded from the information referred to thus far that coated vesicles of the principal class, independent of species and cell type, comprise a structurally homogeneous population which can appear as a membranous enclosure with radiating projections in median sections, or a polygonal network when a whole coated vesicle, or its isolated enclosing structure, is observed *en face*.

Occurrence of coated vesicles
Species

The comparative data which follow are representative of, rather than a complete record of, the occurrence of these organelles. Literature in which coated vesicles are dealt with is scattered and badly indexed. Most papers on ultrastructure do not contain any information on this topic and a complete review of this whole area would be uneconomical. To date I have been able to find publications describing coated vesicles in Protozoa (Rosenbluth & Wissig, 1964), Coelenterata (Westfall, 1973), three species of Mollusca (Chalazonitis, 1969; McKenna & Rosenbluth, 1973; Skelding, 1973) and in many representatives of the Arthropoda – for example the Crustacea (Bunt, 1969; Holtzman, Freeman & Kashner, 1970) and Insecta (Bowers, 1964; Anderson, 1964; Stay, 1965; Roth & Porter, 1965; Porter, Kenyon & Badenhausen, 1967; Lane, 1968). No papers recording the occurrence of coated vesicles in the cells of the Annelida, Nematoda, Platyhelminthes or Porifera were discovered. Coated vesicles have been most extensively studied in the Vertebrata, especially the mammals; it is not surprising that about half the available references concern different cell types in the rat (Table 1).

Cell types
Tables 2, 3 and 4 show that many cell types originating from each of the three primary germ layers of the embryo contain coated vesicles. Data on the neuron, the oocyte, and the cells providing foetal and neonatal support are excluded because these are to be referred to later in more detail.

Variation in the state of cells
Most examples presented in Tables 2 to 4 refer to completely differentiated cells of adult, healthy and unaffected animals. In addition many workers have demonstrated coated vesicles in the cells of embryonic and neonatal or young organisms. They occur in sinusoidal lining cells of embryonic rat liver and bone marrow (Bankston & De Bruyn, 1974) and in differentiating ameloblasts of the kitten (Kallenbach, 1976). Degenerating cells also contain coated vesicles. The anuran lateral motor column neuron (Decker, 1974) is one example worthy of note. The existence of coated vesicles is not incompatible with in-vitro conditions, as evidenced, for example, in cultured rat macrophages (Shirahama & Cohen, 1970) and human fibroblasts (Röhlich & Allison, 1976). Coated vesicles have been shown in tumour cells, for example in mouse melanoma

Table 1. *Distribution of references used for this chapter according to the species whose cells have been shown to contain coated vesicles[a]*

Vertebrata					Invertebrata	
Rat	Man	Mouse	Guinea Pig	Others[b]	Insects[c]	Others[d]
48.0	7.0	5.2	4.7	21.2	8.2	5.2

[a] Relative figures are given as a percentage of all references.
[b] The other mammals, as well as representatives of Pisces, Amphibia, Reptilia and Aves.
[c] Pooled for different species.
[d] Including Crustacea, Mollusca and Coelenterata.

Table 2.   *Coated vesicles in different cell types derived from the ectoderm[a]*

Cell type	Species	Reference
Epithelial cells of tongue filiform papillae	Mouse	Weinstock & Wilgram (1970)
Epithelial cells of mammary gland[b]	Rat	Bargmann & Welsch (1969); Khokhlov (1977)
Columnar cells of sternal gland	Termites, *Zootermopsis nevadensis* and *Z. angusticollus*	Stuart & Satir (1968)
Cells of transitional epithelium of bladder and urethra	Rat	Hicks (1966)
'Intercalated' cells from kidney collecting duct	Rat	Griffith, Bulger & Trump (1968)
Type I cells from carotid body	Cat Rabbit	Biscoe & Stehbens (1966)
Enteroendocrine cells of large intestine	Man	This chapter, Fig. 8
Melanocytes of differentiating feather	Fowl	Maul & Brumbaugh (1971)
Hair cells of the organ of Corti[b]	Cat	Pluzhnikov & Nevorotin (1974)
Hair cells of the papilla basillaris	Caiman	Andrews (1975)

[a] Except the neuron, glial and Schwann cells.
[b] The number of the references has been abridged.

(Novikoff, Albala & Biempica, 1968) and in tissue-cultured human melanoma (Maul & Romsdahl, 1970). Experimental interference of many kinds does not result in destruction of coated vesicles. Cultured exocrine pancreatic cells stimulated by secretogogue administration (Jamieson & Palade, 1971), acinar cells of rat parotid gland after starvation of the animal (Hand, 1972), and cultured rat peritoneal macrophages infected with T_2 bacteriophages (Friend, Rosenau, Winfield & Moon, 1969) all contain coated vesicles. These data can be supplemented by our finding that coated vesicles occur in cells of some human tissues affected by disease (e.g. chronic glomerulonephritis: Plotkin & Nevorotin, 1976; acute bacterial dysentery: Nevorotin & Dobykin, 1978). In our laboratory, coated vesicles have been shown in human blood platelets of both healthy individuals and those suffering from leukaemia (Vashkinel & Petrov, 1974; Petrov & Vashkinel, 1977; see also Plate 1c). The evidence from platelets can be interpreted as an indication that the nucleus does not seem

Table 3. *Coated vesicles in different cell types derived from the endoderm*

Cell type	Species	Reference
Columnar cells of small intestine epithelium[a]	Rat	Bennet (1970)
Columnar cells of midgut	Tick, *Hyalomma asiaticum*	Raikhel (1974)
Columnar cells of large intestine epithelium	Man	Nevorotin & Dobykin (1978)
Clara cells of bronchial mucosa	Rat	Kuhn, Callaway & Askin (1974)
Ameloblasts[a]	Opossum	Lester (1970)
	Cat	Kallenbach (1976)
Acinar cells of parotid and Von Ebner glands[a]	Rat	Hand (1971)
Hepatocytes[a]	Rat	Bruni & Porter (1965)
	Fowl	Roth & Porter (1962)
Cells of fat body	Insect, *Calpodes ethlius*	Locke & Collins (1968)
		Locke & Collins (1968)
Acinar cells of exocrine pancreas	Guinea pig	Jamieson & Palade (1971)
Follicular cells of thyroid gland[a]	Rat	Seljelid (1967)
	Man	Klinick, Oertel & Winship (1970)
Mammotrophs and somatotrophs of anterior pituitary[a]	Rat	Pelletier (1973); This chapter, Plate 1(*d*)–(*f*)
Gonadotrophs of anterior pituitary	Rat	Lubersky-Moore, Poliakoff & Worthington (1975)

[a] The number of these references has been abridged.

to be necessary for the maintenance of coated vesicles. There is apparently no information on the occurrence of coated vesicles in spermatozoa, cells which possess a nucleus but lack certain other cytoplasmic organelles (Novikoff, Novikoff, Davis & Quintana, 1973). Fawcett (1965) has suggested that the formation of coated vesicles in the guinea pig erythroblast is arrested in mitosis. This raises the question of whether coated vesicles are affected by the mitotic cycle in other cell types. This observation is clearly of importance and deserves special investigation. From the reviewed data it would seem reasonable to conclude that the only cell type definitely lacking coated vesicles is the mature erythrocyte of mammals, where many other cytoplasmic organelles are also absent.

Table 4. *Coated vesicles in different cell types derived from the mesoderm*

Cell type	Species	Reference
Muscle cells		
Skeletal	Sheep	Morton & Rowe (1974)
Myocardial[a]	Lizard	Forbes & Sperelakis (1974)
Smooth	Mollusc,	McKenna & Rosenbluth
	Mytilus edulis	(1973)
Kidney proximal tubule	Rat	Maunsbach (1966)
cells[a]	Man	Plotkin & Nevorotin (1976)
Granulosa cells of corpus	Monkey,	Gulgas (1974)
luteum	*Macaca mulatta*	
Endometrial cells	Rabbit	Davies & Hoffman (1975)
Epithelial cells of vas deferens	Rat	Friend & Farquhar (1967)
Epithelial cells of ventral	Rat	Helminen & Ericsson (1970)
prostate lobe		
Odontoblasts	Rat	Weinstock & Leblond (1974)
Chondrocytes		
Elastic cartilage	Mouse	Sanzone & Reith (1976)
Costal and tracheal cartilage	Rat	Dearden (1975)
Osteoclasts	Rat	Kallio, Garant & Minkin
	Mouse	(1971)
	Pike, *Esox lucius*	
Endothelial cells		
Lymph and blood capillaries	Rat; guinea pig	Palade & Bruns (1968)
Bone marrow and spleen	Rat	De Bruyn & Cho (1974)
sinusoids[a]		
Phagocytic pericytes of	Rat	Van Deurs (1976)
brain arterioles, capillaries,		
and venules		
Pericapillary macrophages[b]	Rat	Bruns & Palade (1968)
Kupffer cells of liver	Guinea pig	Fawcett (1965)
erythroblasts		
Eosinophilic leucocytes	Rat	Komiyama & Spicer (1975)
Lymphocytes	Rat	Biberfeld (1971)
Megakaryocytes	Rat	Behnke (1968)
Reticular cells of thymus	Rat	This chapter, Fig. 10

[a] The number of references is abridged.

[b] The examples of cultured macrophages are not given in the table.

Function: the pathways

Soon after coated vesicles were discovered by Bessis & Breton Gorius (1957) and further described by Edward Gray (1961) it became obvious that they function in broadly the same way as smooth cytoplasmic microvesicles (Roth & Porter, 1962, 1965; Wissig, 1962; Palay, 1963; Bowers, 1964). On pinching off from the plasmalemma or the membrane of an organelle a coated vesicle apparently moves through the ground cytoplasm toward some membrane-limited body, or to the plasmalemma again where it fuses with target membrane. As a result, a certain portion of a substance segregated by the coated vesicle at the point of departure will be released at the destination. The biological effect(s) of this process are presumably dependent on the nature of the substance transferred, its amount, and the metabolic situation at the time and place. Questions concerning selectivity in the uptake of substances by coated vesicles, as well as the mechanism of their formation, and the forces directing them to the target organelle will not be considered at this juncture. The aim here is to differentiate between a few major pathways of intracellular transport in which coated vesicles operate as vehicles. In Fig. 1 the known and suggested intracellular pathways of coated vesicles are schematically illustrated to serve as a guide in the following portion of this chapter.

Endocytosis

It is now well established that different substances can be segregated from the extracellular environment by coated vesicles. This fact has been verified by utilising marker substances (tracers) for the investigation of endocytosis. Several examples of this type of experiment are presented in Table 5, which shows the tracer used and the cytoplasmic target organelle to which the tracer molecules are delivered by the endocytic coated vesicle. The pathways from the plasmalemma to multivesicular bodies operating as hetero-phagosomes (pathway 1 in Fig. 1) or to heterophagosomes of different appearances (pathway 2 in Fig. 1) are apparently followed by endocytic coated vesicles in many cell types. In addition to those listed in Table 5 there are other similar examples where these pathways have been demonstrated in tracer experiments. Pathway 1 has been demonstrated in epithelial cells of rat vas deferens (Friend & Farquhar, 1967), rat cerebellar neurons (Becker, Hirano & Zimmerman, 1968) and rat lymphocytes treated with phytohaemag-glutinin (Biberfeld, 1971). Pathway 2 has been shown in rat kidney proximal tubule cells (Maunsbach, 1966) and in oocytes of the cockroach (Anderson, 1964), fish (Droller & Roth, 1966) and rat (Kang, 1974). It is interesting to note that coated vesicles which travel along pathway 2 usually shed their

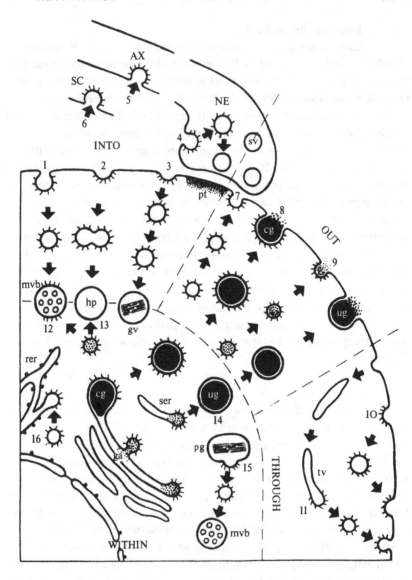

Fig. 1. A schematic representation of coated vesicles pathways. This hypothetical cell is tentatively endowed with some structures pertinent to coated vesicles, and innervated by an axon synaptically attached to it at an axosomatic synapse. The purpose of the illustration is to map the known and suggested routes of coated vesicles. Two dotted radial lines separate arbitrarily delimited areas in which coated vesicles do, or may, operate as endocytic ('INTO'), exocytic ('OUT') and transcytoplasmic ('THROUGH') vehicles. The dashed line parallel to the cellular border delimits the area where coated

enclosing lattices before they fuse with heterophagosomes; sometimes the loss of the polygonal lattice is preceded by fusion of coated vesicles with each other (Roth & Porter, 1962, 1964; Bowers, 1964). That some extracellular substances enter nerve endings and neurosecretory terminals within coated vesicles (pathway 4) is an established fact confirmed by many workers (e.g. Bunt, 1969; Zacks & Saito, 1969; Nagasawa *et al.*, 1971). On the other hand, there are only two additional indications that coated vesicles can participate in pathways 5 and 6 — endocytosis in axons and in the Schwann cells surrounding axons (Holtzman, 1969; Zacks & Saito, 1969). Considering all the evidence presented it follows that endocytosis by coated vesicles is a widespread phenomenon found in many cell types and many animal species. However, it does not seem to be found in all cell types capable of endocytosis. For example, in cultured rat adenohypophysis horseradish peroxidase (HRP) is internalised by smooth endocytic microvesicles of mammotrophs and somatotrophs (Pelletier, 1973). And in our recent tracer experiments with human biopsies of bronchial mucosa cultured in HRP-containing media, the ciliated cells of the epithelium were seen to engulf the tracer by means of relatively large vacuoles lacking polygonal lattice structure (Nevorotin, Chernyakova,

---

vesicles do, or may, move in the cytoplasm without coming into contact with the plasma membrane ('WITHIN'). The following pathways considered in the text are marked: 1, from the plasmalemma to multivesicular bodies (mvb); 2, from the plasmalemma to heterophagosomes (hp); 3, from the plasmalemma to Golgi vesicles and vacuoles (gv); 4, from the synaptolemma of a nerve ending (NE) into the cytoplasm, with the possible loss of polygonal lattice structure to become a plain synaptic vesicle (sv); 5, from the axolemma of an axon (AX) into the axoplasm, possibly eventually fusing with multivesicular bodies; 6, from the plasmalemma of a Schwann cell (SC) surrounding the axon into the cytoplasm (this route may extend to become transcellular); 7, from the Golgi apparatus (ga) to the postsynaptic membrane, with the possible contribution of the shell to a postsynaptic thickening (pt); 8, from the Golgi apparatus to the plasmalemma with the resulting discharge of the granular contents from the coated granule (cg); 9, from the Golgi apparatus, or smooth endoplasmic reticulum (ser) of GERL type to the plasmalemma; 10, across the cell; 11, from the plasmalemma through the tubular vesicles (tv) to the plasma membrane; 12, 13, from the Golgi apparatus and/or GERL to the multivesicular bodies and/or heterophagosomes; 14, from the Golgi apparatus or GERL to the ordinary (uncoated) secretory granules (ug) destined for exocytosis; 15, from procollagen- or mucopolysaccharide-containing secretory granules (pg) to multivesicular bodies (mvb); 16, from the Golgi apparatus to the rough endoplasmic reticulum (rer).

Table 5. *Some tracer-evidenced examples of endocytic coated vesicles*

Cell type	Species	Tracer	Target body (bodies) with which coated vesicles fuse	Reference
Perikarya of spinal ganglion neuron	Toad	Ferritin	Multivesicular body (P1)[a]	Rosenbluth & Wissig (1964)
Oocyte	Moth, *Hyalophora cecropia*	Ferritin	Protein droplets (P2)[a]	Stay (1965)
Follicular cells of thyroid gland	Rat	Ferritin	Colloid droplets (P2)[a]	Seljelid (1967)
Cells of fat body	Insect, *Calpodes ethlius*	HRP[b]	Protein (P2)[a]	Locke & Collins (1968)
Schwann cells surrounding giant axons	Lobster	HRP[b]	Destination not indicated, probably pathway 6 (P6)[a]	Holtzman et al. (1970)
Axons and terminals of posterior pituitary	Rat	HRP[b]	Multivesicular bodies[c] (P4, P5)[a]	Theodosis et al. (1976)

[a] Corresponds to pathway (P) indicated in Fig. 1.
[b] Horseradish peroxidase.
[c] In many other examples of nerve and neruosecretory terminals the endocytic coated vesicles shed their lattices to give rise to plain synaptic microvesicles.

Fedoseyev & Gerasin, 1979). Furthermore, there is some confusion concerning the type of endocytic vesicles employed by the macrophage, a cell renowned for its endocytic capacity (Gordon, 1973). When cultured in media containing heterologous amyloid (Shirahama & Cohen, 1970) or ferritin (Nagura & Asai, 1976), the macrophage does endocytose the marker substances using coated vesicles; in analogous HRP experiments the tracer is internalised by smooth vesicles and vacuoles (Steinman, Brodie & Cohn, 1976). This discrepancy can be accounted for, at least in part, by the difference in the tracers used, if one assumes that the chemical composition of a substance destined for endocytosis can, in the macrophage, influence the type of endocytic vesicle formed. In cell types where endocytosis is carried out simultaneously by both coated vesicles and smooth microvesicles the separate contribution of each kind of vesicular transport can vary greatly, from a situation in which coated vesicles are much the commonest, as in endothelial cells of myeloid sinusoids (De Bruyn,

Michelson & Becker, 1975), to one in which they are rare compared with smooth microvesicles, as, for instance, in axons and neurosecretory terminals of posterior pituitary (Theodosis, Dreifus, Harris & Orci, 1976). Approximately equal numbers of each type of endocytic vesicle are observed in the phagocytic pericytes of rat brain vasa; in the endothelial cells of the same blood vessels smooth microvesicles prevail (Van Deurs, 1976). There is apparently therefore no distinct relationship between cellular capacity for endocytosis as such and the degree of involvement of coated vesicles in the process.

Exocytosis

Some indirect evidence indicates that coated vesicles can transfer the products of intracellular metabolism for export from the cell by exocytosis. Thus, in rat hepatocytes coated vesicles laden with very-low-density lipoprotein (VLDL) can be found fusing with the plasma membrane facing the space of Disse (pathway 9 in Fig. 1) (Ehrenreich, Bergeron, Siekevitz & Palade, 1973). An uncertainty remains as to whether the characteristic shell of coated vesicles is acquired at the Golgi apparatus – which is known to participate in the packaging of VLDL (Wisse, 1970; Farquhar, Bergeron & Palade, 1974; Alexander, Hamilton & Havel, 1976) – or at some other point along the route the VLDL takes from the Golgi apparatus to the plasmalemma. This qualification applies equally to the case of the giant casein-granule-containing coated vesicles in the epithelial cells of rat mammary gland: the release of the casein granules from these coated vesicles into the alveolar lumina has been demonstrated, but their formation has not (Franke *et al.*, 1976). On the other hand in Type 1 cells of rabbit or cat carotid body (Biscoe & Stehbens, 1966), exocrine pancreatic cells of guinea pig (Jamieson & Palade, 1971), or in granule-containing secretory-motor interneurons of *Hydra littoralis* (Westfall, 1973), the involvement of the Golgi apparatus in the formation of granule-containing coated vesicles is evident. In the perikarya of the neurons of embryonic rat superior cervical ganglion observed at different stages of in-vitro synaptogenesis, coated vesicles have been found, both lying free in the cytoplasm between the forming postsynaptic membrane and the Golgi apparatus, and in membranous continuity with the Golgi or postsynaptic membranes; the whole picture implies a centrifugal movement of the coated vesicles characteristic of secretory processes and is illustrated in Fig. 1 as pathway 7 (Rees, Bunge & Bunge, 1976). In these and similar investigations, tracer experiments have apparently not been performed to rule out endocytosis. Fortunately such an alternative can be rejected without additional experimentation in the case of mammotrophs of rat anterior pituitary where a natural tracer, the secretory granule, indicates the only possible pathway

(pathway 8 in Fig. 1) for the granule-containing coated vesicles: from the maternal Golgi cisterna (Plate 1*d*), through the ground cytoplasm (Plate 1*e*) to the plasmalemma, with the final exocytic release of the granule into the extracellular compartment of the gland (Plate 1*f*) (Nevorotin, 1977).

Transcellular routes

In 1968 Palade & Bruns suggested that in guinea pig blood and lymphatic capillaries both coated vesicles and smooth microvesicles can transfer some plasma constituents through the endothelial wall, with endocytosis being involved at the initial stage of the route and exocytosis operating at the end (pathway 10 in Fig. 1). Transcytoplasmic passage of coated vesicles between opposite sides of a cell is also a probable route in the endothelial cells of rat brain vasa (Van Deurs, 1976), in the myeloid sinusoidal lining cells (De Bruyn *et al.*, 1975), in the bodies of ependymal end processes of glial cells of rat brain (Brightman, 1965), and in the Schwann cells surrounding the axons of the lobster (Holtzman *et al.*, 1970), in other words, in situations characterised by an intense vesicular transport through cells with narrow cytoplasm. In the light of recent data (Johansson, 1976) on the occurrence of bidirectional vesicular transport in the wall of skeletal muscle capillaries one can suggest that the transcellular transfer of some substances by coated vesicles may proceed both from the vascular lumen to the perivascular space, and vice versa. In the columnar cells of the epithelium of neonatal rat small intestine, ferritin-labelled antibodies to the immunoglobulins of maternal milk are engulfed by smooth pinocytic microvesicles which are formed at the apical plasmalemma; these vesicles then elongate in the apical cytoplasm of the columnar cells to become so-called tubular vesicles which, in turn, give rise to the typical coated vesicles. These travel in a lateral or basal direction and subsequently release their tracer into the intercellular space (pathway 11 in Fig. 1) (Rodewald, 1973).

Intracellular routes with no plasma membrane contact

There are no as yet completely reliable methods for tracing the passage of a coated vesicle unless it comes into contact with the plasma membrane at some point. Presumably, however, a coated vesicle may tentatively be regarded as a vehicle between two intracellular organelles provided that: (1) all supposed participants in this kind of exchange, i.e. the coated vesicle and the two organelles, have a chemically identical internal composition, so that the same substance can be detected inside the three; (2) it has been established that one of these organelles is, from a biochemical point of view, likely to transfer that substance to the other; and (3) each of the organelles can be found in membranous continuity or close proximity with such coated vesicles. All these conditions appear to be met in the epithelial cells of rat vas deferens where acid phosphatase(AcPase)-positive coated vesicles have been found in membranous continuity with Golgi cisternae and multivesicular bodies. Both kinds of organelle also stain positively for the enzyme (Friend & Farquhar, 1967). The functional interrelationships expected between the structures involved has a strong theoretical basis in the lysosomal concept (De Duve & Wattiaux, 1966; Smith & Farquhar, 1966; Gahan, 1967; De Duve, 1969; Bennet & Leblond, 1971). In addition to pathway 12 (Fig. 1), i.e. from the Golgi apparatus to multivesicular bodies, coated vesicles can also transfer lysosomal hydrolases from the Golgi apparatus and/or Golgi endoplasmic reticulum–lysosomal system (GERL) to heterophagic vacuoles (pathway 13 in Fig. 1) (Friend *et al.*, 1969), or to the autophagic vacuoles, secondary lysosomes and, probably, lipofuscin bodies (Brunk & Ericsson, 1972). The suggested pathways of coated vesicles from some immature secretory granules to multivesicular bodies (pathway 15 in Fig. 1) (Weinstock & Leblond, 1971, 1974), or from the Golgi apparatus to the rough endoplasmic reticulum (pathway 16 in Fig. 1) (Martin & Spicer, 1973) are not substantiated sufficiently to rule out alternative interpretations.

Function: structural and metabolic effects

The data in the previous section indicate that coated vesicles may travel between definite loci within the cell, transferring substances in a way practically indistinguishable from smooth-walled vesicular transport. To demonstrate some functional implications of this process, its structural and metabolic consequences will be considered with special reference to the situations in which coated vesicles operate parallel to their smooth counterparts.

A contribution of the polygonal lattice

It has been suggested that in nerve endings of the CNS the shell lost by endocytic coated vesicles (pathway 4, Fig. 1) can be utilised as building material for the dense projections of the presynaptic bag (Gray & Willis, 1970); in the postsynaptic pole of the developing axosomatic synapse the shell of exocytic coated vesicles probably contributes to the postsynaptic thickening (Rees *et al.*, 1976).

Membrane turnover

As with smooth microvesicles and other membrane-bounded bodies which are capable of fusing with, or pinching off from, membranes in various situations, coated vesicles also participate in membrane turnover by internalising membrane where they form and contributing it to the membrane pool of a target organelle. However, the proportion of membrane turnover owing to coated vesicles rather than to other participants would seem to depend on their relative number. It is evident that coated vesicles in the endothelial cells of myeloid sinusoids (De Bruyn et al., 1975) or oocytes (Roth & Porter, 1964; Droller & Roth, 1966) could play an appreciable role in transferring plasma membrane inward since both these types of cell are abundantly furnished with coated pits presumably ready to give rise to plentiful coated vesicles in response to the appropriate stimuli. On the other hand, the contribution of coated vesicles to membrane turnover in endothelial cells of brain vasa seems likely to be negligible because of their low incidence compared with that of smooth microvesicles in this cell type (Van Deurs, 1976).

Catabolism

It has been shown that coated vesicles in many cell types can participate in the events relating to the ingestion of diverse foreign substances and the lysosomal degradation of both foreign (endocytosed) and intracellular (segregated) material within some specialised compartments of the cell. The following manifestations of the activity of the cellular lysosomal apparatus are worth considering in relation to coated vesicles.

Heterophagy. The involvement of coated vesicles in transport of different substances from the extracellular space to heterophagosomes and multivesicular bodies functioning as heterophagosomes has been considered in the previous section (see Endocytosis, p. 31). A few additional remarks dealing more directly with degradation than with endocytosis are appropriate here. Thus, there does not seem to be any indication that coated and uncoated vesicles might in any way influence the character of lysosomal degradation of the internalised substance within heterophagosomes. The onset of this process can take place immediately after the beginning of endocytosis by coated vesicles in the case of kidney proximal tubule cells (Christensen, 1975; Ottosen, Bode, Madsen & Maunsbach, 1976), or be delayed by evidently extracellular stimuli in the case of fat body cells of insect (Locke & Collins, 1968). Responses of the lysosomal apparatus to the ingestion of extracellular material have been shown to depend on the chemical composition of the material, availability of lysosomal enzymes and the environmental situation during the lysosomal degradation (Gordon, 1973; Mego, 1973; Madsen et al., 1976; Madsen & Christensen, 1976). The rate of the transport of the endocytosed material from outside the cell to heterophagosomes can be high, independent of whether the vesicle is coated, as in proximal tubule cells

(Christensen, 1975), or not, as in the case of HRP-treated macrophages, where endocytosis is carried out by smooth vesicles and vacuoles (Steinman *et al.*, 1976).

Secretion of lysosomal enzymes. In the epithelial cells of rat vas deferens some AcPase-positive coated vesicles may fuse with the plasmalemma thereby discharging the enzyme from the cell (Friend & Farquhar, 1967). Some coated vesicles positive for AcPase are found lying near the plasma membrane of hepatocytes (Plate 2*b*). This admittedly indirect evidence for the exocytic transport of the hydrolase into the bile canalicular lumen was however not supported, because no direct continuity between these coated vesicles and plasmalemma was seen.

Intracellular transfer of lysosomal enzymes. Palay (1963) was the first to suggest that coated vesicles in the neuron might have relevance to lysosomes; in rat hepatocytes a special class of coated vesicles moving from the Golgi apparatus to multivesicular bodies was distinguishable from those probably engaged in endocytosis (Bruni & Porter, 1965). These two classes, one analogous to heterophagosomes and the other actually primary lysosomes, were further characterised by the application of electron microscopic cytochemistry to rat vas deferens (Friend, 1966; Friend & Farquhar, 1967). Examples, other than those described above, of AcPase-positive coated vesicles acting in the capacity of primary lysosomes are numerous (Novikoff *et al.*, 1968; Friend *et al.*, 1969; Maul & Romsdahl, 1970; Novikoff, Novikoff, Quintana & Hauw, 1971; Decker, 1974). In our laboratory AcPase-positive coated vesicles have often been observed in the different types of rat brain neurons; epithelial cells of the mammary gland of virgin, lactating, and litter-weaned rats; in rat thymus macrophages and reticular cells; rat kidney proximal tubule cells; rat hepatocytes; different granule-containing cell types of rat adenohypophysis; in human ciliated and goblet cells of bronchial mucosa; and in human columnar and enteroendocrine cells of large intestine epithelium — in fact in all the cells where we have successfully performed the histochemical reaction for demonstrating acid phosphatase. A few representative micrographs demonstrate these AcPase-positive coated vesicles in different cell types (Plate 2*a*–*g*). At this juncture the following general points should be made. In each cell type observed, coated vesicles were found apparently pinching off from Golgi cisternae, Golgi-related smooth endoplasmic reticulum (SER) of GERL (Novikoff *et al.*, 1971), or the innermost Golgi lamella (Hand, 1971). On rare occasions, some cisternae of AcPase-positive SER located outside the Golgi region displayed patches of lattice coating suggestive of the initial steps in the formation of coated vesicles (Plate 2f). Either the tightly curved tips of the Golgi and SER cisterna, or the flat surfaces, may bulge outward and give rise to AcPase-positive coated vesicles (Plate 2a). Whilst the nascent secretory granule is still connected to the AcPase-filled Golgi cisterna its membrane may contribute to the formation of coated vesicles of the primary lysosomes (Plate 2g). When examined at relatively high magnifications, and using photoprocesses for contrast enhancement (Krug & Weide, 1972), AcPase-positive coated vesicles typically show a heavy lead phosphate precipitate unevenly distributed in the matrix and along the membrane (see Plate 2e and f). It is interesting to compare this with AcPase-positive smooth vesicles of the same size range (60–100 nm), which exhibit a pattern of relatively weak enzyme reaction and a preferentially marginal distribution of the precipitate (cf. Plate 2e and h). Unfortunately, the Gomori-based reactions with lead as a capture ion seem to produce a certain degree of diffusion of the enzymatically produced product (Cornelisse & Van Duijn, 1973; Van Duijn, 1973). We can therefore reach no more definite conclusion than to suggest tentatively that the AcPase-positive coated vesicles probably represent a somewhat separate population of primary lysosomes with still unknown functional differences from the primary lysosomes with smooth membranes. In common with other workers we have not been able to reveal any convincing evidence of AcPase-positive coated vesicles actually fusing with the target hetero- or autophagosomes, although indirect indications of the process, such as coated vesicles 'crowding around' the target, or the occurrence of characteristic structure on its surface (Plate 2i) were not unusual in the cell types examined. One notable fact is that AcPase-positive coated vesicles are generally in the minority compared with those lacking reaction product. To what extent this reflects the functional heterogeneity of coated vesicles, specifically their ability to transfer some other lysosomal hydrolases, remains obscure. At any rate, hydrolases other than acid phosphatase have been demonstrated in some coated vesicles (Decker, 1974).

Crinophagy. In the follicular cells of rat thyroid coated vesicles appear to endocytose the contents of the follicular lumen; a process which probably results in the digestion of the protein moiety of thyroglobulin by lysosomal hydrolases within the cell (Seljelid, 1967; De Duve, 1969). An analogous mechanism is supposed to occur in the epithelial cells of rat mammary gland, where the casein granules previously released into alveolar lumina are apparently transferred from it to multivesicular bodies within coated vesicles (Khokhlov, 1977). Recently these data have been supported by tracer experiments which have demonstrated the pathway using HRP *in vitro*

(S.E. Khokhlov, personal communication). In both these cases the endocytic uptake of the extracellular secretory products in a given number of coated vesicles was accompanied by the formation of many more large smooth endocytic vacuoles when the process was stimulated (by thyrotrophic hormone in the case of the thyroid, and the weaning of the young from the dam in the case of the mammary gland). Unlike the situation in the thyroid and mammary glands, in which the target for the coated-vesicle-mediated crinophagy may be the product previously secreted from the cytoplasm, the process of lysosomal degradation in the rat salivary gland probably starts before granular release, utilising the Golgi-derived AcPase-positive coated vesicles which fuse with some newly formed or previously stored secretory granules (Hand, 1971, 1972). The transfer of lysosomal hydrolases to the secretory granules in the pituitary mammotrophs and somatotrophs seems to be carried out by some different mechanism (Smith & Farquhar, 1966; Farquhar, 1969, 1971) (see also Plate 2*g* of this chapter). Therefore, as far as the involvement of coated vesicles in crinophagy is concerned, no generalised conclusion fitting to all secretory cell types can be derived from the facts available. However these data strongly indicate the importance of coated vesicles in a wide variety of intracellular catabolic processes.

Secretion

Although a detailed analysis of coated vesicles as carriers of secretory products is presented in a later chapter a few remarks have already been made on this subject in passing. Two other points will be mentioned here by way of introduction. First, in rat pituitary mammatrophs the formation, transportation and exocytosis of secretory granules involving participation of coated vesicles appears to be only an accessory method of mammatrophic hormone secretion, because the majority of the secretory granules are uncoated. Most of these coated vesicles lack large areas of surface lattice structure (Plate 1*d* and *e*). In this context it is worth noting a rare occurrence: that granule-containing coated vesicles are found in other hormone-secreting cells of rat adeno-hypophysis. Secondly, coated vesicles may sometimes serve as auxiliary devices which facilitate secretion by transferring the additional portions of Golgi-originated granular material to ordinary (uncoated) granules destined for exocytosis (pathway 14 in Fig. 1). Probably, this is the situation in the epithelial cells of rat ventral prostate lobe (Helminen & Ericsson, 1970), and in the epithelial cells of mouse oral epithelium (Weinstock & Wilgram, 1970).

Some other capacities of coated vesicles

Mucopolysaccharide turnover. It has been suggested that in chondrocytes procollagen peptidase is transferred within coated vesicles from the extracellular space to Golgi saccules and vacuoles filled with the synthesised procollagen molecules (pathway 3 in Fig. 1), which leads to intracellular collagen maturation owing to cleavage of the terminal bonds of procollagen by the hydrolase (Dearden, 1975). In odontoblasts of developing rat teeth the maturation of collagen is probably facilitated by coated vesicles which remove a certain portion of the membrane delimiting the procollagen-containing secretory granules (Weinstock & Leblond, 1974). Analogous structural relationships may occur in the acid glycoprotein-producing ameloblasts (Weinstock & Leblond, 1971). In both these cases the coated vesicles are presumed to fuse with the multivesicular or dense bodies (pathway 15 in Fig. 1). In rat megakaryocytes Golgi-derived coated vesicles filled with acid mucopolysaccharides fuse with demarcation membranes which may promote the formation of the external coat of preformed blood platelets (Behnke, 1968).

Melanogenesis. Coated vesicles of some melanin-producing cells apparently transfer tyrosinase from the Golgi apparatus and/or GERL to premelanosomes. This may lead to their subsequent maturation and melanosome formation (Novikoff, Albala & Biempica, 1968; Maul & Romsdahl, 1970; Maul & Brumbaugh, 1971). In fish pigment cells melanosome maturation appears to proceed in a quite different way without the participation of coated vesicles (Turner, Taylor & Tchen, 1975).

Intracellular synthesis. Coated vesicles isolated from fowl oocytes show a vigorous incorporation of some radioactive precursors of protein, carbohydrate and lipid metabolism (Schjeide & San Lin, 1967; Schjeide et al., 1969). Whether this is due to the shell, or the contents of the coated vesicle remains uncertain. Major questions concerning the contribution of coated vesicles to vitellogenesis, nerve transmission and some other cellular events are not considered here. They will be introduced in the appropriate later chapters.

Conclusion
The data presented here clearly show that coated vesicles are extremely widespread, if not ubiquitous, structures, which appear to be as common in different animal cell types as mitochondria, endoplasmic reticulum or ribosomes. Further analysis indicated that coated vesicles, as do smooth microvesicles, operate within the cytoplasm, transferring substances from one point to another. Not infrequently they use the same intracellular pathways, and have apparently identical biological effects to those of smooth microvesicles.

However a distinct selectivity seemingly related to the chemical composition of the transmitted substance may be observed, as, for example, in the macrophage. In a few cases, however, functionally similar in a general way, a degree of difference between the two kinds of vesicular carriers was experimentally demonstrable either when probes were applied, as in the investigation of the thyrocyte (Seljelid, 1967), and the macrophage cultured in the ferritin-containing medium (Nagura & Asai, 1976), or when more specific methods of processing were used, as in case of the AcPase-positive coated vesicles considered in this paper.

If one assumes that the chemical composition of the membranous wall of coated vesicles is identical to that of smooth microvesicles, then it is the shell that must somehow provide specificity for the coated vesicles with respect to the vesicular contents. Rosenbluth & Wissig (1964) were the first to propose that the surface structure of the coated vesicle might serve as a mechanical device facilitating a local bending of the plasma membrane in the area of formation of the endocytic vesicle. Kanaseki & Kadota (1969) developed the idea on more detailed structural and mechanical premises. In the absence of any ultrastructurally visible connection between the polygonal lattice and the transmitted substance, evidence that the substrate determines the type of vesicle formed is weak. We therefore propose a physical rather than chemically important distinction between coated and uncoated vesicle formation.

Thus if a membrane at the area of formation of an endocytic, or Golgi-derived vesicle is easily 'flexible', then a smooth microvesicle or a secretory granule might be formed by some unspecified but genetically predetermined mechanism. In cases where a higher mechanical resistance of a membranous locus is met with, the mechanical forces which secure the bending of the membrane at the area of formation of a vesicle or a granule might become insufficient to accomplish the process. In this case a polygonal lattice supposedly lying free in the vicinity of the membrane (Kanaseki & Kadota, 1969) might somehow be directed to the spot, and mechanically facilitate the process of vesiculisation. What might influence the proposed differences in the membranous 'flexibility'? Besides probable local differences in the membrane proper, the contribution to the net 'flexibility' of the complex of the substance which is to be segregated should be taken into account. To date, only a few isolated findings lend credence to the suggestion that membrane flexibility is dependent on some physical parameters of the bound substance. It is possible that active areas of bone resorption in the osteoclast are a case in point. Here the material to be endocytosed is bulky and rigid, and the membrane is populated with coated vesicles laden with the material; the major part of the cell surface of this whole area is decorated

with polygonal lattice structure (Kallio, Garant & Minkin, 1971) as if the
membrane were prepared for more efficient formation of endocytic vesicles.
A somewhat analogous situation may be observed in the rat mammotroph of
anterior pituitary, where the secretory granules are of anomalously high
density (Giannattasio, Zanini & Meldolesi, 1975). Since not all granules of
the mammotroph possess the lattice structure it might be speculated that
granules differ in density or that lattices may be lost after vesicle formation.

In the macrophage cultured with ferritin (Nagura & Asai, 1976) the stimuli
governing the formation of coated or smooth endocytic vesicles appear to
depend primarily on some physical conditions in the environment and energy
metabolism.

When both coated and smooth vesicles are competent to handle the same
substance within a single cell, the local deviations in the concentrations of
the substance at sites of future endocytic vesicle formation may be interpreted
as the probable causes of the differing membranous 'flexibility', which on this
hypothesis triggers the appropriate program guiding the formation of either
coated or smooth microvesicles. If this is the case, then, for such a cell,
experiments in which the concentrations of endocytosis-inducing substances
are varied should result in changes of the proportionate contribution made to
endocytosis by each vesicle population. If the variation is straightforward this
may correlate directly with the number of each type of vesicle present.
Experiments of this type may well be of use in the elucidation of the precise
significance behind the presence of a population of both smooth and coated
microvesicles in many cells.

I am indebted to S.E. Khokhlov for valuable suggestions and technical
assistance, to V.K. Vashkinel and A.M. Dobykin for kindly presenting some
preparations and for advice, to Nathali L. Orshanskaya for the illustration,
and to Vera N. Shapiro for her encouragement.

References

Alexander, C.A., Hamilton, R.L. & Havel, R.J. (1976). Subcellular localisation
of β apoprotein of plasma lipoproteins in rat liver. *Journal of Cell
Biology,* **69**, 241–63.
Anderson, E. (1964). Oocyte differentiation and vitellogenesis in the roach
Periplaneta americana. Journal of Cell Biology, **20**, 131–56.
Andrews, K.H. (1975). Morphological criteria for the differentiation of
synapses in vertebrates. *Journal of Neural Transmission, Suppl.* **12**,
1–37.
Bankston, P.W. & De Bruyn, P.P.H. (1974). The permeability to carbon of the
sinusoidal lining cells of the embryonic rat liver and rat bone marrow.
American Journal of Anatomy, **141**, 281–90.

Bargmann, W. & Welsch, U. (1969). On the ultrastructure of the mammary gland. In *Lactogenesis: the Initiation of Milk Secretion at Parturition*, ed. M. Reynolds & S.J. Folley, pp. 43–52. Philadelphia: University of Pennsylvania Press.

Becker, M.D., Hirano, A. & Zimmerman, M.D. (1968). Observation of the distribution of exogenous peroxidase in the rat cerebrum. *Journal of Neuropathology and Experimental Neurology*, 27, 439–52.

Behnke, O. (1968). An electron microscope study of the megacaryocyte of the rat bone marrow. I. The development of the demarcation membrane system and the platelet surface coat. *Journal of Ultrastructure Research*, 24, 412–33.

Bennet, G. (1970). Migration of glycoprotein from Golgi apparatus to cell coat in the columnar cells of the duodenal epithelium. *Journal of Cell Biology*, 45, 668–73.

Bennet, G. & Leblond, C.P. (1971). Passage of fucose-^{3}H label from the Golgi apparatus into dense and multivesicular bodies in the duodenal columnar cells and hepatocytes of the rat. *Journal of Cell Biology*, 51, 875–80.

Bessis, M. & Breton Gorius, J. (1957). Iron particles in normal erythroblasts and normal and pathological erythrocytes. *Journal of Biophysical and Biochemical Cytology*, 3, 503–5.

Biberfeld, P. (1971). Endocytosis and lysosome formation in blood lymphocytes transformed by phytohemagglutinin. *Journal of Ultrastructure Research*, 37, 41–68.

Biscoe, T.J. & Stehbens, W.E. (1966). Ultrastructure of the carotid body. *Journal of Cell Biology*, 30, 563–78.

Blitz, A.L. & Fine, R.E. (1976). Functional and compositional similarities between coated vesicles, synaptic vesicles and sarcoplasmic reticulum fragments. *Journal of Cell Biology*, 70, 204A.

Bowers, B. (1964). Coated vesicles in the pericardial cells of the aphid *Mysus persicae* Sulz. *Protoplasma*, 59, 351–67.

Brightman, M.W. (1965). The distribution within the brain of ferritin injected into cerebrospinal fluid compartments. II. Parenchymal distribution. *American Journal of Anatomy*, 117, 193–220.

Bruni, C. & Porter, K.R. (1965). The fine structure of the parenchymal cell of the normal rat liver. General observations. *American Journal of Pathology*, 46, 691–756.

Brunk, U. & Ericsson, J.L.E. (1972). Electron microscopical studies on rat brain neurons. Localisation of acid phosphatase and mode of formation of lipofuscin bodies. *Journal of Ultrastructure Research*, 38, 1–15.

Bruns, R.R. & Palade, G.E. (1968). Studies on blood capillaries. II. Transport of ferritin molecules across the wall of muscle capillaries. *Journal of Cell Biology*, 37, 277–99.

Bunt, A.H. (1969). Formation of coated and 'synaptic' vesicles within neurosecretory axon terminals of the crustacean sinus gland. *Journal of Ultrastructure Research*, 28, 411–21.

Chalazonitis, N. (1969). Differentiation of membranes in axonal endings in the neuropil of *Helix*. In *Cellular Dynamics of the Neuron*, ed. S.H. Barondes, *Symposium of the International Society for Cell Biology*, pp. 229–43. New York & London: Academic Press.

Christensen, E.I. (1975). Rapid initiation of lysosomal digestion of low molecular weight protein in rat kidney slices. *Journal of Ultrastructure Research*, 50, 373.

Cornelisse, C.J. & Van Duijn, P. (1973). A new method for the investigation of the kinetics of the capture reaction in phosphatase cytochemistry. I. Theoretical aspects of the local formation of the crystalline precipitates. *Journal of Histochemistry and Cytochemistry*, 21, 607–13.

Davies, J. & Hoffman, L.H. (1975). Studies on the progestational endometrium of the rabbit. II. Electron microscopy, day 0 to day 13 of gonadotrophin induced pseudo-pregnancy. *American Journal of Anatomy*, 142, 335–66.

Dearden, L.C. (1975). Periodic fibrillar material in intracellular vesicles and in electron-dense bodies in chondrocytes of rat costal and tracheal cartilage at various ages. *American Journal of Anatomy*, 144, 323–33.

De Bruyn, P.P.H. & Cho, J. (1974). Contractile structures in endothelial cells of splenic sinusoids. *Journal of Ultrastructure Research*, 49, 24–33.

De Bruyn, P.P.H., Michelson, S. & Becker, R.P. (1975). Endocytosis, transfer tubules and lysosomal activity in myeloid sinusoidal endothelium. *Journal of Ultrastructure Research*, 53, 133–51.

Decker, R.S. (1974). Lysosomal packaging in differentiating and degenerating anuran lateral motor column neurons. *Journal of Cell Biology*, 61, 599–612.

De Duve, C. (1969). The lysosome in retrospect. In *Lysosomes in Biology and Pathology*, vol. 1, ed. J.T. Dingle & H.B. Fell, pp. 3–40. Amsterdam: North-Holland.

De Duve, C. & Wattiaux, R. (1966). Function of lysosomes. *Annual Review of Physiology*, 28, 435–93.

Douglas, W.W., Nagasawa, J. & Schulz, R.A. (1971). Coated vesicles in neurosecretory terminals of posterior pituitary glands shed their coats to become smooth 'synaptic' vesicles. *Nature, London*, 232, 340–1.

Droller, M.J. & Roth, T.F. (1966). An electron microscope study of yolk formation during oogenesis in *Lebistes reticulatus* Guppyi. *Journal of Cell Biology*, 28, 209–32.

Ehrenreich, J.H., Bergeron, J.J.M., Siekevitz, P. & Palade, G.E. (1973). Golgi fractions prepared from rat liver homogenates. I. Isolation procedure and morphological characterization. *Journal of Cell Biology*, 59, 45–72.

Farquhar, M.G. (1969). Lysosome function in regulating secretion: disposal of secretory granules in cells of the anterior pituitary gland. In *Lysosomes in Biology and Pathology*, vol. 2, ed. J.T. Dingle & H.B. Fell, pp. 463–82. Amsterdam: North-Holland.

Farquhar, M.G. (1971). Processing of secretory products by cells of the anterior pituitary gland. In *Subcellular Organisation and Function in Endocrine Tissues*, ed. H. Heller & K. Lederis, part I, pp. 79–124. Cambridge University Press.

Farquhar, M.G., Bergeron, J.J.M. & Palade, G.E. (1974). Cytochemistry of Golgi fractions prepared from rat liver. *Journal of Cell Biology*, **60**, 8–25.

Fawcett, D.W. (1965). Surface specialisations of absorbing cells. *Journal of Histochemistry and Cytochemistry*, **13**, 75–91.

Forbes, M.S. & Sperelakis, N. (1974). Spheroidal bodies in the junctional sarcoplasmic reticulum of lizard myocardial cells. *Journal of Cell Biology*, **60**, 602–15.

Franke, W.W., Lüder, M.R., Kartenbeck, J., Zerban, H. & Keenan, T.W. (1976). Involvement of vesicle coat material in casein secretion and surface regeneration. *Journal of Cell Biology*, **69**, 173–95.

Friend, D.S. (1966). Peroxidase absorption in the vas deferens of the rat. *Journal of Cell Biology*, **31**, 37A.

Friend, D.S. & Farquhar, M.G. (1967). Functions of coated vesicles during protein absorption in the rat vas deferens. *Journal of Cell Biology*, **35**, 357–76.

Friend, D.S., Rosenau, W., Winfield, J.C. & Moon, H.B. (1969). Uptake and degradation of T_2 bacteriophage by rat peritoneal macrophage. *Laboratory Investigation*, **20**, 275–82.

Gahan, P.G. (1967). Histochemistry of lysosomes. *International Review of Cytology*, **21**, 1–63.

Giannattasio, G., Zanini, A. & Meldolesi, J. (1975). Molecular organisation of rat prolactin granules. I. In vitro stability of intact and 'membrane-less' granules. *Journal of Cell Biology*, **64**, 246–51.

Gordon, A.H. (1973). The role of lysosomes in protein catabolism. In *Lyso-somes in Biology and Pathology*, vol. 3, ed. J.T. Dingle, pp. 89–137. Amsterdam: North-Holland.

Gray, E.G. (1961). The granule cells, mossy synapses and Purkinje spine synapses of the cerebellum: light and electron microscope observa-tions. *Journal of Anatomy*, **95**, 345–56.

Gray, E.G. & Pearse, H.L. (1971). On understanding the organisation of the retinal receptor synapses. *Brain Research*, **35**, 1–15.

Gray, E.G. & Willis, R.A. (1970). On synaptic vesicles, complex vesicles and dense projections. *Brain Research*, **24**, 149–68.

Griffith, L.D., Bulger, R.E. & Trump, B.F. (1968). Fine structure and staining of mucosubstances on 'intercalated cells' from the rat distal con-voluted tubule and collecting duct. *Anatomical Record*, **160**, 643–62.

Gulgas, B.J. (1974). The corpus luteum of the rhesus monkey (*Macaca mulatta*) during late pregnancy. An electron microscopic study. *American Journal of Anatomy*, **139**, 95–122.

Hand, A.R. (1971). Morphology and cytochemistry of the Golgi apparatus of rat salivary gland acinar cells. *American Journal of Anatomy*, **130**, 141–58.

Hand, A.R. (1972). The effects of acute starvation on parotid acinar cells.

Ultrastructural and cytochemical observations on ad libitum-fed and starved rats. *American Journal of Anatomy*, 135, 71–92.

Helminen, H.J. & Ericsson, J.L.E. (1970). On the mechanism of lysosomal enzyme secretion. Electron microscopic and histochemical studies on the epithelial cells of the rat's ventral prostate lobe. *Journal of Ultrastructure Research*, 33, 528–49.

Hicks, R.M. (1966). The function of the Golgi complex in transitional epithelium. Synthesis of the thick cell membranes. *Journal of Cell Biology*, 30, 623–43.

Holtzman, E. (1969). Lysosomes in the physiology and pathology of neurons. In *Lysosomes in Biology and Pathology*, vol. 1, ed. J.T. Dingle & H.B. Fell, pp. 192–216. Amsterdam: North-Holland.

Holtzman, E., Freeman, A.R. & Kashner, L.A. (1970). A cytochemical and electron microscope study of channels in the Schwann cells surrounding lobster giant axons. *Journal of Cell Biology*, 44, 438–45.

Jamieson, J.D. & Palade, G.E. (1971). Synthesis, intracellular transport, and discharge of secretory proteins in stimulated pancreatic exocrine cells. *Journal of Cell Biology*, 50, 135–58.

Johansson, B.R. (1976). Blood capillary uptake of macromolecules. *Journal of Ultrastructure Research*, 57, 227–8.

Kallenbach, E. (1976). Fine structure of differentiating ameloblasts in the kitten. *American Journal of Anatomy*, 145, 283–318.

Kallio, D.M., Garant, P.R. & Minkin, C. (1971). Evidence of coated membranes in the ruffled border of the osteoclast. *Journal of Ultrastructure Research*, 37, 169–77.

Kanaseki, T. & Kadota, K. (1969). The 'vesicle in a basket'. A morphological study of the coated vesicle isolated from the nerve endings of the guinea pig brain, with special reference to the mechanism of membrane movements. *Journal of Cell Biology*, 42, 202–20.

Kang, Y.H. (1974). Development of the zona pellucida in the rat oocyte. *American Journal of Anatomy*, 139, 535–66.

Karnovsky, M.A. (1965). Formaldehyde–glutaraldehyde fixative of high osmolality for use in electron microscopy. *Journal of Cell Biology*, 27, 137A–138A.

Khokhlov, S.E. (1977). Ultrastructural changes in secreting cells of the albino rat mammary gland during the estrous cycle. *Tzitologia (Russia)*, 19, 604–11.

Klinick, G.H., Oertel, J.E. & Winship, T. (1970). Ultrastructure of normal human thyroid. *Laboratory Investigations*, 22, 2–22.

Komiyama, A. & Spicer, S.S. (1975). Microendocytosis in eosinophilic leucocytes. *Journal of Cell Biology*, 64, 622–35.

Krug, W. & Weide, H.G. (1972). Wissenschaftliche Photographie in der Anwendung. *Wege zur Informationsausschöpfung Photographischer Schwarzweiss-Negative*. Leipzig: Geest & Portig.

Kuhn, C., Callaway, L.A. & Askin, F.B. (1974). The formation of granules in the bronchiolar Clara cell of the rat. I. Electron microscopy. *Journal of Ultrastructure Research*, 49, 387–400.

Lane, N.J. (1968). Distribution of phosphatases in the Golgi region and associated structures of thoracic ganglionic neurons in the grasshopper, *Melanopus differentalis. Journal of Cell Biology,* 37, 89–104.

Lester, K.S. (1970). On the nature of 'fibrils' and tubules in developing enamel of the opossum, *Didelphis marsupialis. Journal of Ultrastructure Research,* 30, 64–77.

Locke, M. & Collins, J.V. (1968). Protein uptake into multivesicular bodies and storage granules in the fat body of an insect. *Journal of Cell Biology,* 36, 453–83.

Lubersky-Moore, J.L., Poliakoff, S.J. & Worthington, W.C. (1975). Ultrastructural observations of anterior pituitary gonadotrophs following hypophysial portal vessel infusion of luteinizing hormone-releasing hormone. *American Journal of Anatomy,* 144, 549–55.

McKanna, J.A. (1973). Cyclic membrane flow in the ingestive–digestive system of peritrich protozoans. II. Cup-shaped coated vesicles. *Journal of Cell Science,* 13, 677–86.

McKenna, O.C. & Rosenbluth, J. (1973). Myoneural and intermuscular junctions in a molluscan smooth muscle. *Journal of Ultrastructure Research,* 42, 434–50.

Madsen, K., Bode, F., Ottosen, P.D., Baumann, K. & Maunsbach, A.B. (1976). Effect of basic amino acids on kidney protein uptake and structure of the proximal tubule. *Journal of Ultrastructure Research,* 57, 221–2.

Madsen, K. & Christensen, E.I. (1976). Protein digestion in kidney slices from rats with chronic mercury intoxication. *Journal of Ultrastructure Research,* 57, 221.

Martin, B.J. & Spicer, S.S. (1973). Ultrastructural features of cellular maturation and ageing in human trophoblast. *Journal of Ultrastructure Research,* 43, 133–49.

Maul, G.G. & Brumbaugh, J.A. (1971). On the possible function of coated vesicles in melanogenesis of the regenerating fowl feather. *Journal of Cell Biology,* 48, 41–8.

Maul, G.G. & Romsdahl, M.M. (1970). Ultrastructural comparison of two human malignant melanoma cell lines. *Cancer Research,* 30, 2782–90.

Maunsbach, A.B. (1966). Absorption of ferritin by rat kidney proximal tubule cells: electron microscopic observations of the initial uptake phase in cells of microperfused single proximal tubules. *Journal of Ultrastructure Research,* 16, 1–12.

Mego, J.L. (1973). Protein digestion in isolated heterolysosomes. In *Lysosomes in Biology and Pathology,* vol. 3, ed. J.T. Dingle, pp. 138–68. Amsterdam: North-Holland.

Morton, D.J. & Rowe, R.W.D. (1974). Development of ovine skeletal muscle: the occurrence of a sarcolemmal process on fetal myotubes. *Journal of Ultrastructure Research,* 47, 142–52.

Nagasawa, J., Douglas, W.W. & Schulz, R.A. (1971). Micropinocytotic origin of coated and smooth microvesicles ('synaptic vesicles') in neurosecretory terminals of posterior pituitary gland demonstrated by incorporation of horseradish peroxidase. *Nature, London*, 232, 341–2.

Nagura, H. & Asai, J. (1976). Pinocytosis by macrophages: kinetics and morphology. *Journal of Cell Biology*, 70, 89A.

Nevorotin, A.J. (1977). Coated vesicles and their functions in the cells of animals. *Tzitologia (Russia)*, 19, 5–14.

Nevorotin, A.J., Chernyakova, D.N., Fedoseyev, G.B. & Gerasin, V.A. (1979). An investigation of bronchial epithelium permeability in chronic bronchitis using an electron dense tracer. *Archives of Pathology (Russia)*, 41, 67–73.

Nevorotin, A.J. & Dobykin, A.M. (1978). Vacuolar apparatus in cellular pathology. An electron cytochemical study. In *Annual Proceedings of the Leningrad Society for Morbid Pathologists*, vol. 19, pp. 146–50. Leningrad.

Novikoff, A.B., Albala, A. & Biempica, L. (1968). Ultrastructural and cytochemical observations on B-16 and Harding–Passey mouse melanomas. The origin of premelanosomes and compound melanosomes. *Journal of Histochemistry and Cytochemistry*, 16, 299–319.

Novikoff, A.B., Novikoff, P.M., Davis, C. & Quintana, N. (1973). Studies of microperoxisomes. V. Are microperoxisomes ubiquitous in mammalian cells? *Journal of Histochemistry and Cytochemistry*, 21, 737–55.

Novikoff, P.M., Novikoff, A.B., Quintana, N. & Hauw, J.-J. (1971). Golgi apparatus, GERL, and lysosomes of neurons in rat dorsal root ganglia studied by thick section and thin section cytochemistry. *Journal of Cell Biology*, 50, 859–86.

Ockleford, C.D. (1976). A three dimensional reconstruction of the polygonal pattern on placental coated-vesicle membranes. *Journal of Cell Science*, 23, 83–91.

Ottosen, P.D., Bode, F., Madsen, K.M. & Maunsbach, A.B. (1976). Handling of low molecular weight protein by the kidney proximal tubule. *Journal of Ultrastructure Research*, 57, 220.

Palade, G.E. & Bruns, R.R. (1968). Structural modulations of plasmalemmal vesicles. *Journal of Cell Biology*, 37, 633–49.

Palay, S.F. (1963). Alveolate vesicles in Purkinje cells of the rat's cerebellum. *Journal of Cell Biology*, 19, 89A–90A.

Pearse, B.M.F. (1975). Coated vesicles from pig brain: purification and biochemical characterisation. *Journal of Molecular Biology*, 97, 93–8.

Pelletier, G. (1973). Secretion and uptake of peroxidase by rat adenohypophyseal cells. *Journal of Ultrastructure Research*, 43, 445–59.

Petrov, M.N. & Vashkinel, V.K. (1977). Peculiarities attending the morphology and the function of platelets in myeloproliferative diseases. *Problems in Hematology (Russia)*, 5, 18–23.

Plotkin, V.Ya. & Nevorotin, A.J. (1976). Protein reabsorption by the cells of the proximal tubules of the kidneys of patients with chronic diffuse glomerulonephritis depending on the degree of proteinuria. *Urologia and Nephrologia (Russia)*, 4, 19–22.

Pluzhnikov, M.S. & Nevorotin, A.J. (1974). Ultrastructure of hair cells of the organ of Corti in cat. *Vestnik of Otorhinolaringologia (Russia)*, 1, 40–50.

Porter, K.R., Kenyon, K. & Badenhausen, S. (1967). Specialisations of the unit membranes. *Protoplasma*, 63, 262–74.

Raikhel, A.S. (1974). An electron microscopic study of endocytosis in the midgut cells of the tick *Hyalomma asiatica*. *Tzitologia (Russia)*, 16, 1499–504.

Rees, R.P., Bunge, M.B. & Bunge, R.P. (1976). Morphological changes in the neuronic growth cone and target neuron during synaptic junction development in culture. *Journal of Cell Biology*, 68, 240–63.

Rodewald, R. (1973). Intestinal transport of antibodies in the newborn rat. *Journal of Cell Biology*, 58, 189–211.

Röhlich, P. & Allison, A.C. (1976). Oriented pattern of membrane-associated vesicles in fibroblasts. *Journal of Ultrastructure Research*, 57, 94–103.

Rosenbluth, J. & Wissig, S.L. (1964). The distribution of exogenous ferritin in toad spinal ganglia and the mechanism of its uptake by neuron. *Journal of Cell Biology*, 23, 307–25.

Roth, T.F. & Porter, K.R. (1962). Specialised sites on the cell surface for protein uptake. In *Electron Microscopy: Fifth International Congress for Electron Microscopy*, vol. 2, *Biology*, ed. S.S. Breese, p. 114. New York & London: Academic Press.

Roth, T.F. & Porter, K.R. (1963). Membrane differentiation for protein uptake. *Federation Proceedings*, 22, 178A.

Roth, T.F. & Porter, K.R. (1964). Yolk protein uptake in the oocyte of the mosquito *Aedes aegypti* L. *Journal of Cell Biology*, 20, 313–32.

Roth, T.F. & Porter, K.R. (1965). Yolk protein uptake in the oocyte of the mosquito *Aedes aegypti* L. In *Molecular and Cellular Aspects of Development*, ed. E. Bell, pp. 391–409. New York: Harper & Row.

Sanzone, C.F. & Reith, E.J. (1976). The development of the elastic cartilage of the mouse pinna. *American Journal of Anatomy*, 146, 31–72.

Schjeide, O.A. & San Lin, R.J. (1967). In vitro synthesis by coated vesicles (synthesomes). *Journal of Cell Biology*, 35, 121A.

Schjeide, O.A., San Lin, R.I., Grellert, E.A., Galey, F.R. & Mead, J.F. (1969). Isolation and preliminary chemical analysis of coated vesicles from chicken oocytes. *Physiological Chemistry and Physics*, 1, 141–63.

Seljelid, R. (1967). Endocytosis in the thyroid follicle cells. II. A microinjection study of the origin of colloid droplets. *Journal of Ultrastructure Research*, 17, 401–20.

Shirahama, T. & Cohen, A.S. (1970). The association of hemidesmosome-like plaque and dense coating with the pinocytic uptake of a heterologous fibrillar protein (amyloid) by macrophages. *Journal of Ultrastructure Research*, 33, 587–97.

Skelding, J.M. (1973). The fine structure of the kidney of *Achatina achatina*

(L.) *Zeitschrift für Zellforschung und Mikroskopische Anatomie*, 147, 1–29.

Slautterback, D.B. (1967). Coated vesicles in absorptive cells of *Hydra*. *Journal of Cell Science*, 2, 563–72.

Smith, R.E. & Farquhar, M.G. (1966). Lysosome function in the regulation of the secretory process in cells of the anterior pituitary gland. *Journal of Cell Biology*, 31, 319–47.

Stay, B. (1965). Protein uptake in the oocytes of the cecropia moth. *Journal of Cell Biology*, 26, 49–62.

Steinman, R.M., Brodie, S.E. & Cohn, Z.A. (1976). Membrane flow during pinocytosis. A stereologic analysis. *Journal of Cell Biology*, 68, 665–87.

Stuart, A.M. & Satir, P. (1968). Morphological and functional aspects of an insect epidermal gland. *Journal of Cell Biology*, 36, 527–49.

Theodosis, D.T., Dreifuss, J.J., Harris, M.C. & Orci, L. (1976). Secretion-related uptake of horseradish peroxidase in neurohypophysial axons. *Journal of Cell Biology*, 70, 294–303.

Turner, W.A., Taylor, J.D. & Tchen, T.T. (1975). Melanosome formation in the goldfish: the role of multivesicular bodies. *Journal of Ultrastructure Research*, 51, 16–31.

Van Deurs, B. (1976). Observations on the blood-brain barrier in hypertensive rats, with particular reference to phagocytic pericytes. *Journal of Ultrastructure Research*, 56, 65–77.

Van Duijn, P. (1973). Fundamental aspects of enzyme cytochemistry. In *Electron Microscopy and Cytochemistry*, ed. E. Wisse, W.Th. Daems, J. Molenaar & P. Van Duijn, pp. 3–23. Amsterdam: North-Holland.

Vashkinel, V.K. & Petrov, M.N. (1974). An electron microscopic study of blood platelets in acute leukemia. In *Annual Proceedings of the Leningrad Society for Morbid Pathologists*, vol. 15, pp. 7–8. Leningrad.

Weinstock, A. & Leblond, C.P. (1971). Elaboration of the matrix glycoprotein of enamel by the secretory ameloblasts of the rat incisor as revealed by radioautography after galactose-^{3}H injection. *Journal of Cell Biology*, 51, 26–51.

Weinstock, M. & Leblond, C.P. (1974). Synthesis, migration, and release of precursor collagen by odontoblasts as visualised by radioautography after ^{3}H-proline administration. *Journal of Cell Biology*, 60, 92–127.

Weinstock, M. & Wilgram, G.F. (1970). Fine structural observations on the formation and enzymatic activity of keratinosomes in mouse tongue filiform papillae. *Journal of Ultrastructure Research*, 30, 262–77.

Westfall, J.A. (1973). Ultrastructural evidence for a granule-containing sensory-motor-interneuron in *Hydra littoralis*. *Journal of Ultrastructure Research*, 42, 268–82.

White, J.G. (1967). A simple method of preservation of fine structure in blood cells. *Thrombosis et Diathesis Haemorrhagica*, 18, 745–53.

Wisse, E. (1970). An electron microscopic study of the fenestrated endothelial lining of rat liver sinusoids. *Journal of Ultrastructure Research*, 31, 125–50.

Wissig, S.L. (1962). Structural differentiation in the plasmalemma and cytoplasmic vesicles of selected epithelial cells. *Anatomical Record,* 142,
292.
Zacks, S.I. & Saito, A. (1969). Uptake of exogenous horseradish peroxidase by
coated vesicles in mouse neuromuscular junctions. *Journal of Histochemistry and Cytochemistry,* 17, 161–70.

Plates

Plate 1. Unless otherwise stated the specimens have been fixed in
Karnovsky's (1965) fixative by immersion, postosmicated, dehydrated
in graded alcohol series, and embedded in Araldite. All specimens
were stained with saturated uranyl acetate overnight in the block,
and after sectioning conventionally double-contrasted with uranium
and lead salts. (*a*) Portion of Golgi area in a rat anterior pituitary
mammotroph. A tangentially cut shell characteristic of coated
vesicles is seen near a secretory granule. The nearly equal dimensions
of these structures indicate that the coated vesicle may contain such
a granule. X 159 300. (*b*) An ordinary-sized coated vesicle and a
giant granule-containing one in a mammotroph show that their
polygonal lattices may be structurally identical. X 159 300. (*c*)
Portion of a human blood platelet from a preparation fixed using
the method of White (1967) shows a smooth microvesicle and a
typical coated vesicle between two alpha-granules. The micrograph
was kindly supplied by Valentina K. Vashkinel. X 94 000. (*d*) A
nascent granule forming in a Golgi cisterna of a mammotroph shows
a few bristles on the cytoplasmic face of the granular membrane.
Another granule also has several slightly blurred projections (arrows).
X 77 400. (*e*) A partially coated granule-containing vesicle in
mammotroph cytoplasm. A shell of either an ordinary, or a granule-
containing coated vesicle is evident nearby (arrow). X 77 400. (*f*) A
granule-containing coated vesicle of a mammotroph, apparently discharging its contents into the extracellular space. X 77 400.

Plate 2. (*a*) The preparations presented in (*a*)–(*h*) were processed for
acid phosphatase (AcPase) demonstration using a Gomori-derived
technique which has cytidine 5′-monophosphate as substrate (Friend
& Farquhar, 1967; Novikoff *et al.*, 1971). The preparation in (*i*) is
conventionally processed without any cytochemical technique. (*a*)
An AcPase-positive coated vesicle apparently pinching off from the
flat surface of a Golgi cisterna of an enteroendocrine cell from human
large intestinal epithelium. The micrograph was kindly supplied by
Alexander M. Dobykin. X 68 100. (*b*) An AcPase-positive coated
vesicle near the plasma membrane facing a bile canaliculus lumen of
a rat hepatocyte. Glutaraldehyde fixation of vascular perfusion
through the portal vein. X 68 100. (*c*) Two AcPase-positive coated
vesicles in the Golgi area of a rat thymus reticular cell. Fixation by
retrograde perfusion through abdominal aorta. X 53 000. (*d*) and (*e*)

Two AcPase-positive coated vesicles near the enzyme-filled Golgi cisterna are seen in a neuron of rat brain cortex after conventional printing of an electron micrograph negative (*d*). The same negative printed at higher magnification after being processed for contrast enhancement (Krug & Weide, 1972) shows heavy and uneven distribution of lead phosphate precipitates within the coated vesicles. Fixed by vascular perfusion. X 68 100; X 218 100. (*f*) An AcPase-filled cisterna of smooth endoplasmic reticulum located beyond the Golgi area in a rat pituitary somatotroph. The slightly dilated tip of the cisternal profile (arrow) has surface structure implying that it is the initial phase in the formation of an AcPase-positive coated vesicle. X 68 100. (*g*) A nascent secretory granule in the course of formation from the innermost AcPase-filled Golgi cisterna of a mammotroph. An AcPase-positive coated vesicle appears to pinch off from the cytoplasmic side of the granule (arrow). There is marginal distribution of the enzyme on the coated vesicle, the nascent granule, and that already budded off from the Golgi apparatus (arrowhead). X 68 100. (*h*) Golgi area in a ciliated cell from human bronchial mucosa. Several smooth, weakly AcPase-positive microvesicles are present, all of which show a pattern of marginal distribution of the enzyme. X 68 100. (*i*) A multivesicular body with two patches of lattice structure on opposite sides, present in a columnar cell of human large intestinal epithelium. The micrograph was kindly supplied by Alexander M. Dobykin. X 68 100.

A.J. Nevorotin   Plate 1

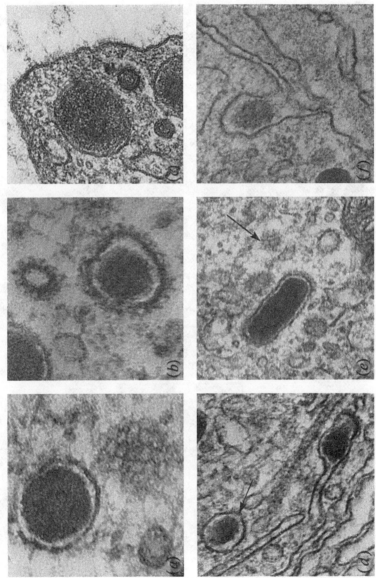

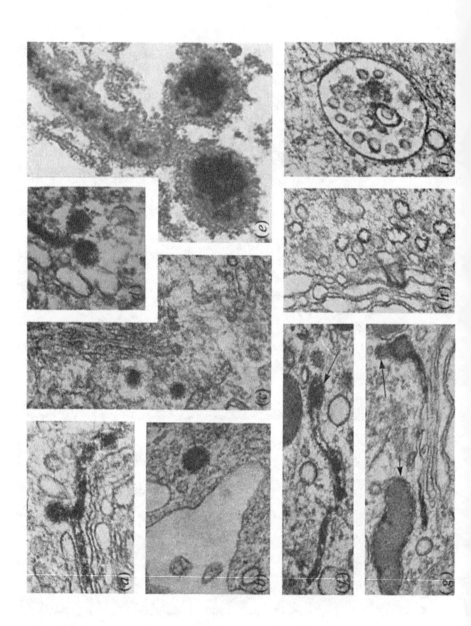

3

Coated vesicles: their occurrence in different plant cell types

ELDON H.NEWCOMB

General characteristics

Alveolate coated vesicles of similar appearance have been observed
in a broad spectrum of plant groups from high to low, and in many different
cell types. However, in only three or four papers have these coated vesicles in
plants been examined or considered in any detail; most of the reports of these
organelles have consisted merely of brief remarks noting their presence and
intracellular location in papers devoted primarily to other subjects. Many
investigators have reported that coated vesicles occur predominantly in two
regions of the plant cell, namely in the vicinity of dictyosomes and beneath
the plasmalemma, and have suggested that they arise from cisternae of the
former and fuse with the latter.

With the exception of the vesicles associated with the contractile vacuoles
of certain algae, and the spiny coated vesicles of limited distribution in flowering plants, coated vesicles have proved to be remarkably uniform in morphology
and size wherever encountered in the plant kingdom. The ordinary coated
vesicles of plants are approximately 85–90 nm in diameter, including coat,
and possess readily observable unit membrane structure surrounded on their
cytoplasmic face by an alveolate or reticulate layer. In median sections the
coat exhibits radiating spokes or columellar projections about 25 nm long
(Plate 1a and b). In tangential sections the coat can be seen to consist of
polygonally packed ridges and in the most favorable views, of what appear to
be pentagonally or hexagonally packed units (Plate 1c). Internally, plant
coated vesicles are rather nondescript; they are of moderate electron opacity,
with a fuzzy or finely fibrous material projecting partway into the interior
from the inner surface of the bounding membrane.

Coated vesicles in lower plants

Some of the earliest reports and best ultrastructural evidence for the
existence of coated vesicles in plants have come from studies of algae. Plant

coated vesicles were first reported by Berkaloff (1963) in a brown alga, and
subsequently by Manton, Leedale and others in a number of other algae.
Comparison of these algal vesicles with those of other plants strongly suggests
that an 'alveolate' or 'reticulate' coat construction is probably common to all
of them. Many of the coated algal vesicles differ from those of higher plants,
however, in being more variable in size and shape, and generally larger, with
the consequence that the coats are constructed of a greater number of units
or compartments and constitute proportionately less of the diameter of the
vesicles than is the case for the coated vesicles of higher plants.

The coated vesicles observed by Berkaloff (1963) in the brown alga
Himanthalia lorea represented both vesicles which were located quite close to
dictyosomes and enlargements at the margins of Golgi cisternae. The coats
were described as 'fine fibres' projecting outward perpendicular to the membrane for a distance of 10–20 nm. Whether the coats were in reality alveolate,
and thus similar to those of higher plants, is difficult to decide from inspection
of the micrographs.

The report of Berkaloff was succeeded in the following year by a paper of
Manton's (1964) in which 'hairy vesicles' were reported in the swimming cells
(zoospores) of the green alga *Stigeoclonium*. The vesicles were all closely
associated with a large contractile vacuole, some lying in its vicinity and others
confluent with it so that they appeared to be either merging with or departing
from it. These algal vesicles are larger than the typical coated vesicles of higher
plants, and tend to be ovoid rather than spherical. They were present in large
numbers in association with the active contractile vacuole, and were assumed
to be discharging directly to the outside through fusion with the plasmalemma
delimiting the vacuole. Since the vesicles disappeared when the contractile
vacuole of the swimming cell finally ceased to function, they were considered
to be an integral part of the water-controlling mechanism of the naked swimming cell.

Leedale, Meeuse & Pringsheim (1965) described coated vesicles similarly
attached to the plasmalemma near the place of discharge of contractile vacuoles
in euglenoids. The authors considered that the vesicles were probably involved
with the contractile vacuole in osmoregulation. Again the vesicles are variable
in size and larger than the common type of coated vesicles, but their coats
are clearly reticulate as is readily ascertained from inspection of the illustrations.

'Hairy vesicles' were repeatedly singled out for attention and were clearly
illustrated in a number of Manton's classic papers on flagellate algae, particularly
marine algae belonging to the classes Haptophyceae and Prasinophyceae
(Manton, 1964, 1966, 1967*a, b*). She found vesicles with alveolate surfaces to

be common in many pigmented flagellates, often attached as 'diverticulae' to the scale-forming Golgi cisternae. In some cases the coated vesicles were seen as secondary structures arising from larger vesicular enlargements at cisternal margins. In the haptophycean alga *Prymnesium parvum* (Manton, 1966), large 'hairy' pits in the plasmalemma were observed lying on one side of each flagellar base. It can be seen that the pits exhibit on the cytoplasmic side of the plasmalemma an alveolate pattern seemingly identical to that of conventional coated vesicles. Manton suggested that the pits could be interpreted as the sites of fusion of scale-containing vesicles, or of the accumulation of remnants of these vesicles after discharge, or as both, and drew attention to the great variability in the size and shape of the pits to support her interpretation. Alveolate coated protuberances of dictyosome cisternae were also present; in Manton's figure 4 for example, the reticulate outlines of a developing vesicle can be seen clearly in face view.

Numerous illustrations of nascent coated vesicles attached to the margins of Golgi cisternae can be seen also in Manton's papers on the alga *Chrysochromulina chiton* (Manton, 1967a, b). The vesicles are seen both budding from and lying near the Golgi cisternae. Clearly alveolate in surface architecture, they are variable in size and generally larger than those of higher plants. Similar vesicles associated with dictyosome cisternae containing early stages in placolith formation were figured by Manton & Leedale (1969) in two marine coccolithophorids. The numerous coated vesicles seen in the vicinity of Golgi structures in green flagellates with scaly flagella (Manton, Rayns & Ettl, 1965) are also alveolate in structure.

Coated vesicles resembling one another closely have also been found in several species of algae in the Xanthophyta (Falk, 1967, 1969), in the alga *Synura* in the Chlorophyta (Schnepf & Deichgräber, 1969), and in the green algae *Chara* (Pickett-Heaps, 1967; Sievers, 1967) and *Micrasterias* (Kiermayer, 1970). In all of these reports, the vesicles appear to be quite uniform in size and structure, and resemble closely the coated vesicles observed elsewhere in the plant kingdom.

In summary, the evidence suggests that there are two closely related types of coated vesicles in algae, both derived from Golgi cisternae and both with alveolate surface architecture. One type resembles closely the coated vesicles commonly observed in other plants and probably migrates to and fuses with the plasmalemma generally, while the other type is larger and more variable in shape, and probably fuses with the plasmalemmas of functioning contractile vacuoles confined to certain algae.

A few papers have noted the presence of coated vesicles in other lower plants. Coated vesicles of conventional appearance have been seen in the slime mould *Physarum* (Aldrich, 1969). The vesicles were especially numerous near the spindle poles during prophase and telophase in dividing myxamoebae. Schnepf (1973) has directed attention to the coated vesicles in dividing leaf cells of the moss *Sphagnum*, where they were observed in association with the dictyosomes and also fusing with the plasmalemma of the phragmoplast.

Coated vesicles in vascular plants

Only a few reports pertain to coated vesicles in vascular plants below the angiosperms. Coated vesicles of conventional appearance have been identified in the egg of the fern *Pteridium* (Bell & Duckett, 1976), in cells of a young fern (*Polypodium vulgare*) gametophyte (Fraser & Smith, 1974), and in young spores from male cones of the gymnosperm *Podocarpus* (Vasil & Aldrich, 1970). They have also been identified in cambial cells of pine (Srivastava, 1966).

Coated vesicles indistinguishable from those reported from lower plants have been widely observed in angiosperms, the frequency with which they have been cited probably being to a large extent simply a measure of the attention that various cell types and developmental stages have received. They have been noted in association with the cell plate both in dividing root tip cells of bean (*Phaseolus vulgaris*) (Hepler & Newcomb, 1967) and in dividing cells of very young leaves of tobacco (*Nicotiana tabacum*) (Cronshaw & Esau, 1968b), and with the phragmoplast in the endosperm of the African blood lily, *Haemanthus katherinae* (Bajer, 1968; Lambert & Bajer, 1972). They have also been found in dividing cells regenerated from protoplasts of soybean (*Glycine max*) (Fowke, Bech-Hansen, Gamborg & Constabel, 1975).

Among the numerous studies of differentiating sieve elements of flowering plants, several have included reference to coated vesicles. They have been recorded in differentiating sieve elements in various monocots, including *Musa velutina* (Behnke, 1969), corn (*Zea mays*) (Singh & Srivastava, 1972), and various palms (Parthasarathy, 1974). In differentiating sieve elements of dicots they have been described in elm (*Ulmus americana*) (Evert & Deshpande, 1969) and broad bean (*Vicia faba*) (Zee, 1969). Their occurrence in other vascular tissue components has also been noted, including cambial cells of white ash (*Fraxinus americana*) (Srivastava, 1966), differentiating vessel elements of corn (Srivastava & Singh, 1972), and transfer cells in minor veins in leaves of a composite, *Senecio vulgaris* (Gunning & Pate, 1969).

The occurrence of coated vesicles in the angiosperm microspore and microgametophyte has also been established. Dickinson & Heslop-Harrison (1971) noted their presence in the cytoplasm of the developing pollen grain (microspore) of *Lilium* during growth of the nexine 2 and intine wall layers.

Coated vesicles were abundant near the plasmalemma and were observed
fusing with it. Coated vesicles in a mature microgametophyte were demonstrated by Franke, Herth, Van der Woude & Morré (1972), who found them
in the cytoplasm of *Clivia* pollen tubes.

Coated vesicles have also been studied in root hairs of radish (*Raphanus
sativus*) (Bonnett & Newcomb, 1966), in lateral root primordial cells of field
bindweed (*Convolvulus arvensis*) (Bonnett, 1969), and in growing cultured
plant cells of *Haplopappus gracilis* (Franke & Herth, 1974). In addition coated
vesicles have been recorded in mucilage-secreting placentary papillae on the
ovary of *Aptenia* (Kristen, 1976, 1978), in epidermal and parenchyma cells
of the flower spur of nasturtium (*Tropaeolum majus*) (Rachmilevitz & Fahn,
1975), in oil gland cells of *Heracleum* (Schnepf, 1969), in guard cells of pea
(*Pisum sativum*) (Singh & Srivastava, 1973), in various tissues of the embryo
of lettuce (*Lactuca sativa*) during germination (Srivastava & Paulson, 1968),
and in mature secretory trichomes on the leaves and stems of *Pharbitis nil*
(Unzelman & Healey, 1974).

Origin and fate of coated vesicles in plants

Numerous investigators have observed a concentration of coated
vesicles in the vicinity of dictyosomes, as well as what appear to be similar
bodies budding from the dictyosome margins, and have concluded that the
vesicles have their origin in this organelle. In one of the earliest papers on the
subject that utilised higher plant material, Cunningham, Morré & Mollenhauer
(1966) described the isolation from onion of dictyosomes bearing on their
tubate peripheries both smooth-membraned vesicles of various sizes and also
rough-membraned vesicles of uniform size. The latter, termed 'rough vesicles'
by the authors, are presumed to have been nascent coated vesicles (cf. also
Mollenhauer & Morré, 1966).

In a few cases, the coated vesicles have been noted arising secondarily from
smooth-surfaced vesicles after the latter have separated from dictyosomes.
For example, in the radish root hair (Bonnett & Newcomb, 1966), the apparent
origin of coated vesicles by evagination from larger vesicles in the neighbourhood
of dictyosomes is a common phenomenon (Plate 1*d*). A similar type of development was noted by Manton (1964, 1966, 1967*a, b*) in her studies of marine
flagellate algae. It appears that the coats develop during budding and mature
after the evaginations have separated from the parent vesicles. Of course it is
generally appreciated, as in the cases just cited, that though it can hardly be
doubted, it has not actually been proved that the vesicles are moving from the
dictyosomes exocytically and not vice versa.

In an early paper on coated vesicles in plant cells, Bonnett & Newcomb

(1966) noted that these vesicles were concentrated near the cell surface in radish root hairs, and were occasionally seen apparently fusing with the plasmalemma. Also, indistinct polygonal configurations were noted in regions where the plasmalemma had been sectioned tangentially. These fields of polygons were interpreted as surface views of coats at points of fusion between vesicles and plasmalemma.

Similarly, Hepler & Newcomb (1967), in a study of cell division in the apical meristem of bean roots, pointed out that the projections of coated vesicles remained visible on the plasmalemma for a brief period after the vesicles fused at the cell plate (Plate 1*e*). They suggested that the coats might be playing a role in effecting vesicle fusion with the plasmalemma of the young plate.

In subsequent work in the author's laboratory, patches of polygons have been noted at or near the inner surface of the plasmalemma in young, rapidly growing cells of several different plant materials (Plate 1*f*). In favourable material these patches are so numerous and extensive and seemingly so closely associated with the plasmalemma and microtubules as to suggest that the coats have remained behind after vesicle fusion with the cell surface, and have coalesced to form an extensive reticulum in the cytoplasm next to the plasmalemma.

Franke & Herth (1974) extended importantly the observations and inter-pretations regarding the fate of the vesicle coats, utilising exponentially growing cells of the composite *Haplopappus gracilis* cultivated in suspension. This is especially favourable material in which the forming cell plates are rich in fusing coated vesicles and in coat-bearing regions of the plasmalemma. A relatively large percentage (as much as 60 %) of the new plasma membrane bore the coat pattern, suggesting that the membrane of the coated vesicles had been incorporated into it. As the phragmoplast matured into the primary cross wall, the ratio of coat-bearing to uncoated regions declined.

These observations strongly suggested (Franke & Herth, 1974) that in a rapidly growing plasma membrane where cell plate formation is completed within a few minutes, membrane material of vesicles identified by their coat markers becomes an integral part of the plasmalemma. Alternative explanations, such as the formation of inwardly migrating coated vesicles over much of the cell plate during rapid growth of the plasmalemma, were considered to be highly unlikely.

In a study of the ultrastructural details of the manner in which the lateral root protrudes through the older tissues in root segments of the field bindweed (*Convolvulus arvensis*), Bonnett (1969) has made some of the most novel and suggestive observations and drawn some of the most specific conclusions yet available about one of the roles that coated vesicles might play in plants.

Briefly, ahead of the growing primordium of the lateral root, the cortical parenchyma cells of the old root are degraded, so that only the cell walls remain. Thus a thick layer representing the walls of numerous collapsed cortical parenchyma cells accrues at the tip of the advancing root primordium. In the outermost cell layer of this primordium, coated vesicle activity is intense, particularly along the plasmalemma bounding the outer wall facing and abutting the layer of collapsed walls. Numerous protuberances of this plasmalemma are covered with projecting spokes characteristic of coated vesicles, and within some of these protuberances there are darkly staining structures (Plate 1g). Similar structures are visible deeper in the wall.

Bonnett (1969) has suggested that these structures within the wall may represent proteins, specifically hydrolases, that have been transported to the plasmalemma by coated vesicles arising from dictyosomes in the cytoplasmic interior. The enzymes so released by fusion of the coated vesicles with the plasmalemma might move across the wall of the outermost primordial cell and then function in degrading the adjacent cortical cells. That coated vesicles might convey hydrolases or other proteins to the wall had been suggested earlier by Bonnett & Newcomb (1966) in their study of coated vesicles in radish root hairs.

Spiny coated vesicles

The 'spiny vesicle' constitutes a distinct variety of coated vesicle; it is covered with numerous radiating tubular projections, and thus is clearly different in surface morphology from the commoner type of coated vesicle encountered in plant cells. Spiny vesicles have been reported in only a few species of higher plants, and only within certain cells of the plant body. Vesicles of similar surface morphology appear not to have been reported from any plants below the angiosperms, nor from animal cells.

Spiny vesicles were first reported from root tips of bean (*Phaseolus vulgaris*), where they were observed in cells adjacent to protophloem sieve elements (Newcomb, 1967). Typically the vesicles occur in clusters or aggregates in the cytoplasm (Plate 2). They are variable in both shape and size, being irregularly spherical to ovoid and ranging in diameter from about 40 to 150 nm, excluding the projections. Occasionally vesicles are seen that appear to be dividing or fusing.

The projections radiating outward from the bounding membrane are about 15 nm apart (centre-to-centre) at their points of origin, and appear to be distributed more or less uniformly over the entire vesicle surface. They are cylindrical, with a length of about 32 nm and a thickness of 8 nm. In favourable longitudinal views and transverse sections they exhibit an electron-opaque cortex and a transparent core, and thus are considered to be tubular in nature (Plate 2, inset).

Spiny vesicles are further distinguished by their tendency to occur in clusters, ranging in size from a few to several hundred individuals. It seems probable that they are embedded in a viscous matrix and remain associated for this reason; it is suggestive that other cytoplasmic components, including free ribosomes, are almost totally excluded from the vesicular aggregate (Plate 2).

Several other features also set spiny vesicles sharply apart from the ordinary alveolate coated vesicles. In the first place, spiny vesicles appear to be of much more limited distribution within the plant kingdom. Similar vesicles were seen also in *Dianthus chinensis* (Newcomb, 1967) and have since been described in *Cucurbita maxima* (Cronshaw & Esau, 1968*a*), *Coleus blumei* (Steer & Newcomb, 1969), *Nicotiana tabacum* (Cronshaw & Esau, 1968*b*), and *Salix babylonica* (Deshpande, 1974). Additionally, a tubular organelle-like component covered with projections quite similar in appearance has been described in *Beta* and *Tetragonia* (Esau & Gill, 1970*a*). In all these cases, the unusual vesicles have been seen only in differentiating cells of the phloem strands. They have, however, been observed in phloem tissue in several organs of the plant body, including roots, stems, petioles and leaf blades. In a few cases, under special circumstances as described below, they have also been encountered in cells other than those of the phloem.

In the original report on spiny vesicles, a close association with developing P-protein was noted, giving rise to the suggestion that the vesicles or their spines might contribute to P-protein formation (Newcomb, 1967). ('P-protein' refers to the massive proteinaceous inclusions which form in differentiating sieve elements and some related cells of the phloem strand, particularly in dicotyledonous angiosperms (Esau, 1965).) While the association of spiny vesicles with P-protein has been observed also by other investigators, puzzling and contradictory aspects have emerged, as described below.

In the bean root P-protein is produced in the procambial cells of the phloem and pericycle bordering on the protophloem elements. The spiny vesicles are restricted in occurrence to these same cells; they appear before and during P-protein formation, and disappear as the P-protein bodies enlarge. Occasionally they are seen in small numbers at the margins of the P-protein (Newcomb, 1967).

Subsequent observations on spiny vesicles and P-protein in the phloem of the shoot of *Coleus blumei* (Steer & Newcomb, 1969) supported the suggestion that the two components might be closely related. Cronshaw & Esau (1968*a*) also found spiny vesicles associated with P-protein in phloem cells in shoots of

Cucurbita maxima. In healthy plants of tobacco, Esau & Gill (1970*b*) found spiny vesicles and P-protein occurring independently and jointly in phloem parenchyma cells in leaf veins and also in young cells around protophloem sieve tubes in root tips. In leaves of plants infected with tobacco mosaic virus, both spiny vesicles and P-protein were found in the phloem parenchyma of the veins, and in a larger number of cells and in larger amounts than in non-infected leaves. Deshpande (1974) observed similar vesicles in parenchyma cells in the primary phloem of young shoots of weeping willow (*Salix babylonica*). Again, in some instances the spiny vesicles occurred near P-protein bodies.

The principal objection to the hypothesis that spiny vesicles play a role in P-protein synthesis is that, with the exception of *Cucurbita*, they have not been demonstrated in the sieve elements themselves, where the accumulation of P-protein is most marked. Since the development of sieve elements has been studied ultrastructurally in a number of species of angiosperms over a period of years in several laboratories, it is highly unlikely that such a distinctive component as the spiny vesicle would have been overlooked. For example, Esau (1971) observed no spiny vesicles in the numerous differentiating sieve elements studied in *Mimosa pudica*. Rough endoplasmic reticulum, dictyosomes producing smooth and coated vesicles, and free ribosomes accompanied the nascent P-protein. Similarly, Palevitz & Newcomb (1971), in a study of P-protein development in several species of legumes, failed to find spiny vesicles in the sieve elements, though encountering them in phloem parenchyma cells.

Spiny vesicles have not been reported in any plants below the angiosperms. Phloem differentiation has been studied in *Welwitschia,* the anomalous gymnosperm from the Namib desert of South-West Africa, by Evert, Bornman, Butler & Gilliland (1973). No spiny vesicles were observed, but segments of endoplasmic reticulum bearing spine-like processes resembling those of spiny vesicles were present in the immature sieve cells. *Welwitschia* does not produce P-protein.

Also detracting from the hypothesis that spiny vesicles are causally related to P-protein formation are several papers reporting the occurrence of these components in cells which have no obvious relationship to phloem tissue. Clusters of spiny vesicles were encountered by Fowke, Bech-Hansen & Gamborg (1974) in some of the cells regenerating from protoplasts isolated from leaves of the dicot *Ammi visnaga*. In an especially interesting finding, Jones (1969) reported that barley aleurone cells treated with gibberellic acid for 24 hours contained spiny vesicles. Paulson & Webster (1970) found spiny vesicles in the greatly enlarged plant cells produced in tomato roots by nematode infection. During the transformation of small, thin-walled root cells into greatly enlarged 'giant cells' which act as nutrient sources for the developing nematodes, crystalline inclusions were formed which were assumed to be storage protein. Spiny vesicles occurred in abundance adjacent to the inclusions but only infrequently elsewhere in the cytoplasm, encouraging the authors to suggest that the vesicles may function in the transport of storage protein, or of enzymes involved in its synthesis. Finally, Franke *et al.* (1972) found spiny vesicles (as well as conventional coated vesicles) in pollen tubes of *Clivia*.

The projecting spines of the vesicles observed in the above four studies are similar in dimensions and spacing to those described for the spiny vesicles of phloem cells. Also, it should be noted that the tubular component described by Esau & Gill (1970a) in young parenchyma cells in roots and leaves of sugar beet (*Beta vulgaris*) is morphologically distinct from spiny vesicles, yet bears 'spines' or tubular projections remarkably similar in appearance to those of spiny vesicles. Bracker & Grove (1971) have described an orderly arrangement of minute tubules projecting perpendicularly outward from the outer mito-chondrial membranes in hyphae of the oomycetous fungus *Pythium ultimum*. As pointed out by the authors, the tubules, although shorter, are similar in diameter to the 'spines' on the spiny vesicles of angiosperms.

References

Aldrich, H.C. (1969). The ultrastructure of mitosis in myxamoebae and plasmodia of *Physarum flavicomum*. *American Journal of Botany*, 56, 290–9.

Bajer, A. (1968). Fine structure studies on phragmoplast and cell plate formation. *Chromosoma*, 24, 383–417.

Behnke, H.-D. (1969). Aspekte der Siebröhren-Differenzierung bei Monocotylden. *Protoplasma*, 68, 289–314.

Bell, P.R. & Duckett, J.G. (1976). Gametogenesis and fertilisation in *Pteridium*. *Botanical Journal of the Linnaean Society*, 73, 47–78.

Berkaloff, C. (1963). Les cellules méristématiques d'*Himanthalia lorea* (L.) S.F. Gray. Étude au microscope électronique. *Journal de Microscopie*, 2, 213–28.

Bonnett, H.T., Jr (1969). Cortical cell death during lateral root formation. *Journal of Cell Biology*, 40, 144–59.

Bonnett, H.T., Jr & Newcomb, E.H. (1966). Coated vesicles and other cytoplasmic components of growing root hairs of radish. *Protoplasma*, 62, 59–75.

Bracker, C.E. & Grove, S.N. (1971). Surface structure on outer mitochondrial membranes of *Pythium ultimum*. *Cytobiologie*, 3, 229–39.

Cronshaw, J. & Esau, K. (1968a). P-protein in the phloem of *Cucurbita*. I. The development of P-protein bodies. *Journal of Cell Biology*, 38, 25–39.

Cronshaw, J. & Esau, K. (1968*b*). Cell division in leaves of *Nicotiana*. *Protoplasma*, 65, 1–24.

Cunningham, W.P., Morré, D.J. & Mollenhauer, H.H. (1966). Structure of isolated plant Golgi apparatus revealed by negative staining. *Journal of Cell Biology*, 28, 169–76.

Deshpande, B.P. (1974). On the occurrence of spiny vesicles in the phloem of *Salix*. *Annals of Botany*, 38, 865–8.

Dickinson, H.G. & Heslop-Harrison, J. (1971). The mode of growth of the inner layer of the pollen-grain exine in *Lilium*. *Cytobios*, 4, 233–43.

Esau, K. (1965). *Plant Anatomy*, 2nd edn. New York: Wiley.

Esau, K. (1971). Development of P-protein in sieve elements of *Mimosa pudica*. *Protoplasma*, 73, 225–38.

Esau, K. & Gill, R.H. (1970*a*). A spiny cell component in the sugar beet. *Journal of Ultrastructure Research*, 31, 444–55.

Esau, K. & Gill, R.H. (1970*b*). Observations on spiny vesicles and P-protein in *Nicotiana tabacum*. *Protoplasma*, 69, 373–88.

Evert, R.F., Bornman, C.H., Butler, V. & Gilliland, M.G. (1973). Structure and development of the sieve-cell protoplast in leaf veins of *Welwitschia*. *Protoplasma*, 76, 1–21.

Evert, R.F. & Deshpande, B.P. (1969). Electron microscope investigation of sieve-element ontogeny and structure in *Ulmus americana*. *Protoplasma*, 68, 403–32.

Falk, H. (1967). Zum Feinbau von *Botrydium granulatum* Grev. (Xanthophyceae). *Archiv für Mikrobiologie*, 58, 212–27.

Falk, H. (1969). Fusiform vesicles in plant cells. *Journal of Cell Biology*, 43, 167–74.

Fowke, L.C., Bech-Hansen, C.W. & Gamborg, O.L. (1974). Electron microscopic observations of cell regeneration from cultured protoplasts of *Ammi visnaga*. *Protoplasma*, 79, 235–48.

Fowke, L.C., Bech-Hansen, C.W., Gamborg, O.L. & Constabel, F. (1975). Electron-microscope observations of mitosis and cytokinesis in multinucleate protoplasts of soybean. *Journal of Cell Science*, 18, 491–507.

Franke, W.W. & Herth, W. (1974). Morphological evidence for de-novo formation of plasma membrane from coated vesicles in exponentially growing cultivated plant cells. *Experimental Cell Research*, 89, 447–51.

Franke, W.W., Herth, W., Van der Woude, W.J. & Morré, D.J. (1972). Tubular and filamentous structures in pollen tubes: possible involvement as guide elements in protoplasmic streaming and vectorial migration of secretory vesicles. *Planta*, 105, 317–41.

Fraser, T.W. & Smith, D.L. (1974). Young gametophytes of the fern *Polypodium vulgare* L. *Protoplasma*, 82, 19–32.

Gunning, B.E.S. & Pate, J.S. (1969). 'Transfer cells.' Plant cells with wall ingrowths, specialised in relation to short distance transport of solutes – their occurrence, structure, and development. *Protoplasma*, 68, 107–33.

Hepler, P.K. & Newcomb, E.H. (1967). Fine structure of cell plate formation in the apical meristem of *Phaseolus* roots. *Journal of Ultrastructure Research*, 19, 498–513.

Jones, R.L. (1969). Gibberellic acid and the fine structure of barley aleurone cells. II. Changes during the synthesis and secretion of α-amylase. *Planta*, 88, 73–86.

Kiermayer, O. (1970). Elektronenmikroskopische Untersuchungen zum Problem der Cytomorphogenese von *Micrasterias denticulata* Bréb. *Protoplasma*, 69, 97–132.

Kristen, U. (1976). Die Morphologie der Schleimsekretion im Fruchtknoten von *Aptenia cordifolia*. *Protoplasma*, 89, 221–33.

Kristen, U. (1978). Ultrastructure and a possible function of the intercisternal elements in dictyosomes. *Planta*, 138, 29–33.

Lambert, A.-M. & Bajer, A.S. (1972). Dynamics of spindle fibers and microtubules during anaphase and phragmoplast formation. *Chromosoma*, 39, 101–44.

Leedale, G.F., Meeuse, B.J.D. & Pringsheim. E.G. (1965). Structure and physiology of *Euglena spirogyra*. I and II. *Archiv für Mikrobiologie*, 50, 68–102.

Manton, I. (1964). Observations on the fine structure of the zoospore and young germling of *Stigeoclonium*. *Journal of Experimental Botany*, 15, 399–411.

Manton, I. (1966). Observations on scale production in *Prymnesium parvum*. *Journal of Cell Science*, 1, 375–80.

Manton, I. (1967a). Further observations on the fine structure of *Chrysochromulina chiton* with special reference to the haptonema, 'peculiar' Golgi structure and scale production. *Journal of Cell Science*, 2, 265–72.

Manton, I. (1967b). Further observations on scale formation in *Chrysochromulina chiton*. *Journal of Cell Science*, 2, 411–18.

Manton, I. & Leedale, G.F. (1969). Observations on the microanatomy of *Coccolithus pelagicus* and *Cricosphaera carterae*, with special reference to the origin and nature of coccoliths and scales. *Journal of the Marine Biological Association of the United Kingdom*, 49, 1–16.

Manton, I., Rayns, D.G. & Ettl, H. (1965). Further observations on green flagellates with scaly flagella: the genus *Heteromastix* Korshikov. *Journal of the Marine Biological Association of the United Kingdom*, 45, 241–55.

Mollenhauer, H.H. & Morré, D.J. (1966). Golgi apparatus and plant secretion. *Annual Review of Plant Physiology*, 17, 27–46.

Newcomb, E.H. (1967). A spiny vesicle in slime-producing cells of the bean root. *Journal of Cell Biology*, 35, C17–22.

Palevitz, B.A. & Newcomb, E.H. (1971). The ultrastructure and development of tubular and crystalline P-protein in the sieve elements of certain papilionaceous legumes. *Protoplasma*, 72, 399–426.

Parthasarathy, M.V. (1974). Ultrastructure of phloem in palms. I. Immature sieve elements and parenchymatic elements. *Protoplasma*, 79, 59–91.

Paulson, R.E. & Webster, J.M. (1970). Giant cell formation in tomato roots caused by *Meloidogyne incognita* and *Meloidogyne hapla* (Nematoda) infection. A light and electron microscope study. *Canadian Journal of Botany*, 48, 271–6.

Pickett-Heaps, J.D. (1967). Ultrastructure and differentiation in *Chara* sp. I. Vegetative cells. *Australian Journal of Biological Sciences*, 20, 539–51.

Rachmilevitz, T. & Fahn, A. (1975). The floral nectary of *Tropaeolum majus* L. The nature of the secretory cells and the manner of nectar secretion. *Annals of Botany*, 39, 721–8.

Schnepf, E. (1969). Über den Feinbau von Öldrüsen. IV. Die Ölgänge von Umbelliferen: *Heracleum sphondylium und Dorema ammoniacum*. *Protoplasma*, 67, 375–90.

Schnepf, E. (1973). Mikrotubus-Anordnung und -Unordnung, Wandbildung und Zellmorphogenese in jungen *Sphagnum*-Blättchen. *Protoplasma*, 78, 145–73.

Schnepf, E. & Deichgräber, G. (1969). Über die Feinstruktur von *Synura petersenii* unter besonderer Berücksichtigung der Morphogenese ihrer Knieselschuppen. *Protoplasma*, 68, 85–106.

Sievers, A. (1967). Elektronenmikroskopische Untersuchungen zur geotropischen Reaktion. II. Die polare Organisation des normalwachsenden Rhizoids von *Chara foetida*. *Protoplasma*, 64, 225–53.

Singh, A.P. & Srivastava, L.M. (1972). The fine structure of corn phloem. *Canadian Journal of Botany*, 50, 839–46.

Singh, A.P. & Srivastava, L.M. (1973). The fine structure of pea stomata. *Protoplasma*, 76, 61–82.

Srivastava, L.M. (1966). On the fine structure of the cambium of *Fraxinus americana* L. *Journal of Cell Biology*, 31, 79–93.

Srivastava, L.M. & Paulson, R.E. (1968). The fine structure of the embryo of *Lactuca sativa*. II. Changes during germination. *Canadian Journal of Botany*, 46, 1447–53.

Srivastava, L.M. & Singh, A.P. (1972). Certain aspects of xylem differentiation in corn. *Canadian Journal of Botany*, 50, 1795–804.

Steer, M.W. & Newcomb, E.H. (1969). Development and dispersal of P-protein in the phloem of *Coleus blumei* Benth. *Journal of Cell Science*, 4, 155–69.

Unzelman, J.M. & Healey, P.L. (1974). Development, structure and occurrence of secretory trichomes in *Pharbitis*. *Protoplasma*, 80, 285–303.

Vasil, I.K. & Aldrich, H.C. (1970). A histochemical and ultrastructural study of the ontogeny and differentiation of pollen in *Podocarpus macrophyllus* D. Don. *Protoplasma*, 71, 1–37.

Zee, S.-Y. (1969). Fine structure of the differentiating sieve elements of *Vicia faba*. *Australian Journal of Botany*, 17, 441–56.

Plates

Plate 1. (*a*) A pair of coated vesicles in a young radish root hair. Scale bar represents 100 nm. (From Bonnett & Newcomb, 1966.) (*b*) Coated vesicle in a root tip of tobacco. Scale bar represents 100 nm. (Micrograph by Dr E.L. Vigil, University of Wisconsin.) (*c*) Tangential section of a coated vesicle in a radish root hair showing polygonal units of the coat (arrow). Scale bar represents 100 nm. (Micrograph by Dr H.T. Bonnett, Jr, University of Wisconsin.) (*d*) Section of a radish root hair showing two dictyosomes surrounded by vesicles. Stages in the formation of mature coated vesicles are represented by indistinctly coated evaginations from smooth-surfaced vesicles and by separated vesicles with relatively distinct coats (see arrows). Scale bar represents 500 nm. (From Bonnett & Newcomb, 1966.) (*e*) Evidence of recent fusion of a coated vesicle (arrow) with the developing cell plate in a dividing root tip cell of bean (*Phaseolus vulgaris*). Scale bar represents 100 nm. (From Hepler & Newcomb, 1967.) (*f*) Patches of polygonal units (arrows) among cortical microtubules beneath the growing wall in a root tip of tobacco. Scale bar represents 100 nm. (Micrograph by Dr E.L. Vigil, University of Wisconsin.) (*g*) Plasmalemma with protuberance (arrow) bearing spokes characteristic of a coated vesicle. Outer primordial cell of a lateral root of *Convolvulus arvensis*. Note darkly staining structure within the protuberance and others deeper in the wall. Scale bar represents 100 nm. (From Bonnett, 1969.)

Plate 2. Section through a large cluster of spiny vesicles in the cytoplasm of a cell adjacent to a protophloem sieve element in a root tip of bean. Many vesicles are represented in the section only by spines. Scale bar represents 100 nm. *Inset.* At higher magnification the transversely sectioned spines of the vesicles appear tubular (arrows). Scale bar represents 100 nm. (From Newcomb, 1967.)

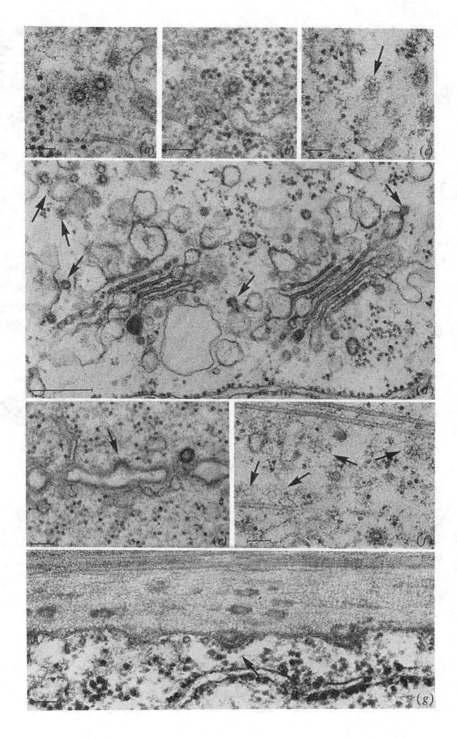

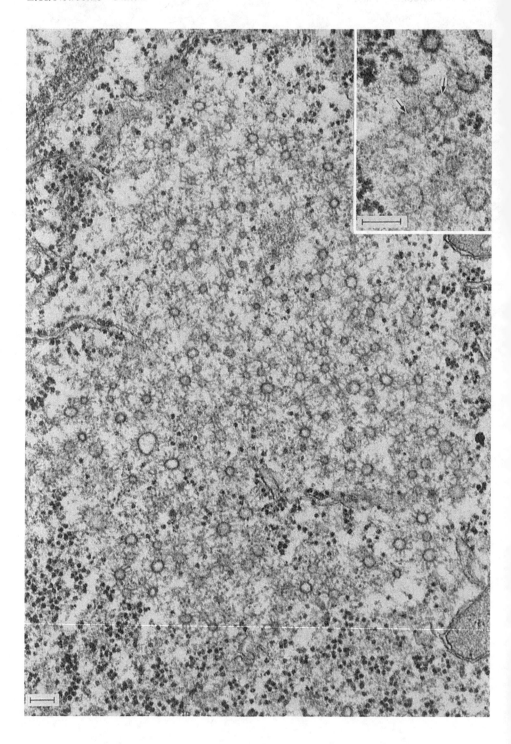

4

Immunoglobulin transmission in mammalian young and the involvement of coated vesicles

RICHARD RODEWALD

The underlying theme in numerous studies on coated vesicles is that these very distinctive and ubiquitous organelles are intimately involved in the selective uptake and transport of exogenous proteins by cells. Yet, even though many recent studies have helped elucidate the structure of these vesicles, it is still not clear what their precise transport functions are in most of the tissues where they have been identified.

Many of the best examples of coated vesicles being implicated in selective transport come from studies on tissues that transfer maternal immunoglobulins to the mammalian young. The tissues involved, depending on the particular species, include the yolk sac and chorioallantoic placenta of the foetus, and the small intestine of the newborn. In each tissue, absorptive cells are present which contain prominent populations of coated vesicles. Furthermore, considerable evidence has established that immunoglobulin transmission across each tissue is selective and, by implication, involves membrane carriers or receptors (Brambell, Halliday & Morris, 1958; Brambell, 1970). However, the exact site of the receptors and the degree of selection within cells are both ambiguous. Much of this uncertainty lies in the fact that all of the tissues serve a second, equally important function for the young, this being the relative non-selective uptake and intracellular digestion of proteins for nutrition. This review considers the physiology of these two functions and the possible roles of coated vesicles in each. It examines the evidence for the presence and locations of specific receptors for immunoglobulins and other proteins, particularly within coated vesicles. It is clear from the outset that our knowledge of this subject is still very incomplete and that many intriguing questions remain to be answered.

Yolk sac transmission

The foetal yolk sac is a major site of transfer of maternal immuno-globulins in several mammalian species (Brambell, 1970). In fact, of those species investigated, only in primates does a different foetal membrane, the chorioallantoic placenta, play an important role in transport. Yolk sac transfer is virtually the only method of transmission in several species, including the rabbit and guinea pig (Brambell *et al.*, 1949; Leissring & Anderson, 1961). This tissue is also important in other species, such as the rat and mouse (Brambell & Halliday, 1956), in which a greater degree of transport takes place after birth across the small intestine. During the latter half of gestation in both groups, the yolk sac splanchnopleur inverts and ruptures at its margin. This causes the absorptive surface, composed of visceral endoderm, to face outward and be bathed by the uterine fluid. The fluid, similar in composition to maternal plasma, is the immediate source of the immunoglobulins which are transported.

Selection

The yolk sac is able to transmit large quantities of both homologous and heterologous IgG, the amount of the latter apparently dependent on the species of origin of the IgG (Brambell, Hemmings & Rowland, 1948; Batty, Brambell, Hemmings & Oakley, 1954; Hemmings, 1956; Barnes, 1959; Koch, Boesman & Gitlin, 1967). All IgG subclasses which have been tested are transferred, but also with different relative efficiencies (Block, Ovary, Kourilsky & Benacerraf, 1963; Hemmings, 1974). In the rabbit, homologous IgM as well as its 7S subunit is transported in significant amounts, but heterologous IgM is not (Hemmings & Jones, 1962; Kaplan, Catsoulis & Franklin, 1965; Hemmings, 1973). In at least the case of IgG, it is the Fc portion of the molecule which is recognised, presumably by receptors in the yolk sac, since this fragment but not the Fab fragment is transmitted efficiently (Brambell, Hemmings, Oakley & Porter, 1960; Kaplan *et al.*, 1965). Further evidence for an IgG receptor comes from the observed saturation of transport at high IgG concentrations (Gitlin & Koch, 1968) and competition between labelled and unlabelled IgG molecules in mixtures (Gitlin & Morphis, 1969).

Although most studies have demonstrated preferential transmission of immunoglobulins by the yolk sac, many transport studies suggest that in at least some species other serum proteins can also be transferred intact to the foetal circulation. In particular, the rabbit yolk sac transports substantial amounts of both homologous and heterologous serum albumin (Winkler, Fitzpatrick & Finnerty, 1958; Hemmings, 1961; Kulangara & Schechtman, 1962; Morgan, 1964; Sonoda & Schlamowitz, 1972*a*) as well as much smaller amounts of other non-immunoglobulin serum proteins (Hemmings, 1961; Morgan, 1964). Yolk sac transfer of albumin has also been documented in the guinea pig (Kulangara & Schechtman, 1963; Koch *et al.*, 1967), rat (Morgan,

1964) and mouse (Gitlin & Koch, 1968; Gitlin & Morphis, 1969). As in the
case of IgG, a large portion of the albumin enters the circulation intact, at
least to the extent that its electrophoretic mobility is unchanged and it can
be precipitated with specific antiserum (Kulangara & Schechtman, 1962;
Sonoda & Schlamowitz, 1972*a*). Serum albumin does not compete with IgG
for transmission across the mouse or rabbit yolk sac (Gitlin & Morphis, 1969;
Sonoda & Schlamowitz, 1972*a*). In the rabbit, kinetic studies suggest that a
carrier, distinct from the IgG receptor, participates in albumin transmission
(Sonoda & Schlamowitz, 1972*a*). No such carrier is apparent in the mouse
(Gitlin & Koch, 1968).

Receptors

The selection exhibited during transmission of IgG and albumin in
the rabbit yolk sac has led Schlamowitz and his colleagues to study in detail
the nature of the possible receptors involved (Schlamowitz, 1976). Sonoda &
Schlamowitz (1972*b*) found that radiolabelled IgG and serum albumin bound
non-competitively to the yolk sac surface *in vitro,* in agreement with their
findings that the two proteins were transported by separate carriers across the
yolk sac *in vivo* (Sonoda & Schlamowitz, 1972*a*). Later studies (Schlamowitz,
Hillman, Lichtiger & Ahearn, 1975; Tsay & Schlamowitz, 1975) have concentrated exclusively on the IgG receptor. Surface membrane vesicles prepared
from yolk sac endoderm were shown to bind either fluorescein-labelled or
radiolabelled rabbit IgG but not similarly labelled bovine IgG, an immunoglobulin
not transmitted by the rabbit yolk sac. Different rabbit IgG fractions isolated
by DEAE—cellulose chromatography were bound by the putative receptor
with affinities between 8.6 and 2.0×10^4 M^{-1}. Treatment of receptor-bearing
membranes with a variety of enzymatic and other chemical reagents indicated
that the receptor was a membrane protein which did not require intact
carbohydrate or divalent cation for binding (Hillman, Schlamowitz & Shaw,
1977). In contrast to this evidence for a specific receptor, Hemmings &
Williams (1974) and Hemmings (1975) used similar membrane preparations
and found no evidence for selective binding of rabbit IgG over bovine
IgG.

Few studies have investigated the nature of the receptors in yolk sacs of
other species. Elson, Jenkinson & Billington (1975) have shown that the
surface of the yolk sac endoderm in the mouse will bind red cells to which
either homologous or heterologous IgG antibodies are absorbed. Binding
requires the presence of the Fc portion of the molecule.

Proteolysis

Quite clearly not all proteins which enter the yolk sac endoderm are transmitted intact to the circulation, as will be apparent from a review of morphological tracer studies. In addition, the detection of cathepsin D (Wild, 1976) and several other lysosomal enzymes within the endoderm (Beck, Lloyd & Griffiths, 1967; Krzyzowska-Gruca & Schiebler, 1967; Christie, 1967) strongly suggests that much of the protein is degraded within the tissue, evidently to provide nutrition for the foetus. Even in the case of IgG transmission, only a fraction of the total protein absorbed by the tissue reaches the circulation in native form (Hemmings, 1957).

Extensive biochemical studies have examined the proteolytic function of the rat yolk sac cultured *in vitro*. This yolk sac was shown to absorb large amounts of both the enzyme horseradish peroxidase and radiolabelled bovine serum albumin (Beck *et al.*, 1967; Parry, Beck & Lloyd, 1968; Williams, Lloyd, Davies & Beck, 1971; Williams, Kidson, Beck & Lloyd, 1975). Both proteins were degraded intracellularly with no evidence for release of intact protein, even though homologous albumin is evidently transmitted intact *in vivo* (Morgan, 1964). The studies suggested a non-selective uptake of proteins. There was selection, however, to the extent that a non-proteinaceous macromolecule, polyvinylpyrrolidone, was absorbed in only small quantities (Williams *et al.*, 1975). Furthermore, the rate of entry of albumin was found to be independent of the protein concentration, which is suggestive of a carrier. Thus, in the rat yolk sac there appears to be a receptor which binds proteins relatively non-selectively and functions in uptake for intracellular digestion. It is not clear, however, in what way this receptor might be related to the specific receptor for albumin evident in the rabbit yolk sac (Sonoda & Schlamowitz, 1972*a, b*).

Morphology of transmission

The endoderm of the visceral yolk sac is composed of a layer of simple columnar epithelial cells. The fine structure of the cells is surprisingly similar in all those species studied in which yolk sac transmission occurs. These species include the rabbit (Petry & Kühnel, 1965; Deren, Padykula & Wilson, 1966; Slade, 1970; Moxon, Wild & Slade, 1976), guinea pig (Petry & Kühnel, 1963; King & Enders, 1970*a, b*) and rat (Padykula, Deren & Wilson, 1966; Lambson, 1966; Krzyzowska-Gruca & Schiebler, 1967; Seibel, 1974). In all cases the cells appear highly specialised for the endocytosis of material at their apical surface (Plate 1). At the base of the microvilli covering this surface are numerous invaginations, often termed caveolae, which appear to give rise to abundant small saccules and tubular vesicles, or canaliculi, within the subjacent cytoplasm (Plate 2). In addition, many vacuoles and larger vesicles are apparent and contain varied amounts of electron-dense material. Frequent membrane connections among the tubules and larger vesicles strongly suggest a pathway of transport between the lumen and large vesicles.

The membranes of the caveolae and small vesicles have a distinctive, fibrillar luminal coat, 30–40 nm thick (Deren *et al.*, 1966; Lambson, 1966; King & Enders, 1970*a*) as well as a frequently discernible cytoplasmic coat approximately 15 nm thick (King & Enders, 1970*a*; Slade, 1975; Moxon *et al.*, 1976). The presence of these coats leaves little doubt that many of these endocytic vesicles are structurally related to coated vesicles similar to those defined by Roth & Porter (1964). However, two features of the vesicles in the yolk sac are important to note. First, only some and not all of the apical vesicles have the cytoplasmic coat (King & Enders, 1970*a*). This has been interpreted to mean either that the cytoplasmic coat is transient and is lost after a vesicle is formed at the cell surface (King & Enders, 1970*a*), or that there are different populations of vesicles which arise separately at the surface, as postulated by Moxon *et al.* (1976) on the basis of other considerations. The second important feature is that those vesicles which do have a distinct cytoplasmic coat may assume a variety of shapes, at least in the guinea pig (King & Enders, 1970*a*), and are not necessarily small spherical vesicles as is most common for coated vesicles in other tissues.

Little is known about the nature of the cytoplasmic coat on yolk sac vesicles. However, cytochemical studies suggest that the luminal coat shares many of the features of the glycocalyx of the apical plasmalemma. In particular, the membranes of both caveolae and microvilli stain similarly with ruthenium red (Jollie & Triche, 1971; King, 1974), alcian blue and concanavalin A (King, 1974). Nevertheless, the coat within the surface invaginations and vesicles appears thicker than the glycocalyx elsewhere and, on the basis of tracer studies, may contain receptors which bind proteins.

Several studies with tracers both for light and electron microscopy have convincingly documented the non-selective endocytosis of proteins into the system of apical vesicles (Lambson, 1966; King & Enders, 1970*b*; Slade, 1970; Wild, 1970; Slade & Wild, 1971; Wild, Stauber & Slade, 1972; Seibel, 1974; Slade, 1975; King, 1977). Large amounts of the proteins evidently pass through the apical tubules and are concentrated within the larger vesicles. The non-selective nature of this pathway, the retention of the bulk of tracer within the large vesicles, and the association of lysosomal enzyme activities with these same vesicles (Krzyzowska-Gruca & Schiebler, 1967; Wild, 1976) all argue forcibly that most proteins transported by this pathway are degraded within the large vacuoles. However, small but perhaps significant amounts of tracer have been identified within the abluminal extracellular spaces (Lambson,

1966; King & Enders, 1970*b*; Slade, 1970; Seibel, 1974). The route by which the tracers gained entry to this compartment was not clear. Although transfer through the cytoplasmic matrix has been suggested (Slade, 1970; Hemmings & Williams, 1974, 1976), a more tenable possibility is that small numbers of tubular or coated vesicles migrate from the apical cytoplasm and fuse with the lateral plasmalemma (King & Enders, 1970*b*; Slade, 1970).

One extremely interesting feature of non-selective endocytosis is that protein tracers which enter the cells tend to bind to the apical plasma membrane (Lambson, 1966; King & Enders, 1970*b*; Slade, 1970). In at least the guinea pig and rat, there is preferential binding to the fibrillar coat found on the luminal surface of the apical invaginations and tubular vesicles (Plate 3) (Lambson, 1966; King & Enders, 1970*b*). In contrast, a non-protein tracer, thorium dioxide, does not bind to the surface coat (King & Enders, 1970*b*), although it too is transported to the apical vacuoles (Carpenter & Ferm, 1966; King & Enders, 1970*b*). This morphological evidence suggests that the apical invaginations and tubular vesicles contain receptors with a broad binding specificity for proteins to provide for rapid non-selective uptake. These morphological results are consistent with the previously mentioned biochemical evidence for a similar non-specific receptor involved in the proteolytic function of the rat yolk sac (Williams *et al.*, 1975).

Ferritin and peroxidase are not suitable tracers for visualising routes of selective transmission of IgG or albumin across cells. Potentially more inform-ative tracers have been employed with varied success. Fluorescein-labelled IgG was used in light microscopic studies on the rabbit yolk sac (Wild, 1970; Slade & Wild, 1971) and radiolabelled IgG was visualised by autoradiography for electron microscopy (Wild *et al.*, 1972; Hemmings & Williams, 1975, 1976). Neither tracer, however, afforded sufficient resolution and sensitivity to enable detection of any differences between intracellular locations of rabbit IgG, some of which was transmitted, and ferritin or bovine IgG which were not. Slade & Wild (1971) and Moxon *et al.* (1976) attempted to use IgG chemically conjugated to ferritin as a more sensitive tracer but found that ferritin when attached to IgG, while not reducing endocytosis, blocked transmission of the IgG to the circulation. They theorised that ferritin, an extremely large protein of 460 000 mol. wt sterically hindered the binding of IgG to its specific receptor.

Slade (1975) and Moxon *et al.* (1976) achieved considerably greater success with the rabbit yolk sac by following transport of either IgG conjugated to peroxidase (40 000 mol. wt), or anti-peroxidase IgG. The latter tracer was particularly appropriate since cells were first exposed to the unmodified anti-body which was then stained specifically with peroxidase following tissue fixation. These tracers, when derived from either rabbit or human IgG were found within the apical canaliculi and their associated large vacuoles, as well as within small coated vesicles which also seemed to form at the apical cell surface (Chapter 1, Plate 2*a*). Several of these latter vesicles, some containing tracer, were fused with the lateral plasmalemma (Plate 4). This finding suggested that the small coated vesicles were responsible for selective transmission to the circulation. Some, but not all, control tracers, including free peroxidase and its conjugate with bovine IgG, were also identified within small coated vesicles. However, no evidence was found for release of these substances from the cells. The investigators proposed that selection of IgG occurs at the apical plasmalemma, in accord with the evidence of Sonoda, Shigematsu & Schlamowitz (1973) for an IgG receptor on this surface. They further postulated that at least two distinct populations of vesicles formed at this surface: the small coated vesicles which were reserved for selective transmission of IgG to the circulation, and the tubular vesicles involved in non-selective endocytosis and transport to lysosomes. This scheme tended to discount the presence within the small coated vesicles of some control tracers which were not transmitted. The investigators speculated that further sub-populations of coated vesicles might be present, including vesicles for selective transmission of serum albumin. Albumin tracers, however, were not studied to test this idea.

King (1977) conducted less extensive experiments with the guinea pig yolk sac, using as tracers either free peroxidase or peroxidase conjugated to homologous IgG. His results were similar to those in the rabbit in that the IgG conjugate could be found within small vesicles, presumably coated vesicles, adjacent to and fused with the lateral membrane. However, King did not describe any differential binding of the two tracers at the apical surface or their segregation into different subpopulations of those coated vesicles which had been previously described (King & Enders, 1970*a*).

Clearly, additional studies on both the guinea pig and rabbit are warranted in order to establish unequivocally whether or not functionally different vesicles form at the apical surface of the endoderm cells. Ideally, these studies should include tracers for serum albumin and IgM, for which there is physiological evidence for selective transmission. Where different vesicle populations occur, careful observations on possible structural differences are needed to resolve, in particular, whether coated vesicles are involved in both selective and non-selective uptake.

Conclusions

Morphological and physiological transport studies both favour the existence of a distinct, highly specific receptor for IgG on the apical plasmalemma of visceral endoderm cells in the yolk sac. This receptor is responsible for selective transmission of IgG to the circulation. In some species receptors for IgM and serum albumin may also be present, as well as a relatively non-specific receptor which functions in endocytosis of proteins for digestion within lysosomes. At least some of these receptors are found within coated vesicles which form at the apical plasmalemma, although not all investigators have been careful to describe the ultrastructure of the endocytic vesicles. Segregation of receptor-bound IgG at the apical surface within a subpopulation of small coated vesicles has been proposed but needs further verification. Regardless of where or how selection occurs, IgG which is transmitted appears to be released from coated vesicles at the lateral plasmalemma of the cells.

Placental transmission

Maternal immunoglobulins are transferred entirely before birth in primates, including humans, by way of the chorioallantoic placenta (Bangham, Hobbs & Terry, 1958a; Dancis et al., 1961; Gitlin, Kumate, Urrusti & Morales, 1964b). The yolk sac in these species remains rudimentary and does not play a significant role in maternal–foetal transport.

The primate placenta is extremely well adapted for maternal–foetal exchange. The fully developed placenta contains a villous trophoblast composed of a single layer of syncytial epithelium which can be as thin as 3–5 μm (Wislocki & Bennett, 1943; Strauss, 1967). The apical surface of the syncytium is in direct contact with maternal blood. Although remnants of a cytotrophoblast are present at the basal surface of this layer, the syncytium in many places rests directly on a thin basement membrane and represents the only significant barrier between the maternal blood and the underlying foetal capillaries.

Selection

As in the case of yolk sac transport in other species, the syncytial trophoblast in primates selectively transports IgG (Bangham et al., 1958; Dancis et al., 1961; Gitlin, Kumate, Urrusti & Morales, 1964a; Kohler & Farr, 1966). However, the extent to which other serum proteins are transported is not entirely clear. Passage of small amounts of other non-immunoglobulin proteins has been detected (Gitlin et al., 1964a, b), and the efficiency of their transfer may be higher than at first apparent in cases where a protein has a short half-life in the circulation. The results of Miller, Zapata, Hutchinson & Gitlin (1973) suggest that some proteins, including albumin, cross the placental barrier by simple diffusion or leakage. In contrast, maternal IgG can reach a higher concentration in the foetus than in the mother and therefore must be transferred by a specific transport mechanism.

The placenta transports all four IgG subclasses (Mellbye, Natvig & Krarstein, 1971), although most studies suggest that IgG2 and IgG4 are transported less efficiently than IgG1 or IgG3 (Wang, Faulk, Struckley & Fudenberg, 1970; Hay, Hull & Torrigians, 1971; Virella, Nures & Tamagnini, 1972; Chandra, 1976). There is no evidence for transfer of any other immunoglobulin classes, including IgM (Gitlin *et al.*, 1964*a*; Mellbye *et al.*, 1971), IgA, IgD (Van Furth, Schuit & Hijmans, 1965; Rowe, Crabbe & Turner, 1968) and IgE (Miller *et al.*, 1973). Of these, only IgM and IgA have been detected in foetal blood and then only after the onset of foetal synthesis (Gitlin & Biasucci, 1969).

Receptors

Matre, Tunder & Enderessen (1975) and Matre (1977) have demonstrated IgG receptors on the apical surface of the human trophoblast by means of IgG adsorbed to red cells. The trophoblast bound cells sensitised with human, rabbit or guinea pig IgG. Binding was strongly inhibited by free IgG or its Fc fragment but not by the $F(ab')_2$ fragment, Fc subfragments or albumin. The relative affinities of different IgG subclasses paralleled their efficiency of transport in that IgG1 and IgG2 bound most strongly. Effects of formaldehyde, periodate and heat treatments suggested that the receptor is a glycoprotein which, unlike the yolk sac IgG receptor, requires intact sugar for binding (Matre, 1977). An IgG receptor has also been detected on trophoblast cells in suspension (Jenkinson, Billington & Elson, 1976) as well as on the trophoblast basement membrane and some other cell types present in the placenta (Faulk, Jeannet, Creighton & Carbonara, 1974; Moskalewski, Ptak & Czannik, 1975; Johnson, Faulk & Wang, 1976).

The nature of the trophoblast receptor has been further elucidated in biochemical studies on protein binding to isolated membranes from trophoblast homogenates. McNabb, Kuh, Dorrington & Painter (1975) demonstrated a single population of membrane receptors which bound all four IgG subclasses but neither IgM nor IgA. IgG1 and IgG2 bound most strongly with an affinity of approximately 4×10^6 M^{-1}. The Fc fragment of IgG1 exhibited a similar affinity. Balfour & Jones (1976) also found selective binding of IgG1, 2 and 3 to a similar membrane preparation, IgG4 not being tested. Binding increased with lowered pH and could be attributed to determinants in the Fc portion of the molecule. Albumin and polyvinylpyrrolidone, in contrast, showed little affinity.

These studies demonstrate with little question that a membrane-bound receptor for IgG, the binding properties of which can account for the overall specificity of IgG transmission, resides within the trophoblast. The studies do not yet provide evidence for any second set of receptors which might function in selective transmission of albumin or non-selective uptake of other proteins.

Morphology of transmission

The fine structure of the epithelial syncytium has been studied by numerous investigators (Wislocki & Dempsey, 1955; Terzakis, 1963; Lister, 1964; Pierce, Midgley & Beales, 1964; Enders, 1965; Strauss, Goldenberg, Hirota & Okudaira, 1965; Strauss, 1967; Tighe, Garrod & Curran, 1967; Dempsey, Lessey & Luse, 1970). It is evident from these studies that the syncytium (Plate 5) shares many of the ultrastructural features of endocytic activity found in the yolk sac. The apical surface is covered by microvilli characterised by frequent branching and bulbous tips. At the bases of the microvilli are large numbers of invaginations and small pinocytic vesicles. Unfortunately, not all investigators have noted specifically the presence of cytoplasmic or luminal coats on the vesicle membranes. Nevertheless, the observations of Lister (1964), Enders (1965), Ockleford (1976) and Ockleford & Whyte (1977) demonstrate that, as in the yolk sac, many of the small vesicles which form at this surface are coated vesicles (Plate 6). Also found in the apical cytoplasm are many small vesicles without cytoplasmic coats and other vesicles similar to the phagolysosomes of the yolk sac, both on the basis of appearance and presence of lysosomal enzymes (Wislocki & Padykula, 1961; Strauss et al., 1965; Christie, 1967; Greene & Spicer, 1970). Similarly, fusion of many apical vesicles with each other (Terzakis, 1963; Dempsey et al., 1970) gives the distinct impression that the lysosomal vesicles arise from the small pinocytic vesicles and contain material from the maternal plasma. However, it is important to note that many of the apical vesicles may instead be secretion granules (Strauss et al., 1965; Strauss, 1967) which reflect the synthesis and export to the apical surface of endogenous materials such as chorionic gonadotrophin (Midgley & Pierce, 1962; Dreskin, Spicer & Greene, 1970).

Ockleford and co-workers have studied in detail the structure and distribution of the prominent coated vesicles of the human syncytium. The vesicles vary in size and shape, with a long axis of 50–130 nm, but all have a well-defined, polygonally patterned coat (Ockleford, 1976; Ockleford, Whyte & Bowyer, 1977) similar to the coat of brain coated vesicles (Kanaseki & Kadota, 1969). On the basis of this striking coat morphology and an analysis of vesicle proteins by gel electrophoresis (Ockleford & Whyte, 1977; Whyte, 1978), the major cytoplasmic coat constituent is probably a protein similar if not identical to the clathrin of other coated vesicles (Pearse, 1975, 1976). Little is known about the luminal coat of these vesicles except that it appears to contain components similar in staining properties to the glycocalyx of the apical plasmalemma (Ockleford & Whyte, 1977).

Other features of the coated vesicles may have a more direct bearing on their transport functions in the placenta. Ockleford *et al.* (1977) have concluded, on the basis of a unimodal size distribution, albeit a strictly morphological criterion, that the vesicles represent a single population. In addition, Ockleford & Whyte (1977) estimated that over 89 % of the coated vesicles lie within 0.54 μm of the apical plasmalemma, a finding which might indicate that most of these vesicles do not transport material across the syncytium. An alternative explanation is that the vesicles lose their cytoplasmic coat shortly after formation and before they move to other regions of the cell (Ockleford & Whyte, 1977). Most investigators have not reported the fusion of coated or other vesicles with the basal surface of the primate syncytium, although the presence of small numbers of coated vesicles in this region has not been specifically ruled out. It is also worth noting that numerous coated vesicles also occur within trophoblast cells of various non-primate chorionic placentae which have not been implicated in immunoglobulin transmission (Enders, 1965; King & Tibbitts, 1976). Obviously, the important question posed by these observations is whether the coated vesicles in the primate placenta are reserved exclusively for selective IgG transfer or have other functions such as non-selective endocytosis, as is apparent in the yolk sac.

The use of IgG tracers and other non-immunoglobulin probes for electron microscopy could help define the functions of coated vesicles in the primate placenta. However, this approach has not yet been widely applied. Reports of Wislocki & Bennett (1943), Muir (1966) and Dempsey *et al.* (1970) indicate that the non-protein tracers colloidal carbon, iron dextran and thorium dioxide enter the trophoblast in only limited quantities. Unfortunately, protein tracers have not been exploited except in studies on the chorioallantoic placentae of the mouse (Robertson, Archer, Papadimitriou & Walters, 1971), rat (Tillack, 1966) and guinea pig (King & Enders, 1971), species in which immunoglobulins are transported across the yolk sac before birth. Each of these studies demonstrated what was presumably non-selective endocytosis of either ferritin or peroxidase by trophoblast cells. The most careful studies have been made on the guinea pig, and here it appears that uptake is similar to non-selective endocytosis by the yolk sac of this species. Protein tracers bound preferentially to membranes of coated pits during endocytosis and were then transported within small vesicles to the lysosome-like vacuoles where they were presumably degraded. Thorium dioxide also entered cells but did not adhere to vesicle membranes. As in the yolk sac, a small amount of the protein tracers reached the intercellular spaces, apparently by release from coated vesicles at the plasmalemma. Transfer across cells has also been reported in the rat (Tillack, 1966), but in the rat and guinea pig release from the cells was considered to be minimal. These studies on non-primate species suggest that coated vesicles in the trophoblast epithelium participate in a carrier-mediated, relatively non-selective endocytosis of protein for intracellular digestion, just as in the yolk sac. By implication, the coated vesicles of the primate placenta may also be involved in non-selective endocytosis, although on the basis of the biochemical binding studies (Balfour & Jones, 1976) a non-specific receptor is probably not involved.

Studies *in vivo* with IgG tracers, such as IgG–peroxidase conjugates and anti-peroxidase IgG, have not yet been reported. Use of these tracers with the primate placenta should be invaluable in determining the role of coated vesicles in selective transmission.

Conclusions

Morphological evidence suggests that coated vesicles are involved in endocytosis of proteins at the apical surface of the primate syncytial trophoblast. However, experiments have not yet established to what extent this uptake is non-selective, as in non-primate chorioallantoic placentae, or is specific for IgG, as would be expected if this uptake represents the first step in transfer of this protein across the cell. Specific IgG receptors are present on the apical surface of the trophoblast, but their distribution on the membrane and, in particular, their relationship to the coated vesicles remain to be established.

Intestinal transmission

The predominant mode of IgG transmission in many mammalian species is by uptake and transfer of colostral and milk immunoglobulins by the absorptive epithelium of the postnatal small intestine (Brambell, 1970). In the horse, pig, ruminants and wallaby, this represents virtually the only means of transfer. In other species, including rodents, a lesser amount of IgG is also transferred by the foetal yolk sac as previously mentioned. Postnatal transmission occurs for only a limited period of time. The duration ranges from as short as 24 hours after birth up to several weeks after birth, depending on the species.

Selection

The degree of selection of IgG from other proteins within the intestinal lumen shows extreme variability from species to species. Transmission in the pig and cow, limited to 24–36 hours after birth, is evidently non-selective. Bangham *et al.* (1958*b*) demonstrated that the calf intestine can transmit all colostral and serum proteins with apparently equal facility. The intestine of each species can transport both homologous and heterologous immunoglobulins and serum albumins (Hansen & Phillips, 1949; Deutsch & Smith, 1957; Balfour & Comline, 1959; Leece, Matrone & Morgan, 1961; Pierce & Smith 1967*a*, *b*; Brown, Smith & Witty, 1968; Porter, 1969, 1972), enzymes (Balconi & Leece, 1966) and polypeptide hormones (Asplund, Grummer & Philips, 1962; Pierce, Risdall & Shaw, 1964), as well as non-protein macromolecules (Balfour & Comline, 1959; Leece *et al.*, 1961; Hardy, 1965). Curiously, however, Staley, Corley, Bush & Jones (1972) have reported that the calf intestine is unable to transmit horse ferritin.

Intestinal transmission in the rat and mouse, in contrast, is highly specific for IgG. Other milk and serum proteins, including albumin, are transferred in the rat much less efficiently if at all (Bangham & Terry, 1957; Jordan & Morgan, 1968; Jones & Waldmann, 1972). Homologous and many heterologous IgG immunoglobulins of all subclasses are transported (Halliday, 1955*a*, 1958; Morris, 1964, 1969, 1976; Jones & Waldmann, 1972; Guyer, Koshland & Knopf, 1976). Other immunoglobulin classes are not selectively transferred (Morris, 1965, 1967; Jones & Waldmann, 1972; Guyer *et al.*, 1976). There is both saturation of transport at high IgG concentrations and competition between different IgG samples in mixtures, as would be predicted if transport were receptor-mediated (Halliday, 1958; Brambell *et al.*, 1958; Morris, 1964; Jones & Waldmann, 1972; Guyer *et al.*, 1976). On the basis of careful kinetic studies (Morris, 1964, 1976; Guyer *et al.*, 1976) there appears to be a single class of specific receptors which recognise Fc determinants.

Site of transfer

It has now been firmly established that selective IgG transmission in the rat, and presumably mouse, is limited to the duodenum and proximal jejunum (Rodewald, 1970; Mackenzie, 1972; Morris & Morris, 1974, 1976*a*; Morris, 1975; Jones, 1976; Waldmann & Jones, 1976). At best, only a very small quantity may be transported by the distal jejunum or ileum, although this has been disputed by one laboratory (Hemmings & Williams, 1977). Morphological and biochemical binding studies (Rodewald, 1976*a*, 1979; Borthistle, Kubo, Brown & Grey, 1977) have confirmed the presence within the proximal intestine of an IgG receptor. This receptor selectively binds the Fc region of all IgG subclasses but neither IgA nor IgM (Borthistle *et al.*, 1977). The receptor is sensitive to treatment with trypsin and disappears at three weeks after birth, the age at which IgG transfer ceases within this region of

the intestine (Morris & Morris, 1974, 1976*b*) as well as within the intact
animal (Halliday, 1955*b*). The IgG receptor appears similar to the receptor
detected earlier in crude membrane fractions and homogenates from whole
intestine (Waldmann & Jones, 1976).

In many studies on those species in which protein transfer is non-specific,
little consideration has been given to the possibility of regional differences in
transport efficiency along the length of the small intestine (Sibalin & Bjorkman,
1966; Mattisson & Karlsson, 1966; Staley, Jones & Marshall, 1968; Kraehenbuhl
& Campiche, 1969; Staley, Jones & Corley, 1969). However, there is some
indication that the jejunum has the highest efficiency on the basis of both
physiological and morphological transport studies (El-Nageh, 1967; Pierce &
Smith, 1967*b*; Staley *et al.*, 1972; Staley, 1977). No evidence for non-specific
receptors has been found in these species.

Proteolysis

There is substantial proteolysis of all milk proteins within the small
intestine of neonatal rats and mice. It has been estimated that even in the case
of IgG, at most only 7–12 % of this protein can be transferred intact in the
rat (Bangham & Terry, 1957; Jones, 1972). However, the method of digestion
differs dramatically in form from luminal digestion in the adult intestine. As
shown by Jones (1972), luminal washes from the neonatal intestine exhibit
only low levels of proteolytic activity within the physiological pH range of
the gut contents (Rodewald, 1976*b*). Instead, considerable proteolytic activity
as well as the presence of lysosomal enzymes are found within the epithelial
cells, with the greatest activities in the cells of the distal small intestine (Noack
et al., 1966; Cornell & Padykula, 1969; Williams & Beck, 1969; Baintner &
Juhász, 1971; Jones, 1972). This has suggested that as material passes down
the intestine, IgG is first removed for transmission by cells in the proximal
region; the remaining proteins are then later absorbed by distal cells and
degraded intracellularly (Rodewald, 1973).

Morris & Morris (1976*a*, 1977*a*, *b*, 1978) have compared quantitatively the
transport and digestive functions of cells in both proximal and distal segments
of the neonatal rat small intestine. They found that proximal cells transfer
intact up to 40 % of that IgG which these cells are able to remove from the
lumen, and digest the remainder. Distal cells, on the other hand, degrade over
98 % of absorbed IgG. Evidently the proximal cells in the rat, while able to
transport IgG functionally intact to the circulation, also digest significant
amounts of this protein in a manner which may be analogous to the situation
found in the foetal yolk sac of this and other species.

Luminal and intracellular proteolysis both appear to be strongly suppressed within the small intestine of the pig and ruminants during the period of IgG transmission (Hardy, 1969a; Kraehenbuhl & Campiche, 1969). Although the studies of Pierce & Smith (1967a) and Hardy (1969a, b) suggest that some fragmentation of IgG may occur during transmission, there is little apparent degradation of transmitted albumin (Brown *et al.*, 1968), lactate dehydrogenase (Balconi & Leece, 1966) or insulin (Asplund *et al.*, 1962; Pierce, Risdall & Shaw, 1964). Concomitant with the loss of transfer function is a rapid rise in luminal protease activity beginning at the second day after birth (Hill, 1956; Huber, Jacobson, Allen & Hartmann, 1961; Hardy, 1969a, b; Baintner, 1973).

Morphology of selective transmission in the rat

As might be expected from the foregoing transport studies, the cells responsible for selective transmission in the proximal region of the rat intestine appear functionally related by their ultrastructure to other IgG-transporting tissues. The columnar absorptive cells (Plate 7) have a well-developed brush border, contain large amounts of transported lipid, and in other ways appear similar in general morphology to adult cells (Clark, 1959; Rodewald, 1973). The neonatal cells alone, however, contain throughout their apical cytoplasm large numbers of irregularly shaped, often tubular vesicles and small coated vesicles 100–200 nm in diameter, similar in appearance to those of the yolk sac and placenta (Rodewald, 1970, 1973). Both vesicle types have a cytoplasmic membrane coat which in the case of the tubular vesicles is only evident during their formation at the microvillar surface (Plates 8 and 9). Larger lysosomal vesicles are present but in strictly reduced numbers, reflecting, no doubt, the minimal digestive function in this region of the intestine.

The vesicular pathway of IgG transport across proximal cells has been studied extensively with anti-peroxidase IgG and IgG–tracer conjugates (Rodewald, 1970, 1973, 1976c). According to the sequence of steps revealed by these tracers, luminal IgG first binds selectively to tubular invaginations at the apical surface (Plate 10). The IgG enters the cell within the tubular vesicles and then is transferred to the small coated vesicles, which migrate to the lateral plasmalemma and release their contents to the extracellular space (Plates 11 and 12). These experiments demonstrated that selection occurs at the apical surface: only those immunoglobulins known to be transferred to the circulation adhered to the surface and entered the cells in appreciable quantities. In addition, Fc but not Fab fragments were selected and crossed the cells in the same manner as intact IgG (Rodewald, 1976c). Transport was rapid, with release of IgG evident at the lateral surface between 15 and 30 minutes after uptake. Selective binding at the apical surface and the system of endocytic vesicles disappeared at three weeks after birth, as predicted by prior physiological studies. A similar pathway may also function in the foetal intestine to a limited degree (Lev & Orlic, 1972; Orlic & Lev, 1973, 1976).

Novel to this scheme of transport is the sequential transfer of IgG from the tubular to smaller coated vesicles. The possibility was proposed that this transfer might allow further selection of molecules after initial endocytosis (Rodewald, 1973). This notion is consistent with observations that a portion of the IgG tracers and all of the non-IgG tracers which were observed to enter the cells do not leave by way of the small coated vesicles but instead enter lysosome-like vesicles (Rodewald, 1973, 1976c; Worthington & Graney, 1973b). This agrees with biochemical evidence for some digestive capability within the cells (Morris & Morris, 1977a, b). The possibility that proteins to be transmitted and those to be digested are segregated within functionally distinct vesicles at the apical plasmalemma (Moxon *et al.*, 1976; Morris & Morris, 1977b) has not been ruled out and needs to be tested. However, two morphologically different classes of vesicles were not readily apparent on that surface. Furthermore, at least many of the small coated vesicles clearly appeared to form on the surface of the tubular vesicles.

Receptors

Recent morphological studies on the rat intestine (Rodewald, 1976a, 1979) have focussed on the distribution and possible movement of immunoglobulin receptors within the jejunal epithelial cells. This work was stimulated in large part by the observation of Jones & Waldmann (1972) that IgG binds efficiently to receptors at pH values below 6.5 but not at pH 7.4 or higher. This meant that IgG might bind at a low pH at the luminal cell surface, and be released at the abluminal surface upon exposure to the extracellular plasma at pH 7.4 (Waldmann & Jones, 1973, 1976). Rodewald (1976a, b, 1979), using IgG–peroxidase conjugates, provided direct morphological evidence for selective binding of IgG to the apical surface of proximal absorptive cells at pH 6.0 which was consistent with the measured pH of the luminal contents. This observation has been confirmed by Nagura, Nakane & Brown (1978). By studying binding of tracers to isolated cells, Rodewald (1976a, 1979) further demonstrated the presence on the abluminal cell surface of IgG receptors which shared the same pH sensitivity of binding. Receptors were found over the entire plasma membrane and were not restricted to the coated vesicles attached to the surface. To explain the presence of abluminal receptors, it was suggested that receptors and receptor–IgG complexes were transported across the cells within several distinct membrane compartments, from the apical membrane to the tubular and coated vesicles, and then to the abluminal membrane. A shuttling of IgG receptors in both directions was proposed which, because of the effect of pH on binding, could allow net transport of IgG in the abluminal direction without net movement of receptors (Rodewald, 1979).

Non-selective transport in the rat

Absorptive epithelial cells in the distal regions of the neonatal rat intestine exhibit a very striking morphology which easily distinguishes them from the proximal cells. As noted by many (Clark, 1959; Graney, 1968; Kraehenbuhl, Gloor & Blanc 1966, 1967; Kraehenbuhl & Campiche, 1969; Rodewald, 1973), the distal cells are characterised by extremely large supra-nuclear vacuoles and an intercommunicating system of apical canaliculi. The canaliculi are not identical to the tubular endocytic vesicles of proximal cells. They have a distinctly different membrane structure in which the luminal membrane leaflet is composed of small, regularly arrayed particles, 7 nm in diameter (Wissig & Graney, 1968; Rodewald, 1973; Limbrick & Robertson, 1974). A cytoplasmic coat on these membranes has not been described.

The well-documented ability of the distinctive canaliculi and supranuclear vacuoles to concentrate luminal IgG and other proteins by endocytosis was originally interpreted as part of a pathway for selective IgG transport (Clark, 1959; Kraehenbuhl *et al.,* 1967; Kraehenbuhl & Campiche, 1969). Almost all recent evidence with ultrastructural tracers, however, points to a strictly non-selective phagocytic function for these vesicles (Graney, 1968; Hugon, 1971; Rodewald, 1973; Worthington & Graney, 1973*a*). The transport of all tracers exclusively to the supranuclear vacuoles and the presence of lysosomal enzymes within the vacuoles (Cornell & Padykula, 1969) support the bio-chemical evidence that virtually all proteins which enter the cells are digested (Morris & Morris, 1977*a*, *b*, 1978).

Non-selective transmission in other species

As pointed out previously, proteins including IgG are transported non-selectively and intact to the circulation by the small intestine in the pig and cow. The fine structure of the absorptive cells within the small intestine of these species has not been studied to the same extent as in the rat. However, on the basis of reported studies (Sibalin & Bjorkman, 1966; Staley *et al.*, 1968, 1972; Kraehenbuhl & Campiche, 1969; Munn & Smith, 1974), the jejunal cells, believed responsible for transmission, appear similar in important respects to the distal cells in the rat. They contain an extensive system of endocytic vesicles in their apical cytoplasm and abundant large vacuoles, although these vacuoles are even larger in more distal cells. Most importantly, the endocytic vesicles in the jejunum of both the pig and cow display the striking beaded coat on the luminal membrane surface as seen in rat ileal cells (Staley *et al.*, 1968, 1972). It is interesting to note that some published micrographs (Staley *et al.*, 1972: figures 7 and 10) give the impression of an additional cytoplasmic coat on at least some of these beaded membrane vesicles.

The route of non-selective transfer has been studied using ferritin, peroxidase (Staley *et al.*, 1972; Staley, 1977), anti-peroxidase IgG (Kraehenbuhl & Campiche, 1969) and ferritin—IgG conjugates (Staley *et al.*, 1972), with generally very similar results. In both the cow and pig, tracers in the intestinal lumen are transferred within beaded vesicles to the large vacuoles where they are concentrated. The vacuoles evidently do not contain active lysosomal enzymes (Kraehenbuhl & Campiche, 1969). Instead, the presence of large tracer droplets in the extracellular spaces has suggested that the vacuoles discharge their undigested contents *en masse* at the basal cell surface (Kraehenbuhl & Campiche, 1969; Staley, 1977) as had been postulated previously on the basis of light microscopic observations (Payne & Marsh, 1962; El-Nageh, 1967). However, actual fusion of apical vacuoles with the abluminal plasmalemma has not been documented. In addition, it is difficult to explain the observations of Staley *et al.* (1972) that a ferritin—IgG conjugate, but not free ferritin is able to enter the jejunal cell at the luminal cell surface. No evidence has been presented in any of these studies for the involvement of spherical coated vesicles, in contrast to the findings for the selective transport systems of other tissues already discussed.

Conclusions

Intestinal transmission of IgG may be either selective or non-selective, depending on the species. In the neonatal rat, selective transmission occurs across cells in the duodenum and proximal jejunum. A highly specific IgG receptor is present on the apical cell surface and within endocytic tubular vesicles which at the time of formation appear to have a cytoplasmic coat. Receptor-bound IgG is transported from the tubular vesicles to spherical coated vesicles which, aided by a shift in pH, discharge the IgG at the lateral plasmalemma. Proteolysis of IgG is largely suppressed both within these proximal cells and within the gut lumen. Instead, most other proteins are absorbed non-selectively by ileal cells and degraded intracellularly within large vacuoles.

In newborn pigs and cows, proteins are transmitted non-selectively to the circulation within the jejunum. Cells responsible for transport appear most closely related to ileal cells in the rat on the basis of their distinctive beaded vesicles. Proteins are apparently transferred across the cells by way of these vesicles without the involvement of small coated vesicles. Very little digestion of proteins occurs anywhere within the intestine during the short period of transmission.

Overview

An intriguing common problem faced by all of the tissues which have been considered in this review is the separation of IgG transmission from the antithetical function of protein digestion. The solution to this problem as proposed by Brambell (1966, 1970) was that specific receptors existed which could not only account for selection of IgG but would also protect the bound IgG from proteolytic enzymes during transport. From the evidence reviewed here, it is clear that, indeed, receptors do exist when selection occurs but that the receptors need not have any immediate protective function. Instead, the different tissues have dissociated the digestive and transport functions from each other, albeit in apparently three very different ways. In the small intestine of the pig and cow, where transmission is non-selective, the two functions occur at different times after birth. In the case of intestinal transport in the rat, digestion and selective transmission occur simultaneously but for the most part in different regions of the intestine. As evident in the yolk sac and probably in the placenta, the two functions are segregated at the subcellular level by what may be completely separate endocytic mechanisms.

A compelling case can be made that coated vesicles are involved in transport in all of those tissues in which IgG is transmitted selectively. However, it is equally evident that these vesicles do not have a unique structure or one simple function. The vesicles, even within one cell type, can assume different shapes, such as tubular or spherical, and their characteristic cytoplasmic coat may in some instances be transient. Many of the coated vesicles in the different tissues no doubt contain receptors, but these receptors may be specific in their binding, as for IgG, or relatively non-specific, as in cases in which they function in non-selective endocytosis of proteins. The fascinating possibility exists that different receptors within a cell may be segregated within different subpopulations of the coated vesicles.

Future research on these systems will certainly focus on the structure of the receptors and the chemical nature of their interactions with transported proteins. It is highly likely that an additional goal of these studies will be to determine specifically how receptors interact with the molecular components of coated vesicle membranes which are now being characterised in several laboratories.

The author thanks Dale Abrahamson, David Begg, Avon Hudson and Steve Ifshin for their valuable assistance in the preparation of this manuscript. The work was supported by a grant from NIH (AI-11937).

References

Asplund, J.M., Grummer, R.H. & Philips, P.H. (1962). Absorption of colostral γ-globulins and insulin by the newborn pig. *Journal of Animal Science*, 21, 412–13.

Baintner, K. (1973). The physiological role of colostral trypsin inhibitor: experiments with piglets and kittens. *Acta Vet. Acad. Scient. Hung.* 23, 247–60.

Baintner, K. & Juhász, S. (1971). Mucosal proteolytic activity in the small intestine of suckling rats. *Acta Physiol. Acad. Scient. Hung.* 40, 179–86.

Balconi, I.R. & Leece, J.G. (1966). Intestinal absorption of homologous lactic dehydrogenase isoenzymes by the neonatal pig. *Journal of Nutrition*, 88, 233–8.

Balfour, A. & Jones, E.A. (1976). The binding of IgG to human placental membranes. In *Maternofoetal Transmission of Immunoglobulins*, ed. W.A. Hemmings, pp. 61–75. Cambridge University Press.

Balfour, W.E. & Comline, R.S. (1959). The specificity of the intestinal absorption of large molecules by the newborn calf. *Journal of Physiology*, 148, 77–8P.

Bangham, D.R., Hobbs, K.R. & Terry, R.J. (1958a). Selective placental transfer of serum-proteins in the rhesus. *Lancet*, ii, 351–4.

Bangham, D.R., Ingram, P.L., Roy, J.H., Shillam, K.W. & Terry, R.J. (1958b). The absorption of ^{131}I-labelled serum and colostral proteins from the gut of the young calf. *Proceedings of the Royal Society of London, Series B*, 149, 184–91.

Bangham, D.R. & Terry, R.J. (1957). The absorption of I^{131}-labelled homologous and heterologous serum proteins fed orally to young rats. *Biochemical Journal*, 66, 579–83.

Barnes, J.M. (1959). Antitoxin transfer from mother to foetus in the guinea pig. *Journal of Pathology and Bacteriology*, 77, 371–80.

Batty, I., Brambell, R.W.R., Hemmings, W.A. & Oakley, C.L. (1954). Selection of antitoxins by the foetal membranes of rabbits. *Proceedings of the Royal Society of London, Series B*, 142, 452–71.

Beck, F., Lloyd, J.B. & Griffiths, A. (1967). A histochemical and biochemical study of some aspects of placental function in the rat using maternal injection of horseradish peroxidase. *Journal of Anatomy*, 101, 461–78.

Block, K.J., Ovary, Z., Kourilsky, F.M. & Benacerraf, B. (1963). Properties of guinea pig 7S antibodies. VI. Transmission of antibodies from maternal to fetal circulation. *Proceedings of the Society for Experimental Biology and Medicine*, 114, 79–82.

Borthistle, B.K., Kubo, R.T., Brown, W.R. & Grey, H.M. (1977). Studies on receptors for IgG on epithelial cells of the rat intestine. *Journal of Immunology*, 119, 471–6.

Brambell, F.W.R. (1966). The transmission of immunity from mothers to young and the catabolism of immunoglobulins. *Lancet*, ii, 1087–93.

Brambell, F.W.R. (1970). *The Transmission of Passive Immunity from Mother to Young*. Amsterdam: North-Holland.

Brambell, F.W.R. & Halliday, R. (1956). The route by which passive immunity is transmitted from mother to foetus in the rat. *Proceedings of the Royal Society of London, Series B*, 145, 170–8.

Brambell, F.W.R., Halliday, R. & Morris, I.G. (1958). Interference by human and bovine serum and serum protein fractions with the absorption of antibodies by suckling rats and mice. *Proceedings of the Royal Society of London, Series B*, 149, 1–11.

Brambell, F.W.R., Hemmings, W.A., Henderson, M., Parry, H.J. & Rowlands, W.T. (1949). The route of antibodies passing from the maternal to the foetal circulation in rabbits. *Proceedings of the Royal Society of London, Series B*, 136, 131–44.

Brambell, F.W.R., Hemmings, W.A., Oakley, C.L. & Porter, R.R. (1960). The relative transmission of the fractions of papain hydrolyzed homologous γ-globulin from the uterine cavity to the foetal circulation in the rabbit. *Proceedings of the Royal Society of London, Series B*, 151, 478–82.

Brambell, F.W.R., Hemmings, W.A. & Rowlands, W.T. (1948). The passage of antibodies from the maternal circulation into the embryo in rabbits. *Proceedings of the Royal Society of London, Series B*, 135, 390–403.

Brown, P., Smith, M.W. & Witty, R. (1968). Interdependence of albumin and sodium transport in the foetal and new-born pig intestine. *Journal of Physiology*, 198, 365–81.

Carpenter, S.J. & Ferm, V.H. (1966). Electron microscopic observations on the uptake and storage of Thorotrast by rodent yolk sac epithelial cells. *Anatomical Record*, 154, 327.

Chandra, R.K. (1976). Levels of IgG subclasses, IgA, IgM, and tetanus antitoxin in paired maternal and foetal sera: findings in healthy pregnancy and placental insufficiency. In *Maternofoetal Transmission of Immunoglobulins*, ed. W.A. Hemmings, pp. 77–90. Cambridge University Press.

Christie, G.A. (1967). Comparative histochemical distribution of acid phosphatase, non-specific esterase, and β-glucuronidase in the placenta and foetal membranes. *Histochemie*, 12, 189–207.

Clark, S.L. (1959). The ingestion of proteins and colloidal materials by columnar absorptive cells of the small intestine in suckling rats and mice. *Journal of Biophysical and Biochemical Cytology*, 5, 41–50.

Cornell, R. & Padykula, H.A. (1969). A cytological study of intestinal absorption in the suckling rat. *American Journal of Anatomy*, 125, 291–316.

Dancis, J., Lind, J., Oratz, M., Smolens, J. & Vara, P. (1961). Placental transfer of proteins in human gestation. *American Journal of Obstetrics and Gynecology*, 82, 167–71.

Dempsey, E.W., Lessey, R.A. & Luse, S.A. (1970). Electron microscopic observations on fibrinoid and histiotroph in the junctional zone and villi of the human placenta. *American Journal of Anatomy*, 128, 463–84.

Deren, J.J., Padykula, H.A. & Wilson, T.H. (1966). Development of structure and function in the mammalian yolk sac. II. Vitamin B_{12} uptake by rabbit yolk sacs. *Developmental Biology*, 13, 349–69.

Deutsch, H.F. & Smith, V.R. (1957). Intestinal permeability to proteins in the newborn herbivore. *American Journal of Physiology*, 191, 271–6.

Dreskin, R.B., Spicer, S.S. & Greene, W.B. (1970). Ultrastructural localisation of chorionic gonadotropin in human term placenta. *Journal of Histochemistry and Cytochemistry*, 18, 862–74.

El-Nageh, M.M. (1967). Voies d'absorption des gamma globulines du colostrum au niveau de l'intestine grêle du veau nouveau-né. *Annals of Veterinary Medicine*, 11, 384–90.

Elson, J., Jenkinson, E.J. & Billington, W.D. (1975). Fc receptors on mouse placenta and yolk sac cells. *Nature, London*, 255, 412–14.

Enders, A.C. (1965). A comparative study of the fine structure of the trophoblast in several hemochorial placentas. *American Journal of Anatomy*, 116, 29–68.

Faulk, W.P., Jeannet, M., Creighton, W.D. & Carbonara, A. (1974). Immunological studies of human placentae: characterisation of immunoglobulins on trophoblastic basement membranes. *Journal of Clinical Investigation*, 54, 1011–19.

Gitlin, D. & Biasucci, A. (1969). Development of IgG, IgA, IgM, beta 1_c, beta 1_a, C_1 esterase inhibitor, ceruloplasmin, transferrin, hemopexin, haptoglobin, fibrinogen, plasminogen, alpha-1-antitrypsin, mucoid beta-lipoprotein, alpha-2-macroglobulin and prealbumin in the human conceptus. *Journal of Clinical Investigation*, 48, 1433–46.

Gitlin, D. & Koch, C. (1968). On the mechanisms of maternofetal transfer of human albumin and γG globulin in the mouse. *Journal of Clinical Investigation*, 47, 1204–9.

Gitlin, D., Kumate, J., Urrusti, J. & Morales, C. (1964a). The selectivity of the human placenta in the transfer of plasma proteins from mother to fetus. *Journal of Clinical Investigation*, 43, 1938–51.

Gitlin, D., Kumate, J., Urrusti, J. & Morales, C. (1964b). Selective and directional transfer of 7S γ-globulin across the human placenta. *Nature, London*, 203, 86.

Gitlin, D. & Morphis, L.G. (1969). Systems of materno-foetal transport of γG immunoglobulin in the mouse. *Nature, London*, 223, 195–6.

Graney, D.O. (1968). The uptake of ferritin by ileal absorptive cells in suckling rats. An electron microscope study. *American Journal of Anatomy*, 123, 227–54.

Greene, W.B. & Spicer, S.S. (1970). Cytochemistry of multivesicular bodies in human term placenta. *Journal of Histochemistry and Cytochemistry*, 18, 687.

Guyer, R.L., Koshland, M.E. & Knopf, P.M. (1976). Immunoglobulin binding by mouse intestinal epithelial cell receptors. *Journal of Immunology*, 117, 587–93.

Halliday, R. (1955a). The absorption of antibodies from immune sera by the gut of the young rat. *Proceedings of the Royal Society of London, Series B*, 143, 408–13.

Halliday, R. (1955b). Prenatal and postnatal transmission of passive immunity to young rats. *Proceedings of the Royal Society of London, Series B*, 144, 427–30.

Halliday, R. (1958). The absorption of antibody from immune sera and from mixtures of sera by the gut of the young rat. *Proceedings of the Royal Society of London, Series B,* 148, 92–103.

Hansen, R.G. & Phillips, P.H. (1949). Studies on proteins from bovine colostrum. III. The homologous and heterologous transfer of ingested protein to the blood stream of the young animal. *Journal of Biological Chemistry,* 179, 523–7.

Hardy, R.N. (1965). Intestinal absorption of macromolecules in the new-born pig. *Journal of Physiology,* 176, 19–20P.

Hardy, R.N. (1969*a*). The break-down of [^{131}I] γ-globulin in the digestive tract of the new-born pig. *Journal of Physiology,* 205, 435–51.

Hardy, R.N. (1969*b*). Proteolytic activity during the absorption of [^{131}I] γ-globulin in the new-born calf. *Journal of Physiology,* 205, 453–70.

Hardy, R.N., Hockaday, A.R. & Tapp, R.L. (1971). Observations on the structure of the small intestine in foetal, neo-natal and suckling pigs. *Philosophical Transactions of the Royal Society of London, Series B,* 259, 517–31.

Hay, F.C., Hull, M.S. & Torrigïans, G. (1971). The transfer of human IgG subclasses from mother to foetus. *Clinical and Experimental Immunology,* 9, 355–8.

Hemmings, W.A. (1956). Protein selection in the yolk sac splanchnopleur of the rabbit: the distribution of isotope following injection of ^{131}I-labelled serum globulin into the uterine cavity. *Proceedings of the Royal Society of London, Series B,* 145, 186–95.

Hemmings, W.A. (1957). Protein selection in the yolk sac splanchnopleur of the rabbit: the total uptake estimated as loss from the uterus. *Proceedings of the Royal Society of London, Series B,* 148, 76–83.

Hemmings, W.A. (1961). Protein transfer and selection. In *First International Conference on Congenital Malformations,* ed. M. Fishbein, pp. 223–9. Philadelphia: J.B. Lippincott.

Hemmings, W.A. (1973). Transport of IgM antibody to the rabbit foetus. *Immunology,* 25, 165–6.

Hemmings, W.A. (1974). Transport of the subclasses of human IgG across the yolk-sac of the foetal rabbit. *Immunology,* 27, 693–7.

Hemmings, W.A. (1975). Attachment of rabbit and bovine IgG to Schlamowitz vesicles prepared from rabbit foetal yolk-sacs. I. The ratio of the two proteins. *IRCS Medical Science,* 3, 466.

Hemmings, W.A. & Jones, R.E. (1962). The occurrence of macroglobulin antibodies in maternal and foetal sera of rabbits as determined by gradient centrifugation. *Proceedings of the Royal Society of London, Series B,* 157, 27–32.

Hemmings, W.A. & Williams, E.W. (1974). The attachment of IgG to cell components: a reconsideration of Brambell's receptor hypothesis of protein transmission. *Proceedings of the Royal Society of London, Series B,* 187, 209–19.

Hemmings, W.A. & Williams, E.W. (1975). The use of direct deposition electron microscope autoradiography in studies of protein transport. *Journal of Microscopy,* 106, 131–43.

Hemmings, W.A. & Williams, E.W. (1976). The attachment of IgG to cell components of transporting membranes. In *Maternofoetal Transmission of Immunoglobulins*, ed. W.A. Hemmings, pp. 91–111. Cambridge University Press.

Hemmings, W.A. & Williams, E.W. (1977). Quantitative and visualisation studies of the transport of rat and bovine IgG and ferritin across the segments of the small intestine of the suckling rat. *Proceedings of the Royal Society of London, Series B*, 197, 400–25.

Hill, K.J. (1956). Gastric development and antibody transference in the lamb, with some observations on the rat and guinea-pig. *Quarterly Journal of Experimental Physiology*, 41, 421–32.

Hillman, K., Schlamowitz, M. & Shaw, A.R. (1977). Characterisation of IgG receptors of the fetal rabbit yolk sac membrane: localisation to subcellular fraction and effects of chemical agents and enzymes on binding. *Journal of Immunology*, 118, 782–8.

Huber, J.T., Jacobson, N.L. Allen, R.S. & Hartmann, P.A. (1961). Digestive enzyme activities in the young calf. *Journal of Dairy Science*, 44, 1494–501.

Hugon, J.S. (1971). Absorption of horseradish peroxidase by the mucosal cells of the duodenum of mouse. II. The newborn mouse. *Histochemie*, 26, 19–27.

Jenkinson, E.J., Billington, W.D. & Elson, J. (1976). Detection of receptors for immunoglobulin on human placenta by EA rosette formation. *Clinical and Experimental Immunology*, 23, 456–61.

Johnson, P.M., Faulk, P.W. & Wang, A.C. (1976). Immunological studies of human placentae: subclass and fragment specificity of binding of aggregated IgG by placental endothelial cells. *Immunology*, 31, 659–64.

Jollie, W.P. & Triche, T.J. (1971). Ruthenium labeling of micropinocytotic activity in the rat visceral yolk-sac placenta. *Journal of Ultrastructure Research*, 35, 541–53.

Jones, E.A. & Waldmann, T.A. (1972). The mechanism of intestinal uptake and transcellular transport of IgG in the neonatal rat. *Journal of Clinical Investigation*, 51, 2916–27.

Jones, R.E. (1972). Intestinal absorption and degradation of rat and bovine globulins in the suckling rat. *Biochimica et Biophysica Acta*, 255, 530–8.

Jones, R.E. (1976). Studies on the transmission of bovine IgG across the intestine of the young rat. In *Maternofoetal Transmission of Immunoglobulins*, ed. W.A. Hemmings, pp. 325–39. Cambridge University Press.

Jordan, S.M. & Morgan, E.H. (1968). The development of selectivity of protein absorption from the intestine during suckling in the rat. *Australian Journal of Experimental Biology and Medical Science*, 46, 465–72.

Kanaseki, T. & Kadota, K. (1969). The 'vesicle in a basket'. A morphological study of the coated vesicle isolated from the nerve endings of the guinea pig brain, with special reference to the mechanism of membrane movements. *Journal of Cell Biology*, 42, 202.

Kaplan, K.C., Catsoulis, E.A. & Franklin, E.C. (1965). Maternal–foetal transfer of human immunoglobulins and fragments in rabbits. *Immunology*, 8, 354.

Kenworthy, R., Stubbs, J.M. & Syme, G. (1967). Ultrastructure of small intestine epithelium in weaned and unweaned pigs and pigs with post-weaning diarrhoea. *Journal of Pathological Bacteriology*, 93, 493–8.

King, B.F. (1974). An electron microscopic investigation of the surface coat of visceral yolk sac endoderm cells in the guinea pig. *Anatomical Record*, 180, 299–308.

King, B.F. (1977). An electron microscopic study of absorption of peroxidase-conjugated immunoglobulin G by guinea pig visceral yolk-sac *in vitro*. *American Journal of Anatomy*, 148, 447–55.

King, B.F. & Enders, A.C. (1970a). The fine structure of the guinea pig visceral yolk sac placenta. *American Journal of Anatomy*, 127, 397–414.

King, B.F. & Enders, A.C. (1970b). Protein absorption and transport by the guinea pig visceral yolk sac placenta. *American Journal of Anatomy*, 129, 261–88.

King, B.F. & Enders, A.C. (1971). Protein absorption by the guinea pig chorio-allantoic placenta. *American Journal of Anatomy*, 130, 409–30.

King, B.F. & Tibbitts, F.D. (1976). The fine structure of the chinchilla placenta. *American Journal of Anatomy*, 145, 33–56.

Knutton, S., Limbrick, A.R. & Robertson, J.D. (1974). Regular structures in membranes. I. Membranes in the endocytic complex of ileal epithelial cells. *Journal of Cell Biology*, 62, 679–94.

Koch, C., Boesman, M. & Gitlin, D. (1967). Maternofoetal transfer of γG immunoglobulins. *Nature, London*, 216, 1116–17.

Kohler, P.F. & Farr, R.S. (1966). Elevation of cord over maternal IgG immunoglobulin: evidence for an active placental IgG transport. *Nature, London*, 210, 1070–1.

Kraehenbuhl, J.P. & Campiche, M.A. (1969). Early stages of intestinal absorption of specific antibodies in the newborn. An ultrastructural, cytochemical and immunological study in the pig, rat, and rabbit. *Journal of Cell Biology*, 42, 345–65.

Kraehenbuhl, J.P., Gloor, E. & Blanc, B. (1966). Morphologie comparée de la muqueuse intestinale de deux espèces animales aux possibilités d'absorption protéique néonatale différentes. *Zeitschrift für Zellforschung und Mikroskopische Anatomie*, 70, 209–19.

Kraehenbuhl, J.P., Gloor, E. & Blanc, B. (1967). Résorption intestinale de la ferritine chez deux espèces animals aux possibilités d'absorption protéique néonatale différentes. *Zeitschrift für Zellforschung und Mikroskopische Anatomie*, 76, 170–86.

Krzyzowska-Gruca, S. & Schiebler, T.H. (1967). Experimentelle Untersuchungen am Dottersackepithel der Ratte. *Zeitschrift für Zellforschung und Mikroskopischt Anatomie*, 79, 157–71.

Kulungara, A.C. & Schechtman, A.M. (1962). Passage of heterologous serum proteins from mother into foetal compartments in the rabbit. *American Journal of Physiology*, **203**, 1071–80.

Kulangara, A.C. & Schechtman, A.M. (1963). Do heterologous proteins pass from mother to fetus in cow, cat and guinea pig? *Proceedings of the Society for Experimental Biology and Medicine*, **112**, 220–2.

Lambson, R.O. (1966). An electron microscopic visualisation of transport across rat visceral yolk sac. *American Journal of Anatomy*, **118**, 21–51.

Leece, J.C., Matrone, G. & Morgan, D.O. (1961). Porcine neonatal nutrition: absorption of unaltered nonporcine proteins and polyvinylpyrrolidone from the gut of piglets and the subsequent effect on the maturation of the serum protein profile. *Journal of Nutrition*, **73**, 158–66.

Leissring, J.C. & Anderson, J.W. (1961). The transfer of serum proteins from mother to young in the guinea pig. I. Prenatal rates and routes. *American Journal of Anatomy*, **109**, 149–55.

Lev, R. & Orlic, D. (1972). Protein absorption by the intestine of the fetal rat *in utero*. *Science*, **177**, 522–4.

Lister, U.M. (1964). Ultrastructure of the early human placenta. *Journal of Obstetrics and Gynaecology of the British Commonwealth*, **71**, 21–32.

Mackenzie, D.D.S. (1972). Selective uptake of immunoglobulins by the proximal intestine of suckling rats. *American Journal of Physiology*, **223**, 1286–95.

McNabb, T., Kuh, T.Y., Dorrington, K.J. & Painter, R.H. (1975). Structure and function of immunoglobulin domains. V. Binding of immunoglobulin G and fragments to placental membrane preparations. *Journal of Immunology*, **117**, 882–8.

Matre, R. (1977). Similarities of Fcγ receptors on trophoblasts and placental endothelial cells. *Scandinavian Journal of Immunology*, **6**, 953–8.

Matre, R., Tunder, O. & Enderessen, C. (1975). Fc receptors in human placenta. *Scandinavian Journal of Immunology*, **4**, 741.

Mattisson, A.G. & Karlsson, B.W. (1966). Electron microscopic and immuno-chemical studies on the small intestine of newborn piglets. *Arkiv. Zool.* **18**, 575–89.

Mellbye, O.J., Natvig, J.B. & Krarstein, B. (1971). Presence of IgG subclasses and C1q in human cord serum. In *Protides of the Biological Fluids*, ed. H. Peeters, pp. 127–31. Oxford: Pergamon Press.

Midgley, A.R. & Pierce, G.P. (1962). Immunohistochemical localisation of human chorionic gonadotropin. *Journal of Experimental Medicine*, **115**, 289–94.

Miller, D.L., Zapata, R., Hutchinson, D.L. & Gitlin, D. (1973). Maternofetal passage of human IgE in the pregnant monkey, mouse, rat and guinea pig. *Federation Proceedings*, **32**, 1013A.

Morgan, E.H. (1964). Passage of transferrin, albumin, and gamma globulin from maternal plasma to foetus in the rat and rabbit. *Journal of Physiology*, **171**, 26–41.

Morris, B. (1975). The transmission of ^{125}I-labelled immunoglobulin G by proximal and distal regions of the small intestine of 16-day-old rats. *Journal of Physiology*, **245**, 249–59.

Morris, B. & Morris, R. (1974). The absorption of ^{125}I-labelled immunoglobulin G by different regions of the gut in young rats. *Journal of Physiology*, **241**, 761–70.

Morris, B. & Morris, R. (1976a). Quantitative assessment of the transmission of labelled protein by the proximal and distal regions of the small intestine of young rats. *Journal of Physiology*, **255**, 619–34.

Morris, B. & Morris, R. (1976b). The effects of corticosterone and cortisone on the transmission of IgG and the uptake of polyvinylpyrrolidone by the small intestine in young rats. *Journal of Physiology*, **254**, 389–403.

Morris, B. & Morris, R. (1977a). Fractionation studies on the absorption of labelled immunoglobulin G by the gut of young rats. *Journal of Physiology*, **265**, 429–42.

Morris, B. & Morris, R. (1977b). The digestion and transmission of labelled immunoglobulin G by enterocytes of the proximal and distal regions of the small intestine of young rats. *Journal of Physiology*, **273**, 427–42.

Morris, B. & Morris, R. (1978). Immunoglobulin transmission in the neonatal rat. *Journal of Physiology*, **276**, 59P.

Morris, I.G. (1964). The transmission of antibodies and normal γ-globulins across the young mouse gut. *Proceedings of the Royal Society of London, Series B*, **160**, 276–92.

Morris, I.G. (1965). The transmission of anti-*Brucella abortus* agglutinins across the gut in young rats. *Proceedings of the Royal Society of London, Series B*, **163**, 402–16.

Morris, I.G. (1967). The transmission of bovine anti-*Brucella abortus* agglutinins across the gut of suckling rats. *Immunology*, **13**, 49–61.

Morris, I.G. (1969). The selective transmission of bovine γG-globulins across the gut of suckling rodents. *Immunology*, **17**, 139–49.

Morris, I.G. (1976). Intestinal transmission of IgG subclasses in suckling rats. In *Maternofoetal Transmission of Immunoglobulins*, ed. W.A. Hemmings, pp. 341–57. Cambridge University Press.

Moskalewski, S., Ptak, W. & Czannik, Z. (1975). Demonstration of cells with IgG receptors in human placenta. *Biology of the Neonate*, **26**, 268.

Moxon, L.A., Wild, A.E. & Slade, B.S. (1976). Localisation of proteins in coated micropinocytotic vesicles during transport across rabbit yolk sac endoderm. *Cell and Tissue Research*, **171**, 175–93.

Muir, A.R. (1966). On the phagocytosis of iron-dextran by the human plasmoditrophoblast. *Journal of Obstetrics and Gynaecology of the British Commonwealth*, **73**, 966–72.

Munn, E.A. & Smith, M.W. (1974). Uptake of albumin by neonatal pig ileum incubated *in vitro*. *Journal of Physiology*, **242**, 30–2P.

Nagura, R., Nakane, P.K. & Brown, W.R. (1978). Breast milk IgA binds to jejunal epithelium in suckling rats. *Journal of Immunology*, **120**, 1333–9.

Noack, R., Koldovský, O., Friedrich, M., Heringová, A., Jirsová, V. & Schenk, G. (1966). Proteolytic and peptidase activities of the jejunum and ileum of the rat during postnatal development. *Biochemical Journal*, 100, 775–8.

Ockleford, C.D. (1976). A three-dimensional reconstruction of the polygonal pattern on placental coated-vesicle membranes. *Journal of Cell Science*, 21, 83–91.

Ockleford, C.D. & Menon, G. (1977). Differentiated regions of human placental cell surface associated with exchange of materials between maternal and foetal blood: a new organelle and the binding of iron. *Journal of Cell Science*, 25, 279–91.

Ockleford, C.D. & Whyte, A. (1977). Differentiated regions of human placental cell surface associated with exchange of materials between maternal and foetal blood: coated vesicles. *Journal of Cell Science*, 25, 293–312.

Ockleford, C.D., Whyte, A. & Bowyer, D.E. (1977). Variation in the volume of coated vesicles isolated from human placenta. *Cell. Biol. Int. Rep.* 1, 137–46.

Orlic, D. & Lev, R. (1973). Fetal rat intestinal absorption of horseradish peroxidase from swallowed amniotic fluid. *Journal of Cell Biology*, 56, 106–19.

Orlic, D. & Lev, R. (1976). Foetal intestinal absorption of protein and radio-iron *in utero*. In *Maternofoetal Transmission of Immunoglobulins*, ed. W.A. Hemmings, pp. 417–30. Cambridge University Press.

Padykula, H.A., Deren, J.J. & Wilson, T.H. (1966). Development of structure and function in the mammalian yolk sac. I. Developmental morphology and vitamin B_{12} uptake of the rat yolk sac. *Developmental Biology*, 13, 311–48.

Parry, L.M., Beck, F. & Lloyd, J.B. (1968). *In vitro* measurements of intracellular digestion. *Journal of Anatomy*, 103, 393.

Payne, L.B. & Marsh, C.L. (1962). Gamma globulin absorption in the baby pig: the nonselective absorption of heterologous globulins and factors influencing absorption time. *Journal of Nutrition*, 76, 151–8.

Pearse, B.M.F. (1975). Coated vesicles from pig brain: purification and biochemical characterisation. *Journal of Molecular Biology*, 97, 93–8.

Pearse, B.M.F. (1976). Clathrin: a unique protein associated with intracellular transfer of membrane by coated vesicles. *Proceedings of the National Academy of Sciences*, 73, 1255–9.

Petry, G. & Kühnel, W. (1963). Histotopographische und cytologische Studien an den Embryonalhüllen des Meerschweinchens. *Zeitschrift für Zellforschung und Mikroskopische Anatomie*, 59, 625–62.

Petry, G. & Kühnel, W. (1965). Der Feinbau des Dottersackepithels und dessen Beziehung zur Eiweissresorption (Kaninchen). *Zeitschrift für Zellforschung und Mikroskopische Anatomie*, 65, 27–46.

Pierce, A.E., Risdall, P.C. & Shaw, B. (1964). Absorption of orally administered insulin by the newly born calf. *Journal of Physiology*, 171, 203–15.

Pierce, A.E. & Smith, M.W. (1967a). The intestinal absorption of pig and bovine

immune lactoglobulin and human serum albumin by the newborn
pig. *Journal of Physiology*, 190, 1–18.

Pierce, A.E. & Smith, M.W. (1967*b*). The *in vitro* transfer of bovine immune
lactoglobulin across the intestine of newborn pigs. *Journal of
Physiology*, 190, 19–34.

Pierce, G.B., Midgley, A.R. & Beales, T.F. (1964). An ultrastructural study of
differentiation and maturation of trophoblast of the monkey.
Laboratory Investigation, 13, 451–64.

Porter, P. (1969). Transfer of immunoglobulins IgG, IgA and IgM to lacteal
secretions in the parturient sow and their absorption by the neonatal
piglet. *Biochimica et Biophysica Acta*, 181, 381–92.

Porter, P. (1972). Immunoglobulins in bovine mammary secretions. Quantitative
changes in early lactation and absorption by the neonatal calf. *Immunology*, 23, 225–38.

Robertson, T.A., Archer, J.M., Papadimitriou, J.M. & Walters, M.N.I. (1971).
Transport of horseradish peroxidase in the murine placenta. *Journal
of Pathology*, 103, 141–7.

Rodewald, R. (1970). Selective antibody transport in the proximal small
intestine of the neonatal rat. *Journal of Cell Biology*, 45, 635–40.

Rodewald, R. (1973). Intestinal transport of antibodies in the newborn rat.
Journal of Cell Biology, 58, 198–211.

Rodewald, R. (1976*a*). Distribution of immunoglobulin receptors in the small
intestine of the neonatal rat. *Journal of Cell Biology*, 70, 165*a*.

Rodewald, R. (1976*b*). pH-dependent binding of immunoglobulins to intestinal
cells of the neonatal rat. *Journal of Cell Biology*, 71, 666–70.

Rodewald, R. (1976*c*). Intestinal transport of peroxidase-conjugated IgG
fragments in the neonatal rat. In *Maternofoetal Transmission of
Immunoglobulins*, ed. W.A. Hemmings, pp. 137–49. Cambridge
University Press.

Rodewald, R. (1979). Distribution of immunoglobulin G receptors in the
small intestine of the young rat. *Journal of Cell Biology*, in press.

Roth, T.F. & Porter, K.R. (1964). Yolk protein uptake in the oocyte of the
mosquito. *Journal of Cell Biology*, 20, 313–32.

Rowe, D.S., Crabbe, P.A. & Turner, M.W. (1968). Immunoglobulin D in serum,
body fluids and lymphoid tissues. *Clinical and Experimental Immunology*, 3, 477–90.

Schlamowitz, M. (1976). Maternofoetal transmission of protein in the rabbit:
transfer *in vivo* and binding *in vitro* to the yolk sac membrane. In
Maternofoetal Transmission of Immunoglobulins, ed. W.A. Hemmings,
pp. 179–200. Cambridge University Press.

Schlamowitz, M., Hillman, K., Lichtiger, P. & Ahearn, J.J. (1975). Preparation
of IgG binding membrane vesicles from the microvillar brush border
of the fetal rabbit yolk sac. *Journal of Immunology*, 115, 296–302.

Seibel, W. (1974). An ultrastructural comparison of the uptake and transport
of horseradish peroxidase by the rat visceral yolk sac placenta
during mid- and late gestation. *American Journal of Anatomy*, 140,
213–36.

Sibalin, M. & Bjorkman, N. (1966). On the fine structure and absorptive function of the porcine jejunal villi during the early suckling period. *Experimental Cell Research*, 44, 165–74.

Slade, B.S. (1970). An attempt to visualise protein transmission across the rabbit visceral yolk sac. *Journal of Anatomy*, 107, 531–45.

Slade, B.S. (1975). The role of coated micropinocytotic vesicles in the transport of homologous immunoglobulins across the rabbit yolk sac endoderm. *IRCS Medical Science*, 3, 235.

Slade, B.S. & Wild, A.E. (1971). Transmission of human γ-globulin to rabbit foetus and its inhibition by conjugation with ferritin. *Immunology*, 20, 217–23.

Sonoda, S. & Schlamowitz, M. (1972*a*). Kinetics and specificity of transfer of immunoglobulin G and serum albumin across rabbit yolk sac *in utero*. *Journal of Immunology*, 108, 807–18.

Sonoda, S. & Schlamowitz, M. (1972*b*). Specificity in the *in vitro* binding of IgG and rabbit serum albumin (RSA) to rabbit yolk sac membrane. *Journal of Immunology*, 108, 1345–52.

Sonoda, S., Shigematsu, T. & Schlamowitz, M. (1973). Binding and vesiculation of rabbit IgG by rabbit yolk sac membrane. *Journal of Immunology*, 110, 1682–92.

Staley, T.E. (1977). Absorption of horseradish peroxidase by neonatal pig intestinal epithelium: effect of *Escherichia coli* (055B5) on absorption. *American Journal of Veterinary Research*, 38, 1307–14.

Staley, T.E., Corley, L.D., Bush, L.J. & Jones, E.W. (1972). The ultrastructure of neonatal calf intestine and absorption of heterologous proteins. *Anatomical Record*, 172, 559–80.

Staley, T.E., Jones, E.W. & Corley, L.D. (1969). Fine structure of duodenal absorptive cells in the newborn pig before and after feeding of colostrum. *American Journal of Veterinary Research*, 30, 567–81.

Staley, T.E., Jones, E.W. & Marshall, A.E. (1968). The jejunal absorptive cell of the newborn pig: an electron microscopic study. *Anatomical Record*, 161, 497–516.

Strauss, F. (1967). Die normale Anatomie der menschlichen Plazenta. In *Handbuch der Speziellen Pathologischen Anatomie und Histologie*, vol. 7, pp. 1–96. Berlin: Springer-Verlag.

Strauss, L., Goldenberg, N., Hirota, K. & Okudaira, Y. (1965). Structure of the human placenta with observations on ultrastructure of the terminal chorionic villus. *Birth Defects Orig. Art. Ser.* 1, 13–25.

Terzakis, J.A. (1963). The ultrastructure of normal human first trimester placenta. *Journal of Ultrastructure Research*, 9, 268–84.

Tighe, J.R., Garrod, P.R. & Curran, R.C. (1967). The trophoblast of the human chorionic villus. *Journal of Pathological Bacteriology*, 93, 559–67.

Tillack, T.W. (1966). The transport of ferritin across the placenta of the rat. *Laboratory Investigation*, 15, 896–909.

Tsay, D.D. & Schlamowitz, M. (1975). Comparison of the binding affinities of rabbit IgG fractions to the rabbit fetal yolk sac membrane. Use of

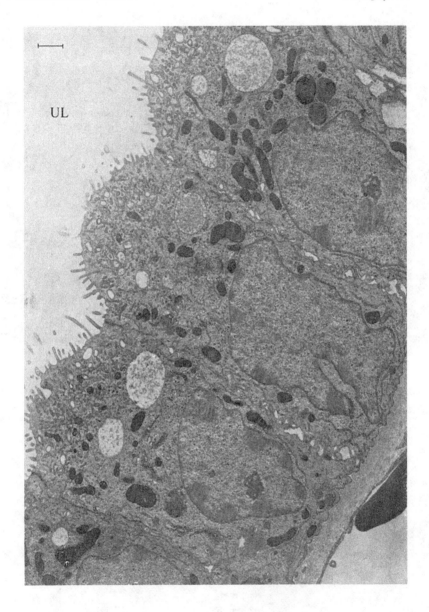

UL

For explanation of plates see p. 100

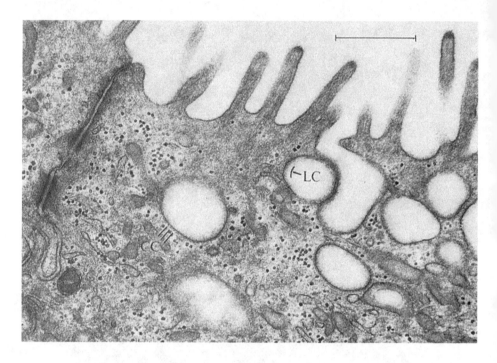

Plate 3

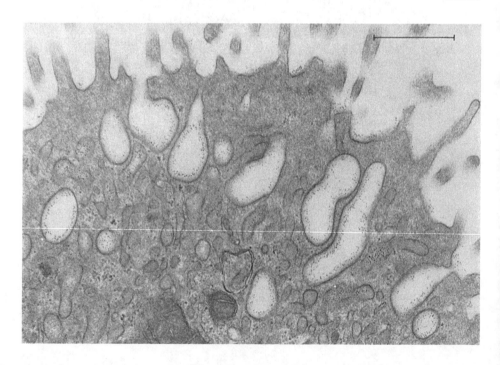

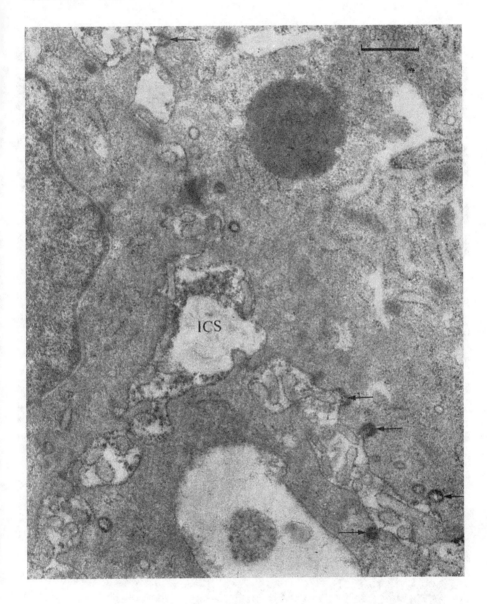

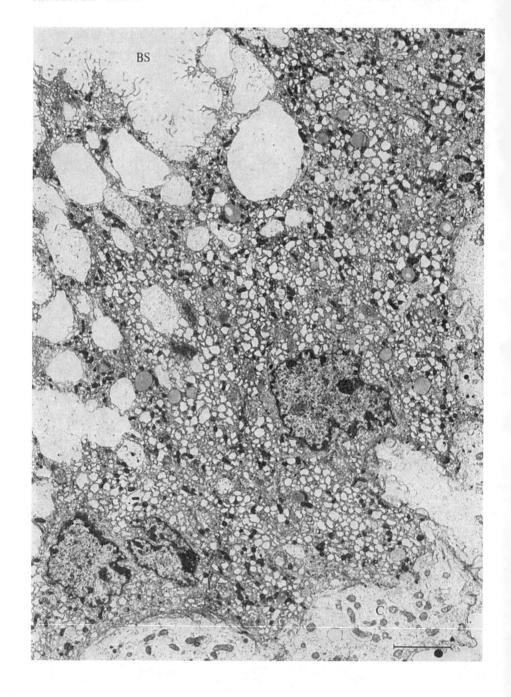

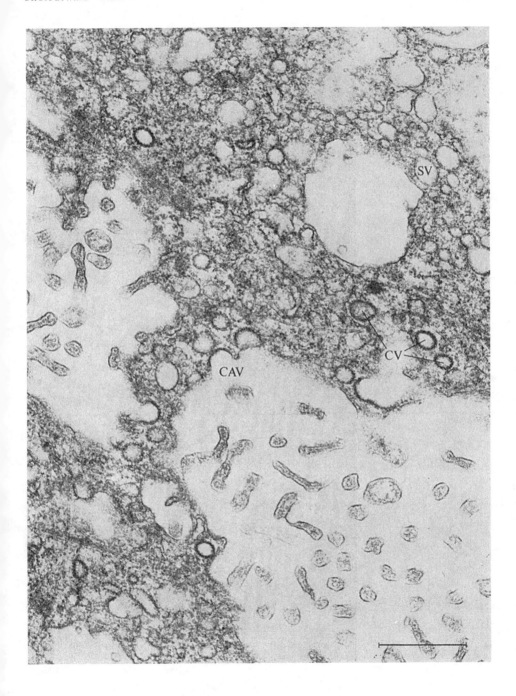

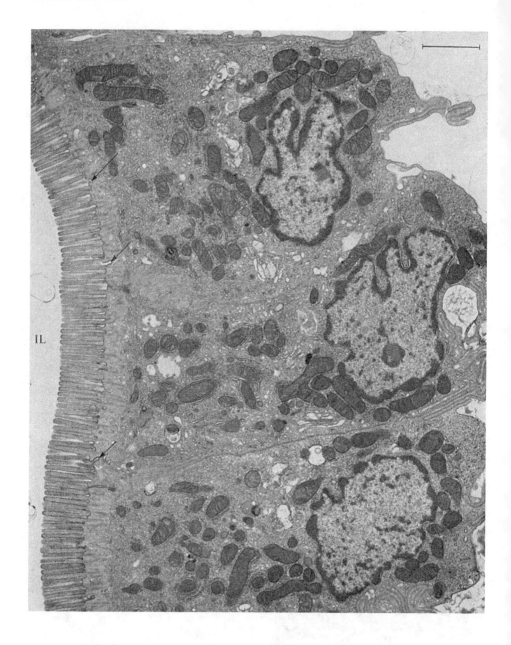

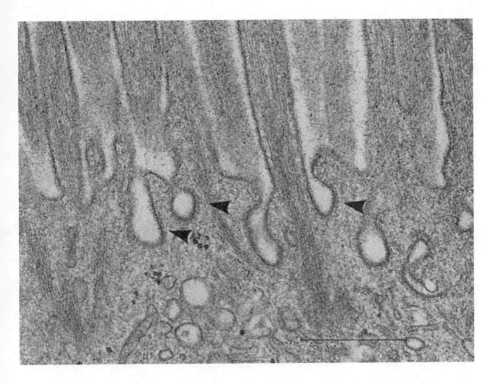

Plate 9

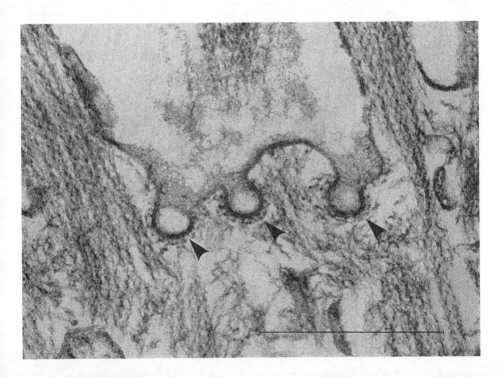

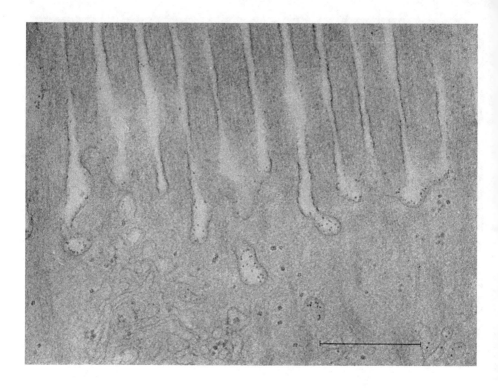

Plate 11

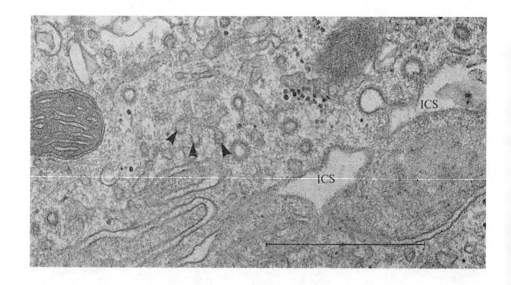

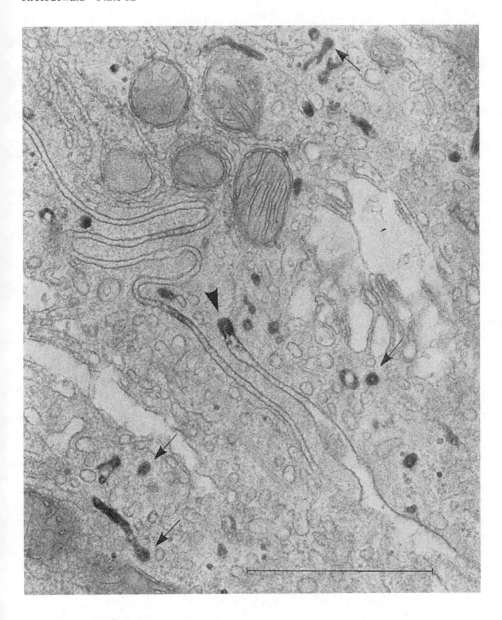

^{22}Na to facilitate quantitation of ^{125}I-IgG binding. *Journal of Immunology*, 115, 939–42.

Van Furth, R., Schuit, H.R.E. & Hijmans, W. (1965). The immunological development of the human fetus. *Journal of Experimental Medicine*, 122, 1173–88.

Virella, G., Nures, M.A. & Tamagnini, G. (1972). Placental transfer of human IgG subclasses. *Clinical and Experimental Immunology*, 10, 475–8.

Waldmann, T.A. & Jones, E.A. (1973). The role of cell surface receptors in the transport and catabolism of immunoglobulins. In *Protein Turnover*, ed. G.E.W. Wolstenholme & M.O'Connor, *Ciba Foundation Symposium*, 9, pp. 5–18. Amsterdam: Elsevier/North-Holland.

Waldmann, T.A. & Jones, E.A. (1976). The role of IgG-specific cell surface receptors in IgG transport and catabolism. In *Maternofoetal Transmission of Immunoglobulins*, ed. W.A. Hemmings, pp. 123–36. Cambridge University Press.

Wang, A.C., Faulk, W.P., Sruckley, M.A. & Fudenberg, H.H. (1970). Chemical differences of adult, foetal and hypogammaglobulinemic IgG immunoglobulins. *Immunochemistry*, 7, 703–8.

Whyte, A. (1978). Proteins of coated micropinocytic vesicles isolated from human placentae. *Biochemical Society Transactions*, 6, 299–301.

Wild, A.E. (1970). Protein transmission across the rabbit foetal membranes. *Journal of Embryology and Experimental Morphology*, 24, 313–30.

Wild, A.E. (1976). Mechanism of protein transport across the rabbit yolk sac endoderm. In *Maternofoetal Transmission of Immunoglobulins*, ed. W.A. Hemmings, pp. 155–67. Cambridge University Press.

Wild, A.E., Stauber, V.V. & Slade, B.S. (1972). Simultaneous localisation of human γ-globulin I^{125} and ferritin during transport across the rabbit yolk sac splanchnopleur. *Zeitschrift für Zellforschung und Mikroskopische Anatomie*, 123, 168–77.

Williams, K.E., Kidson, E.M., Beck, F. & Lloyd, J.B. (1975). Quantitative studies of pinocytosis. II. Kinetics of protein uptake and digestion by rat yolk sac cultured *in vitro*. *Journal of Cell Biology*, 64, 123–34.

Williams, K.E., Lloyd, J.B., Davies, M. & Beck, F. (1971). Digestion of an exogenous protein by rat yolk sac cultured *in vitro*. *Biochemical Journal*, 125, 303–8.

Williams, R.M. & Beck, F. (1969). Intracellular association of pinocytotic capacity and lysosomal enzyme activity in the mammalian small intestine. *Journal of Anatomy*, 104, 175.

Winkler, E.G., Fitzpatrick, J.G. & Finnerty, J.J. (1958). The permeability of the rabbit placenta to homologous albumin. *American Journal of Obstetrics and Gynecology*, 76, 1209–13.

Wislocki, G.B. & Bennett, H.S. (1943). The histology and cytology of the human and monkey placenta, with special reference to the trophoblast. *American Journal of Anatomy*, 73, 345–50.

Wislocki, G.B. & Dempsey, E.W. (1955). Electron microscopy of the human placenta. *Anatomical Record*, 123, 133–68.

Wislocki, G.B. & Padykula, H. (1961). Histochemistry and electron microscopy of the placenta. In *Sex and Internal Secretions,* ed. W.C. Young, vol. 2, pp. 883–957. Baltimore: Williams & Wilkins.

Wissig, S.F. & Graney, D.O. (1968). Membrane modifications in the apical endocytic complex of ileal epithelial cells. *Journal of Cell Biology,* 39, 564–79.

Worthington, B. & Graney, D.O. (1973*a*). Uptake of adenovirus by intestinal absorptive cells of the suckling rat. I. The neonatal ileum. *Anatomical Record,* 175, 37–62.

Worthington, B. & Graney, D.O. (1973*b*). Uptake of adenovirus by intestinal absorptive cells of the suckling rat. II. The neonatal jejunum. *Anatomical Record,* 175, 63–76.

Plates
Plates 1 to 12 are between pp. 98 and 99.

Plate 1. Guinea pig yolk sac endoderm. The prominent tubular vesicles and large vacuoles are seen in the apical regions of the cells next to the uterine lumen (UL). Bar represents 1 μm. (From King & Enders, 1970*a*.)

Plate 2. Apical cytoplasm of a guinea pig yolk sac endoderm cell. The endocytic vesicles which form at the luminal surface are depicted. Both luminal (LC) and cytoplasmic coats (CC) can be resolved on the membranes of several vesicles. Bar represents 0.5 μm. (From King & Enders, 1970*a*.)

Plate 3. Apical surface of a guinea pig yolk sac endoderm cell exposed to ferritin. The ferritin adheres preferentially to the luminal surface coat of the membrane invaginations and apical vesicles. Bar represents 0.5 μm. (From King & Enders, 1970*b*.)

Plate 4. Lateral surfaces of rabbit yolk sac endoderm cells exposed to homologous anti-peroxidase. Specific reaction product appears within the intercellular space (ICS) and within small coated vesicles attached to the plasmalemma (arrows). Bar represents 0.5 μm. (From Moxon *et al.,* 1976.)

Plate 5. Syncytial epithelium from the human placenta. Below the microvillous surface, which projects into the maternal blood space (BS), are numerous vesicles and vacuoles of varied sizes. Cells of the cytotrophoblast (C) are seen near the basal surface of the syncytium. Bar represents 5 nm. (From Ockleford & Menon, 1977.)

Plate 6. Oblique section through the microvillous surface of the syncytium from the human placenta. Numerous caveolae (CAV), small coated vesicles (CV), and smooth surfaced vesicles (SV) can all be identified within the apical cytoplasm. Bar represents 1 μm. (From Ockleford & Whyte, 1977.)

Plate 7. Absorptive epithelial cells from the proximal jejunum of the neonatal rat. The cells have a well-developed brush border which faces the intestinal lumen (IL). Small invaginations (arrows) and endocytic vesicles can be seen near the bases of the microvilli. Compared to other transporting cells, the intestinal cells have fewer apical vacuoles in their cytoplasm. Bar represents 2 μm.

Plate 8. Apical cell surface of a jejunal cell from the neonatal rat. Several small endocytic vesicles of various shapes are seen at the bases of the microvilli. The vesicles often exhibit a cytoplasmic membrane coat (arrowheads) in addition to a luminal coat. Bar represents 0.5 μm. (From Rodewald, 1973.)

Plate 9. Apical cell surface of a jejunal cell. The cell has been glycerinated and treated with a fixative containing tannic acid to reveal the cytoplasmic membrane coat (arrowheads) of the endocytic invaginations. Actin filaments seen in the adjacent cytoplasm have been treated with myosin S1 fragments. Bar represents 0.5 μm. (From the work of Begg, Rodewald & Rebhun, unpublished observations.)

Plate 10. Apical cell surface of a jejunal cell exposed to rat IgG conjugated to ferritin. The IgG conjugate binds preferentially to the membrane invaginations at the bases of the microvilli. Free ferritin and ferritin conjugated to proteins which are not transmitted do not bind. Bar represents 0.5 μm. (From Rodewald, 1973.)

Plate 11. Lateral cell surfaces of adjacent jejunal absorptive cells. Spherical coated vesicles are abundant in the surrounding cytoplasm and vary in size from less than 100 nm to about 200 nm overall diameter. One tubular vesicle is shown with coated vesicles attached to its membrane (arrowheads). ICS, intercellular space. Bar represents 1 μm. (From Rodewald, 1973.)

Plate 12. Lateral cell surfaces of jejunal cells exposed to an Fc–peroxidase conjugate. Small coated vesicles which contain tracer are apparent both free in the cytoplasm and attached to tubular vesicles (arrows). One coated vesicle (arrowhead) appears to be discharging tracer into the intercellular space. Bar represents 1 μm. (From Rodewald, 1976c.)

5

Coated vesicles in neurons

ALAN L. BLITZ & RICHARD E. FINE

The existence of coated vesicles (CVs) in neurons has been known since 1961 when Gray reported 'complex vesicles' in mossy fibre endings of the rat cerebellar cortex. He described them as spheres, 60–80 nm in diameter, surrounded by shells consisting of closely packed 15–20 nm vesicular bodies. This unusual structural organisation was probably only an apparent one. Today the structure assigned to CVs in neurons and other cells, which is based upon detailed analyses of isolated CVs, consists of a spherical lipid-bilayer vesicle enclosed by a protein coat composed of pentagonal and hexagonal subunits (Kanaseki & Kadota, 1969; Kadota & Kadota, 1973a, b; Pearse, 1975; Crowther, Finch & Pearse, 1976; Woods, Woodward & Roth, 1978). The comparison of CVs isolated from different tissues (Pearse, 1975, 1976; Woods et al., 1978) and in thin sections of isolated tissues (Nickel, Vogel & Waser, 1967) revealed no significant differences between the structure and chemical properties of CVs in neurons and in other cells.

For a time there was some doubt whether neuronal CVs were discrete organelles. Difficulty in resolving the three-dimensional organisation of the coat by goniometry led to the idea that the CVs seen in thin sections were fixation artifacts arising from the denaturation of microtubules or of the neuronal cytonet (Gray, 1972, 1975; Westrum & Gray, 1977). This idea, however, was not supported by the studies of CVs isolated from unfixed brain tissue cited above. It was particularly clear from Markham rotations performed on micrographs of negatively stained suspensions of CVs that the coat was a highly regular structure quite unlike aggregates of denatured protein (Woods et al., 1978).

The possibility still remains that the CVs isolated from neural tissue constitute a subset of coated organelles, and that not all of the CVs seen in thin sections of neurons represent the same structural and chemical entities. A contributory factor in this lingering doubt is the fact that the number of CVs which could be counted in micrographs of neurons depended to some extent upon the method by which the tissue was fixed (Ceccarelli, Pensa, DeGiuli & Pelosi, 1967; Paula-Barbosa & Gray, 1974). Those who wish to use the data of others to compare the numbers of CVs in different tissues should bear this fact in mind.

The function of coated vesicles in neurons: pioneering studies

Palay (1963) was the first to speculate about the function of CVs in neurons. He described 'alveolate vesicles' in the Purkinje cells of rat cerebellum in the vicinity of the Golgi complex. He also observed them in continuity with the internal face of the plasma membrane, the cisternae of the smooth endoplasmic reticulum, and other tubular elements. Because of their association with different cellular membranes, he postulated that CVs were participating in the circulation of membrane components between the surface membrane and internal membrane compartments. This idea, although based on little evidence at the time, corresponded closely to a current viewpoint.

While Palay reported that CVs were largely confined to the perikaryon, Andres (1964) observed them in the terminals of axons as well. He noted invaginations of the plasma membrane in the region adjacent to the synapse. These invaginations, which were coated on their cytoplasmic face, appeared to give rise to CVs by budding inward and pinching off – the process which is encountered in micropinocytosis. Andres felt that the CVs which formed in this fashion subsequently fused with intracellular membranous cisternae. These cisternae, which were coated on portions of their surface, eventually gave rise, Andres felt, to uncoated synaptic vesicles (SVs). This scheme for the origin of SVs, with CVs and cisternae as intermediates, was virtually identical to a scheme proposed nearly a decade later (Heuser & Reese, 1972, 1973) and which is discussed below.

Coated vesicles in micropinocytosis

Andres did not regard this scheme as a mechanism for producing SVs, but rather as a mechanism for the re-uptake of neurotransmitters and their breakdown products. In this opinion he was strongly influenced by the findings of Roth & Porter (1964) concerning the uptake by CVs of yolk protein in mosquito oocytes. Following the appearance of this influential article, a number of publications appeared which purported to show that CVs in neurons participated in the uptake by micropinocytosis of extracellular fluid and solutes.

Some of the evidence for this was circumstantial. It was based on observations of coated pits and invaginations of the plasma membranes of a variety of neurons together with the existence of CVs located nearby in the cytoplasm (Evans, 1966; Waxman & Pappas, 1969; Gray & Willis, 1970; Bunge, 1973). Conclusions drawn from these observations were ambiguous, however, because the same images were also consistent with the fusion of CVs with the plasma membrane, the directly opposite situation from micropinocytosis.

The use of extracellularly applied tracers helped to dispel this ambiguity. In a variety of reports, extracellular tracers such as ferritin, thorium dioxide and horseradish peroxidase (HRP) were observed being taken up within CVs by neurons, neurosecretory cells and other brain cells (Rosenbluth & Wissig, 1963, 1964; Becker, Novikoff & Zimmerman, 1967; Holtzman & Dominitz, 1968; Bunt, 1969; Holtzman & Peterson, 1969; Zacks & Saito, 1969; Holtzman, Freeman & Kashner, 1970; Holtzman, 1971; Holtzman *et al.*, 1973; LaVail & LaVail, 1974; Turner & Harris, 1974; Wessells *et al.*, 1974). What is more, in some of these studies, the tracer appeared in CVs shortly after its application and was confined at early times to CVs located near the periphery of the cell. These facts strengthened the contention that the tracers were being taken up when CVs formed from the plasmalemma, rather than being expelled after the fusion of internally derived CVs with the plasmalemma.

Another matter was the quantitative significance of this route of uptake. In few studies did CVs appear to be the only route by which tracers were internalised. Uncoated vesicles bearing tracer arose from the plasmalemma in a similar fashion and in fact appeared to be more numerous (Brightman, 1965; Bunge, 1977). In extreme cases only 2–6 % of the labelled vesicles were coated (Birks, Mackey & Weldon, 1972; Weldon, 1975). During uptake of colloidal thorium dioxide by cultured sensory cells, labelled CVs which were continuous with the plasmalemma were rare (Becker, Hirano & Zimmerman, 1968). It is possible, of course, that at least some of the uncoated vesicles bearing tracers were originally coated and that the coats were labile.

The term micropinocytosis refers to the internalisation by cells of small portions of extracellular fluid and solutes dissolved in that fluid. It may be that CVs in neurons represent a minor uptake route for fluid. However, micropinocytosis also involves the internalisation of a portion of plasma membrane. It is just as likely that this is the specific function of CVs. The incorporation of tracers into CVs may well be incidental to this second type of activity. In fact, when the specific function of the neuron is considered — the evoked release of neurotransmitters and neurosecretory products — this second function of CVs assumes importance. We now examine studies on the functioning of neurons which support strongly the early idea (Palay, 1963) that CVs participate in the recycling of neuronal membrane components.

Neurotransmitter release: exocytosis and recycling

Two discoveries made during the 1950s provided the basic outlines
for the most popular current hypothesis which seeks to explain how neurons
release neurotransmitter substances. The first of these was the discovery of
spontaneous release of acetylcholine by motor nerve endings and the realisa-
tion that this release was in the form of packages or 'quanta' of acetylcholine
consisting of several thousand molecules (Fatt & Katz, 1952). The second
discovery was the SVs which were seen in early electron micrographs of nerve
terminals (Robertson, 1954; DeRobertis & Bennett, 1955). These two findings,
when considered together, led to the 'vesicle hypothesis'. This formulation,
stated briefly, holds that the SVs are the storage sites for acetylcholine, each
SV containing one quantum of transmitter. When the nerve terminal receives
impulses conducted by the axon, the membranes of the SVs fuse or in some
other way become continuous with the presynaptic membrane of the nerve
terminals, thereby releasing their content of acetylcholine into the synaptic
cleft (DeRobertis & Bennett, 1955; Del Castillo & Katz, 1956; DeRobertis, 1958).

It is not the purpose of this discussion to review in depth the evidence for
and against the vesicle hypothesis. It is sufficient to state that this hypothesis
is the only widely held idea which can explain the quantal release of neurotransmitter. Most of the recent experiments designed to test the vesicle
hypothesis have in fact supported it. With current refinements in the hypothesis,
several of the earlier findings which tended to conflict with it can now be
incorporated into its general outlines in a reasonable fashion.

We may discover the relationship of CVs to this mechanism of neurotransmitter release if we consider two predictions of the vesicle hypothesis.
First, during rapid stimulation of nerve terminals, when the membranes of
SVs are presumably fusing with the presynaptic membrane at the region of
the synaptic cleft, the number of SVs in the terminal should progressively
decline. Second, the total surface area of the presynaptic membrane should
increase. Clearly this is an undesirable situation which if allowed to proceed
unchecked would severely impair the ability of nerve terminals to release
transmitter over any appreciable period of time.

For continued nerve function, assuming that the vesicle hypothesis is
correct, there must be a way of retrieving vesicle membrane components from
the external membrane of the neuron. There must also be some means of
replenishing the supply of SVs. If both of these needs are met simultaneously,
that is, if SVs re-form from the synaptic plasma membrane continuously
over time, we have a process known as 'recycling' It is in the retrieval and
recycling of vesicle membrane components that CVs are believed to play their
specific role in nerve endings.

The depletion of SVs during transmitter release has not been seen in every attempt to demonstrate it. Failures have been reported, for example, by Birks, Huxley & Katz (1960) and by Jones & Kwanbunbumpen (1970). These failures were not necessarily refutations of the vesicle hypothesis; they may in fact have been examples of the successful replenishment of SVs.

Support for this explanation was provided by the first study of this kind (DeRobertis, 1958). Stimulation of the rabbit splanchnic nerve at 400 impulses per second led to a decrease in the number of SVs in the nerve terminals. Stimulation at only 100 impulses per second not only failed to produce depletion of SVs but actually produced a slight increase in their number near the cleft region (this also occurred in the experiments of Jones & Kwanbunbumpen cited above). DeRobertis suggested that at the higher rate of stimulation the ability of nerve terminals to replenish the supply of SVs was exceeded, whereas at the lower rate of stimulation the population of SVs was being replenished at a rate equal to or slightly greater than the rate at which they were discharged.

In most cases where depletion of SVs was reported, conditions were chosen in which the activity of the nerve terminals in releasing transmitter was extremely high, or where recycling was presumably blocked. In cholinergic endings rapid stimulation produced a reduction in the number of SVs (Heuser & Reese, 1972, 1973; Pysh & Wiley, 1972, 1974). Stimulation of the hatchet fish giant fibre at low temperature also had this effect (Bennet, Model & Highstein, 1975; Model, Highstein & Bennet, 1975; Bennet, Highstein & Model, 1976). A similar result was obtained by Atwood, Lang & Morin (1972), who stimulated crayfish excitatory and inhibitory axons in the presence of 2, 4-dinitrophenol, a metabolic inhibitor. These last results suggested that the replenishment of SVs depended on metabolic energy.

Application of certain agents which greatly stimulated the spontaneous release of acetylcholine also caused depletion of SVs. When lanthanum chloride was applied to the frog sartorius neuromuscular junction preparation, there was a 10 000-fold increase in the frequency of release of quanta, but if the preparation was fixed during this time, no evidence for depletion of synaptic vesicles could be seen in electron micrographs of the nerve terminals. Later, when the frequency of release gradually subsided, complete depletion of vesicles could eventually be seen (Heuser & Miledi, 1971). Black widow spider venom, which caused a burst of quantal release, also caused marked depletion of SVs (Clark, Mauro, Longenecker & Hurlbut, 1970; Clark, Hurlbut & Mauro, 1972).

Accompanying these changes were dramatic alterations in the structure of the membranes associated with nerve terminals. Rapid preganglionic stimulation of the cat superior cervical ganglion resulted not only in a decrease in the number of SVs, but also in an increase in the circumference of the terminals (Pysh & Wiley, 1972, 1974). After lanthanum treatment, frog motor terminals developed membrane-bound tubes and cisternae. Cisternae were also seen after tetanic stimulation of this preparation (Heuser & Reese, 1972, 1973) and after stimulation of the hatchet fish giant fibre at room temperature (Bennet *et al.*, 1975, 1976; Model *et al.*, 1975). The changes after nerve stimulation were reversible; the terminals regained normal appearance after a period of rest. However, when the hatchet fish preparation was stimulated at 12 °C, whorls of membrane appeared within the terminal which were continuous with the plasma membrane. This change was not reversible as long as the preparation was chilled. Another irreversible change occurred when the frog preparation was treated with black widow spider venom, when infoldings of the plasma membrane were seen (Clark *et al.*, 1972).

When these changes were reversible, the replenishment of the SVs and the return of synaptic plasma membrane to normal following the cessation of stimulation and a period of rest suggested that SVs could form either from the plasma membrane or from the membranous elements which appeared after stimulation. The results further implied that if this process occurred throughout the period of nerve activity, the SVs could be continually recycled during transmitter release.

There are a number of considerations which suggest indirectly that this was the case. In frog motor terminals (Heuser & Reese, 1973) and in preganglionic terminals of the rat superior cervical ganglion (Perri, Sacchi, Raviola & Raviola, 1972), the number of quanta of transmitter released during the period from the onset of stimulation to fatigue was greater than the total number of SVs initially present. In the frog cutaneous pectoris nerve—muscle preparation (Ceccarelli, Hurlbut & Mauro, 1972), stimulation at two impulses per second for 6–8 hours resulted in the release of 80 % of the available acetylcholine without noticeable depletion of the SVs. During the release of the residual acetylcholine, the SVs declined in number. While there is evidence that depletion of transmitter stores hastened the depletion of SVs (Párducz & Fehér, 1970; Párducz, Fehér & Joó, 1971), there is also evidence against this (Pysh & Wiley, 1972; Heuser & Reese, 1973).

None of the observations discussed so far rules out the possibility that SVs are replaced by the de-novo synthesis and assembly of their components. However, this possibility seems unlikely. Bittner & Kennedy (1970), who studied the opener—stretcher neuron of the crayfish, estimated that at an impulse rate of 20 per second the contents of about 4.5×10^9 SVs were discharged per hour. If each SV were used only once and later replaced by a newly assembled vesicle, this would require the biosynthesis of 24 mm^2 of membrane per hour. They estimated that for mammalian neurons this number would be lower but only by a factor of about 10. They concluded that the incorporation of vesicle membrane into the presynaptic membrane must be balanced by retrieval and re-formation of SVs.

Chemical studies of the rate of turnover of the proteins associated with SVs have also supported the notion that they are recycled. When [^{3}H]lysine was used to label SVs in avian brain, the measured half-time of the label in vesicle-associated proteins was 17 hours. When [^{3}H]fucose was the label, the computed half-time was ten days (Koenig, Giamberardino & Bennett, 1973). Employing [^{3}H]leucine as the source of radioactivity, the much longer half-time of 21 days was determined for proteins associated with rat SVs (von Hungen, Mahler & Moore, 1968). While these results were not in quantitative agreement, they all suggested that at least the protein- and glycoprotein-components of SVs were reutilised rather than continuously resynthesised.

Coated vesicles in recycling

The first hint that CVs may play a role in the recycling of SVs comes from the work of Heuser & Miledi (1971) on lanthanum-treated frog motor terminals. In these terminals the number of CVs rose, and a series of membrane-bound tubes and cisternae appeared which were coated on a portion of their surface. After 12 hours no SVs were present, but CVs were prominent. A similar increase in the number of CVs and the appearance of partially coated membrane compartments were seen in the hatchet fish endings after stimulation in the cold (Bennet *et al.*, 1975, 1976). Coated convoluted tubules were seen in endings of the goldfish saccular macula (Hama & Saito, 1977) under normal conditions.

Stronger evidence that SVs originated from the plasma membrane during recycling, and that CVs were involved in this process, came from the use of HRP and other tracers in order to follow the path of membrane movement during nerve activity. The most extensive tracer studies were carried out by Heuser & Reese (1972, 1973) on the frog sartorius preparation. They found that after tetanic stimulation the number of SVs fell and the number of CVs rose in a reversible fashion. When HRP was present in the bath during the stimulation, the tracer first appeared in CVs which budded from the plasma membrane. It next appeared in membrane-bound sacs or cisternae which were partially coated. Finally it appeared in SVs. The CVs arose not from the region of the release sites but from the area adjacent to these sites near the region of contact with the Schwann-cell processes.

They proposed that during exocytosis the membrane of SVs was incorporated into the presynaptic membrane. There followed a translocation of this path of vesicle membrane to the area adjacent to the Schwann-cell process, where it was internalised in the form of CVs. This implied the presence of coat material on the cytoplasmic surface of the plasma membrane where internalisation occurred. The CVs which formed in this manner, they further proposed, fused in the axoplasm to form the partially coated cisternae. These cisternae were then supposed to give rise to SVs by budding; this last event could not be captured.

A number of their findings in a sense comprised a list of criteria which could be applied when proposing that SVs are recycled. We have already discussed the stimulation-dependent nature of uptake of tracer, a finding which suggested that something other than ordinary micropinocytosis was occurring. A related finding was the rise in the number of CVs after stimulation.

Equally important was the observation that the total membrane surface area remained nearly constant during recycling. The relative distribution of membrane surface area between CVs, cisternae, SVs and plasma membrane was different when the recycling process was interrupted at different times, but nearly all of the membrane could be accounted for at each stage.

Also crucial was the observation that the HRP tracer which was incorporated into SVs was expelled from the nerve terminal during a second period of stimulation. This result suggested that the uncoated organelles which contained the tracer were functional SVs and not simply micropinocytic organelles. It was unlikely that during the short time periods employed the tracer was degraded.

In many cases where tracers were internalised in coated and uncoated membrane-bound structures, these organelles ultimately fused with lysosome-like bodies and were presumably degraded (Rosenbluth & Wissig, 1964; Becker *et al.*, 1967; Holtzman, Freeman & Kashner, 1971; Korneliussen, 1972; Abrahams & Holtzman, 1973; LaVail & LaVail, 1974; Schacher, Holtzman & Hood, 1976; Bunge, 1977). In these systems, true recycling could not have occurred. In this regard it is important to distinguish between membrane retrieval and membrane recycling.

Portions of the findings of Heuser & Reese (1972, 1973) were reported for other systems, but there were discrepancies as well. Stimulus-dependent incorporation of HRP into SVs was reported in the cutaneous pectoris nerve—muscle preparations of the frog (Ceccarelli *et al.*, 1972) and in photoreceptors of various animals (Ripps, Shakib & MacDonald, 1976; Schacher *et al.*, 1976; Schaeffer & Raviola, 1977). In photoreceptors, greater incorporation occurred in the dark when the terminals were actively releasing transmitter than in the light when transmitter release was inhibited. Litchy (1973) reported a significant increase in HRP in the axons of frog motor nerves after stimulation but gave evidence that its entry into the axons depended on retrograde axonal transport. Further, the amount of tracer was estimated to be several orders of magnitude greater than the amount which could be contained in SVs. The results implied that there were alternative routes for internalisation of the tracer, and that not all of the tracer was retained in functional SVs.

Many authors have suggested that SVs arose directly from the presynaptic plasma membrane of nerve terminals (Evans, 1966; Holtzman *et al.*, 1971; Ceccarelli *et al.*, 1972, 1973; Jorgensen & Mellerup, 1974; Pysh & Wiley, 1974; Van Herreveld & Trubatch, 1975; Ripps *et al.*, 1976; Schacher *et al.*, 1976). There were also reports that SVs formed directly from CVs as the latter shed their coats (Westrum, 1965; Gray & Willis, 1970; Gray & Pease, 1971). Recycling via CVs and coated cisternae was implied by the work on the hatchet fish synapse and lanthanum-treated frog motor nerves cited earlier. However, Korneliussen (1972) failed to see an increase in the number of CVs after stimulation of rat phrenic nerves.

It is quite possible that there are variations among different synapses in the extent to which recycling occurs, and in the mechanisms by which this is accomplished (Teichberg, Holtzman, Crain & Peterson, 1975). It is not the case that the recycling systems involving coated intermediates are a feature largely confined to cholinergic terminals. CVs are important at least in membrane retrieval, if not in true recycling, in neurosecretory cells and in the chromaffin cells of the adrenal medulla which secrete catecholamines. In these cells the mechanism of retrieval is thought to be somewhat different from the one seen in neuromuscular junctions.

Membrane retrieval in neurosecretory and chromaffin cells

In motor nerve terminals and at certain other synapses, the CVs active in membrane retrieval frequently arise not at the sites of exocytosis, but rather at locations adjacent to these areas (Andres, 1964; Ceccarelli *et al.*, 1972; Heuser & Reese, 1972, 1973; Turner & Harris, 1973; Hama & Saito, 1977). In neurosecretory endings and chromaffin cells, retrieval occurs near the release sites themselves.

In these cells the substances destined for secretion are contained in relatively large, electron-dense secretory granules. When these granules fuse with the limiting membrane of the secretory cell, depressions in this membrane are created. In stimulated cells, these depressions persist for a period of time sufficiently long for them to be detected in electron micrographs. Frequently the electron-dense cores of these granules can then be seen adjacent to the external surface of these depressions, presumably because the cores are in the process of expulsion during exocytosis.

On the cytoplasmic face of these depressions one or more smaller, inwardly directed, depressions were situated. These were coated on their cytoplasmic surface (Diner, 1967; Douglas & Nagasawa, 1971; Grynszpan-Winograd, 1971; Benedeczky & Smith, 1972; Douglas, 1973). Adjacent to this coated membrane region in the cytoplasm of the cells, CVs and uncoated 'microvesicles' or 'synaptic vesicles' were seen. These were approximately 50 nm in diameter, the diameter of the smaller membrane depressions, but considerably smaller than either the secretory granules or the larger membrane depressions, which ranged from 100 to 200 nm in diameter. In some cases the coated microvesicles appeared to be in the process of losing their coats and free coat material was sometimes apparent.

The appearance of the coated and uncoated microvesicles was apparently related to the process of exocytosis, since they increased in number when secretory cells were stimulated to release their secretory products (Nagasawa, Douglas & Schulz, 1970; Douglas & Nagasawa, 1971; Smith, 1971; Douglas, 1973). This was evidence that these structures were participating in the retrieval of secretory granule membrane from the plasma membrane of these cells during or following exocytosis.

This type of membrane retrieval could not properly be considered recycling since the microvesicles were not themselves secretory organelles. In studies with externally applied tracers most of the label which was incorporated into microvesicles by these types of cells eventually entered lysosome-like bodies and was presumably degraded (Holtzman & Dominitz, 1968; Nagasawa, Douglas & Schulz, 1971; Abrahams & Holtzman, 1973; Holtzman *et al.*, 1973; Holtzman, Schacher, Evans & Teichberg, 1977). There was, however, a report of CVs fusing with prosecretory granules in adrenal chromaffin cells (Benedeczky & Smith, 1972).

Coated vesicles in recycling: results from quick-freezing and freeze-cleaving

It is appropriate to ask at this point what the role of CVs in the retrieval of vesicle and secretory membrane is. Or, to restate the question more precisely, what the function of clathrin, the coat protein, is in endocytosis. Kanaseki & Kadota (1969) suggested that vesiculisation occurred at coated regions of plasma membrane when the coat subunits underwent conformational changes. While this may be true, it is quite clear that clathrin is not obligatory for endocytosis. It has been known for many years that uncoated vesicles participate in the internalisation of extracellular fluid and solutes.

The examination of nerve terminals viewed in thin sections did not provide answers to these questions. It was not until the application of the freeze-cleaving technique that hints about a special role for clathrin in nerve terminals were obtained. Furthermore, to obtain a clear picture of the events which transpired during vesicle recycling, it was necessary to devise a method for halting the process within extremely brief time intervals after the onset of nerve stimulation. When this was done it became apparent that the scheme of vesicle recycling in cholinergic nerve terminals presented in the preceding paragraphs had to be modified. The modification was not one of peripheral details alone; the revised scheme provided a powerful rationale for postulating an obligatory role for CVs and for clathrin in the recycling of neuronal sub-cellular membrane components.

Freeze-cleaving studies revealed that in a variety of synapses membrane specialisations occurred on both the presynaptic and postsynaptic membranes which bordered the synaptic cleft. In the case of the postsynaptic membrane, this specialisation took the form of an aggregate of 8–13 nm particles (Sandri, Akert, Livingston & Moor, 1972; Heuser, Reese & Landis, 1974, 1975). Within the presynaptic membrane of motor nerve terminals, two parallel double rows of intramembranous particles could be seen in the cytoplasmic leaflet (Dreyer *et al.*, 1973; Heuser *et al.*, 1974, 1975). In synapses in the central nervous system, in the spinal cord, and in photoreceptor endings, the presynaptic specialisation appeared as a hexagonal array of particles (Pfenninger, Akert, Moor & Sandri, 1971, 1972; Raviola & Gilula, 1975).

After stimulation of frog motor nerves under certain conditions – when the preparation was bathed in fixatives or after treatment with 4-aminopyridine or other agents – rows of 'vesicle openings' could be seen adjacent to the rows of particles at the release site. In untreated terminals these openings were too rare and too short-lived to be detected (Heuser & Reese, 1975a; Heuser, 1976, 1977; Pumplin & Reese, 1977). These openings corresponded in location to the double row of SVs seen in thin sections which defined the so-called 'active zones' of the nerve terminal (Couteaux & Pécot-Dechavassine, 1970).

In addition to these openings at the active zone, another group of larger openings appeared near the Schwann-cell processes and in other regions of the terminal away from the release sites. The position of these openings – the 'coated vesicle openings' – corresponded to the site of origin of CVs seen in thin sections (Landis & Reese, 1974; Heuser, 1976, 1977).

In images of cross-fractured terminals another set of particles, larger than the ones at the release sites, could be seen in association with the membranes of the SVs within the terminals. We will refer to these larger particles as vesicle particles. According to Heuser & Reese (Heuser, 1976), each SV contained two or three of these vesicle particles. Akert, Sandri, Cuénod & Moor (1976) also saw the vesicle particles, but reported only a single particle per vesicle.

During stimulation of the nerve terminal, the vesicle particles were incorporated into the presynaptic plasma membrane in the region of the active zones, and then appeared to move laterally away from the release sites in a direction roughly parallel to the synaptic cleft. At the sites of endocytosis, within the coated vesicle openings, the particles formed clusters or mounds (Heuser & Reese, 1975a).

The evidence suggested that the vesicle particles were actually recycled during transmitter release. When terminals were stimulated in solutions containing fixatives, the total number of larger particles increased in the cytoplasmic leaflet of the plasma membrane (Heuser & Reese, 1975a). An excess of vesicle particles could also be seen when this preparation was treated with brown widow spider venom, which caused massive exocytosis and depletion of SVs (Pumplin & Reese, 1977). Apparently the vesicle particles which were incorporated into the plasma membrane during exocytosis were normally retrieved via CVs. Under conditions when recycling of the particles was presumably inhibited, the vesicle particles accumulated in the plasma membrane.

A crucial piece of information was obtained by the use of a rapid-freezing technique which enabled the frog neuromuscular junction preparation to be frozen to the temperature of liquid nitrogen within milliseconds of the onset of stimulation (Heuser et al., 1975). It became apparent, when this technique was combined with the use of externally applied ferritin as a tracer, that labelled cisternae which had previously been seen in thin sections appeared much more rapidly than labelled CVs. Within 50 milliseconds the ferritin entered the cisternae, which were therefore continuous with the plasma membrane after these brief intervals. The label did not appear within CVs until about one second had elapsed (Heuser, 1977).

These facts implied that the cisternae which seemed to form during the retrieval of vesicle membrane, derived directly from the plasma membrane, and not from the fusion of CVs as had previously been thought. The role of the CVs was apparently confined to the retrieval of the vesicle particles.

The cisternae themselves appeared relatively depleted of vesicle particles, as if the fusion of SVs occurred in a particle-free region of membrane (Damassa et al., 1976; Heuser, 1977). In synapses at the electric organ of *Torpedo*, bulges the size of SVs were seen on the inner leaflet of the presynaptic membrane. These bulges were also depleted of particles (Heuser & Reese, 1975b).

Studies on membrane fusion supported the reasonableness of this scheme. There are several examples in which the fusion of cellular membranes appeared to be confined to regions from which intramembranous particles had been excluded (Satir, Schooley & Satir, 1973; Ashkong, Fisher, Tampion & Lucy, 1975; Chi, Lagunoff & Koehler, 1976; Dahl, Gratzl & Ekerdt, 1976). If the exclusion of intramembranous particles was a precondition for the fusion of SVs with the presynaptic membrane, and if the retrieval of vesicle membrane occurred in the form of particle-poor cisternae, it is reasonable that some independent means of retrieving the particles would be necessary.

Furthermore, what is known about the nature of intramembranous particles also fits this scheme. It is likely that intramembranous particles are aggregates of intrinsic membrane proteins some of which span the lipid bilayer (Singer & Nicolson, 1972). If proteins associated with vesicle particles in fact extend into the axoplasm, they might interact with clathrin which is presumably associated with the cytoplasmic face of the plasma membrane. Clathrin could be induced to assemble at the sites of endocytosis at which the particles appear to cluster. Alternatively, clathrin could specifically recognise the particles and cause them to cluster.

The idea that intramembranous particles were internalised at coated-membrane sites was supported by the results of freeze-cleaving studies of neurosecretory endings in the neurohypophysis (Dreifuss, Akert, Sandri & Moor, 1973). Pits were seen in the plasma membrane of stimulated cells which corresponded in size to the coated invaginations seen in thin sections. These pits contained a higher density of particles than the surrounding areas of membrane. The microvesicles, it will be recalled, appeared to participate in the retrieval of secretory granule membrane after exocytosis.

It may be that in neurosecretory terminals the retrieval of membrane and the retrieval of particles occur together, in distinction to the situation in motor nerve terminals. It may be significant that structures analogous to the cisternae of cholinergic endings have not been reported either in neurosecretory cells or in chromaffin cells. It is conceivable that the formation of cisternae becomes necessary when membrane and particle retrieval occur independently. It is presumably at the level of the cisternae that particles are reincorporated into the membranes giving rise to SVs.

Speculations on the nature of the vesicle particle and its role in neurotransmitter release

If the scheme of particle recycling in cholinergic endings presented in the last section is correct, it follows that the particles found in CVs and those found in SVs should be identical. Whereas the isolation from mammalian brain of homogeneous preparations of SVs in useful yield is probably not currently feasible, it is relatively simple to obtain large quantities of pure or nearly pure CVs. In our own laboratory we undertook the biochemical characterisation of CV membrane in the hope of indirectly gaining information about the nature of SV membrane. In the course of this work we obtained rather surprising results, which may bear directly upon the nature of the vesicle particle.

These results revealed that the lipid bilayer associated with brain CVs was strikingly similar to the membrane of sarcoplasmic reticulum (SR) fragments isolated from skeletal muscle (Blitz, Fine & Toselli, 1977). As is true of SR fragments, preparations of brain CVs displayed calcium-activated ATPase (CaATPase) activity and the ability to take up calcium ions from the medium in the presence of ATP. With both SR and CVs, the uptake was enhanced by potassium oxalate, a calcium-trapping agent.

Polyacrylamide gels of brain CVs contained a 100 000 mol. wt protein which co-migrated with the 100 000 mol. wt CaATPase of SR and which constituted at least 50 % of the membrane-associated protein. In the presence of $[\gamma^{32}P]$ ATP, counts were incorporated into this protein as was previously shown with the SR CaATPase (MacLennan, Yip, Iles & Seeman, 1972).

In SR the CaATPase, which comprises about 70 % of the protein, is organised into membrane-associated aggregates which form 9 nm particles each consisting of about three or four CaATPase molecules (Scales & Inesi, 1976). It is tempting to propose that the intramembranous particles associated with CVs, and therefore also with SVs, are aggregates of CaATPase molecules. Two possible functions for calcium pump sites in CVs and SVs come to mind. One might be catalysis of the transport of calcium ions into the interior of these vesicles, thus in some way promoting their fusion with the synaptic plasma membrane or with internal membranes. A second function of this catalysis might be the removal of calcium ions from the terminal axoplasm thereby helping to maintain the extremely low free-calcium level in this region. These ideas will, we hope, become clearer as we briefly discuss what is known about the mechanism of action of calcium ions in promoting neurotransmitter release and the mechanisms which are thought to control intraterminal calcium concentration.

Calcium ions and neurotransmitter release

Although it has long been known that transmitter release evoked by nerve stimulation is dependent on external calcium ions (Del Castillo & Stark, 1952; Katz & Miledi, 1965), the mode of action of calcium has remained an enigma. The evidence is strong that, at least in squid stellate ganglion cells, external calcium ions enter the terminal during depolarisation prior to release (Miledi & Slater, 1966; Katz & Miledi, 1969; Llinás, Blinks & Nicholson, 1972;

Miledi, 1973; Llinás & Nicholson, 1975). However, only about 100 micro-seconds are available for calcium diffusion (Llinás, Steinberg & Walton, 1976; Llinás, Walton & Hess, 1976). Given the extremely limited mobility of calcium ions in squid axoplasm (Hodgkin & Keynes, 1957; Baker & Crawford, 1972), calcium ions entering the terminal could only diffuse during this time over a distance approximately equal to the diameter of one SV (Gray & Parsegian, 1977). The reasonable conclusion is that calcium must act at or near the release sites.

Perhaps the site of action of calcium resides on the surface of the SV or within its interior. There is some evidence that calcium ions act from the inside of closed lipid bilayers in promoting their fusion (Zakai, Kulka & Loyter, 1976; Ingolia & Koshland, 1978). There is also evidence that stimulation of the adrenal medulla with acetylcholine results in net movement of calcium ions into the secretory granules (Borowitz, 1969). If this is true for the fusion of SVs with the presynaptic membrane, then there is an obvious role for an inwardly directed calcium pump associated with SVs. This reasoning applies also to the fusion of CVs with internal cisternae.

There is also evidence for the existence of calcium-binding sites within SVs. With the frog neuromuscular junction preparation, when the nerve endings of *Narcine* electric organ or isolated SVs were fixed in calcium-containing solutions, electron-dense calcium-containing 'dots' appeared within the vesicles in close association with the inner face of their membranes (Bohan *et al.*, 1973; Boyne, Bohan & Williams, 1974; Politoff, Rose & Pappas, 1974; Pappas & Rose, 1976). It was suggested by Akert *et al.* (1976) that the single particle they observed in cross-fractured SVs corresponded to this calcium-containing dot. Recall, however, that more than one particle per vesicle was seen by others (Heuser, 1976).

Equally plausible is the idea that CVs, SVs and secretory granules take up calcium from nerve terminals as part of a system for regulating the resting free-calcium level. This ability to sequester calcium ions could explain the presence of large amounts of calcium associated with secretory granules (Borowitz, 1967; Thorn, Russell & Villardt, 1975; Normann & Hall, 1978) and the function of calcium-binding sites within cholinergic vesicles.

Various membrane fractions have been isolated from whole brain which sequester calcium in a fashion dependent on ATP (Robinson & Lust, 1968; Ohtsuki, 1969; DeMeis, Rubin-Altschul & Machado, 1970; Duggan & Kelleher, 1975; Trotta & DeMeis, 1975; Duncan, 1976; Tanaka, Takeda & Jaimovich, 1976; O'Driscoll & Duggan, 1977). In some cases these preparations have been shown to contain CaATPase activity as do chromaffin granules isolated from adrenal medullae (Banks, 1965). In most cases these activities were associated with the brain 'microsome' fraction – a heterogeneous fraction composed of fragments of smooth endoplasmic reticulum, plasma membrane and probably other membranes.

Most relevant to our results was the discovery of ATP-dependent calcium uptake associated with synaptosomes previously treated with hypotonic buffers (Kendrick, Blaustein, Fried & Ratzlaff, 1977). This uptake system was insensitive to mitochondrial inhibitors and in addition was stimulated by oxalate, as was our brain CV preparation. In most cases the brain microsome preparations mentioned above did not respond to oxalate.

Nevertheless it must be admitted that these calcium-sequestering fractions were potential sources of contamination of our brain CVs. We have not as yet provided definitive proof that the CaATPase activity was associated specifically with CVs rather than with contaminating membranes, although this preparation appeared to contain few contaminants when examined with electron microscopy (Blitz *et al.*, 1977).

It should also be mentioned that in a recent report polyacrylamide gels of supposedly highly purified SVs from electric organ of *Narcine brasiliensis* did not display a significant band corresponding to a polypeptide of 100 000 mol. wt – the size of the CaATPase presumed to be associated with brain CVs (Wagner, Carlson & Kelly, 1978). Puszkin *et al.* (1976) did report the presence of a protein of similar size (95 000 mol. wt) associated with a brain fraction enriched in SVs. They presented evidence, however, that this protein was antigenically similar to another muscle protein, α-actinin.

The mechanisms for the regulation of the free-calcium-ion concentration in resting nerve terminals are efficient ones. In squid axons the free-calcium concentration was estimated to lie between 20×10^{-9} and 50×10^{-9} M when the external calcium concentration was 1×10^{-2} M (Dipolo *et al.*, 1976). This low level of ionised calcium in axoplasm is thought to be maintained by several mechanisms: (1) calcium-binding molecules which act as calcium buffers (Alema, Calissano, Rusca & Giuditta, 1973; Hillman & Llinás, 1974; Baker & Schlaepfer, 1975, 1978; Mullins, 1976; Dipolo *et al.*, 1976); (2) energy-dependent calcium uptake by mitochondria (Rahamimoff & Alnaes, 1973; Alnaes & Rahamimoff, 1974, 1975; Baker & Schlaepfer, 1975, 1978; Rahamimoff *et al.*, 1975); (3) energy-dependent non-mitochondrial calcium-sequestering systems mentioned above; (4) external sodium/internal calcium exchange (Blaustein & Hodgkin, 1969; Blaustein & Oborn, 1975; Mullins & Brinley, 1975; Blaustein & Ector, 1976; Mullins, 1976; Dipolo *et al.*, 1976; Baker & McNaughton, 1978); and (5) uncoupled energy-dependent calcium efflux which may depend on plasma membrane calcium pumps (Mullins, 1976; Baker & McNaughton, 1978).

As pointed out by Mullins (1976), mechanisms similar to the ones described in (1)–(3) above could regulate the instantaneous free-calcium-ion concentration in nerve terminals, but only mechanisms such as (4) and (5) could maintain long-term calcium balance. These last two represent mechanisms for translocation of calcium out of nerve terminals against a concentration gradient, a process which requires metabolic energy. (Mechanism (4), sodium/calcium exchange, theoretically depends upon the energy-dependent production of a sodium gradient.)

The sequestration of calcium ions within some sort of intracellular sac – the smooth endoplasmic reticulum, for example – could not serve to maintain calcium balance unless this sac were at some time continuous with the plasma membrane and the extracellular environment. Furthermore, there would need to be some mechanism for expelling the contents of the sac.

Theoretically speaking, SVs and secretory granules are sacs of this nature. If in fact the intramembranous particles associated with SVs, CVs and secretory granules are part of a calcium-sequestering system, these organelles could function as calcium-expelling devices by sequestering calcium ions from the nerve terminal and expelling them during exocytosis. This expulsion of calcium would occur during periods of active release of neurotransmitter when there is a transient increase in the calcium concentration within the cell, probably at the very site where release is occurring.

Coated vesicle/neurofilament interactions

It remains to explain how endocytosis by coated vesicles occurs. Heuser & Reese (1975*a*) reported seeing filaments attached to the vesicle particles in cross-fractured images of the endocytosis sites. Perhaps these filaments act in some way to draw this region of membrane inward.

We have recently found evidence for an in-vitro association between neurofilaments and brain CVs. When a purified preparation of CVs was stored at 4 °C a precipitate formed which, when examined in the electron microscope, contained a large number of CVs attached to filaments in the regular array shown in Plate 1. These filaments were 8–10 nm in diameter, the size of neurofilaments (Sack, Fine & Blitz, 1978).

By a variety of biochemical, morphological and immunological techniques, we have demonstrated that these filaments closely resembled neurofilaments isolated from brain. Approximately 10 % of the protein associated with brain CVs co-migrated with the 55 000 mol. wt subunit of neurofilaments. These two proteins also cross-reacted immunologically. Surprisingly, the coat subunit protein cross-reacted with the neurofilament subunit as well.

Even after the suspension of CVs was totally depleted of the neurofilament-containing complex by centrifugation, the 55 000 mol. wt component remained. This finding suggested that the neurofilament subunits were still present, tightly bound to CVs. Perhaps this protein constitutes part of the coat structure, as suggested by Woods *et al.* (1978).

So far a complex of this nature has not been reported in intact neurons. It is therefore difficult to speculate on the physiological significance of this in-vitro interaction.

Coated vesicles in synaptogenesis

We now turn to the other side of the synapse and examine events which occur in postsynaptic neurons during synaptogenesis. An early event in the laying down of synaptic contacts between developing neurites and target cells is the appearance of the subsynaptic web or postsynaptic density, a filamentous thickening of the postsynaptic membrane adjacent to the synaptic cleft. This structure in isolated preparations has been described as a lattice-like array of 20 nm, densely staining subunits (Matus, Walters & Jones, 1975). There is presumptive evidence that during the development of the nervous system, CVs convey to the synaptic region some component or components of this postsynaptic membrane specialisation.

Altman (1971), in studying the development of the rat cerebellar cortex, noted that CVs were more numerous at sites of synaptogenesis than elsewhere in young neurons. Prior to the onset of synapse formation most of the CVs were found in the vicinity of the Golgi apparatus. Also, coated regions were seen adjacent to areas of synapse formation.

Stelzner, Martin & Scott (1973) observed what they considered to be the fusions of CVs with the postsynaptic membrane at developing synapses in the chick cervical spinal cord. However, images of CVs fusing with the plasma membrane were not confined to synaptic regions and occasionally occurred at the presynaptic membrane as well. They suggested that CVs provided either transient adhesion membranes or lasting dense membranes. They regarded the Golgi apparatus as the site of origin of the CVs.

Rees, Bunge & Bunge (1976) made similar findings. They noted that after neuritic growth cones made contact with target neurons in culture, the Golgi apparatus of the target cell hypertrophied and gave rise to an increased number of CVs. There then ensued what they described as a traffic of CVs from the Golgi apparatus to the region of the postsynaptic membrane adjacent to developing synapses. They suggested that the CVs were responsible for the laying down of a precursor of the postsynaptic density.

The straightforward prediction based on these results is that clathrin, the protein comprising the coat of CVs, should be a component of the postsynaptic density. This idea, however, does not seem correct. Polyacrylamide gels of brain fractions enriched in synaptic membranes, synaptic junction complexes or postsynaptic densities did not contain prominent bands corresponding to a polypeptide of 180 000 mol. wt, the molecular weight of the clathrin subunit (Banker, Churchill & Cotman, 1974; Walters & Matus, 1975*a, b*; Therien & Mushynski, 1976; Kelly & Cotman, 1977). The only major protein associated with the postsynaptic density which has been identified with any certainty is tubulin, the subunit of microtubules (Matus, Walters & Mughal, 1975; Walters & Matus, 1975*b*; Therien & Mushynski, 1976).

We would like to propose an alternate function for CVs in synaptogenesis. It is possible that the array of postsynaptic intramembranous particles which appear in images of freeze-fractured synapses are carried as part of the membrane of CVs and are incorporated into the postsynaptic membrane when CVs fuse with it during synaptogenesis. The function of the CV as a carrier of membrane-associated protein aggregates would be consistent with the proposed function of CVs in presynaptic terminals. The function of the subsynaptic web may be to prevent the lateral diffusion of these particles within the fluid lipid bilayer. The function of the particles themselves is a subject for future research.

Conclusion

The point of view implicit in the foregoing discussion is that neuronal CVs have, for the most part, one well-defined function: they mediate the traffic of membrane-associated protein assemblies between different neuronal membrane compartments. This point of view may well be an oversimplification, but it is certainly attractive in view of the conservatism of nature. If this view is adopted, there is no reason to believe that the function of CVs in neurons is in any way different from their general function in other types of cells.

One need not look far for examples. The exciting work of Brown, Goldstein and co-workers (Anderson, Brown & Goldstein, 1977) on internalisation of low-density lipoprotein receptors by fibroblasts has drawn sharp attention to the role of CVs in carrying plasma-membrane-associated protein assemblies.

Nor do neurons have a monopoly on membrane recycling. Tulkens, Schneider & Trouet (1977) found evidence for membrane recycling by rat embryo fibroblasts. It will be no surprise if CVs, which are plentiful in fibroblasts, are shown to participate in this phenomenon.

The rapid increase in knowledge in this decade about the detailed nature of membrane traffic in neurons has mostly resulted from the development of sophisticated technology by the morphologists. The challenge now rests with the biochemists to devise ultramicromethods for isolating and characterising subcellular membrane fractions. When this is accomplished, it may be possible to learn more about what it is that drives and controls the rather bewildering array of membrane transformations which are so much a part of the life of neurons and most other cells. As these mechanisms are uncovered it is likely that CVs and clathrin will be found to play major roles.

References

Abrahams, S.J. & Holtzman, E. (1973). Secretion and endocytosis in insulin-stimulated rat adrenal medulla cells. *Journal of Cell Biology*, 56, 540–58.

Akert, K., Sandri, C., Cuénod, M. & Moor, H. (1976). Solitary membrane associated particles in synaptic vesicles as possible calcium binding sites. *Experientia*, 32, 784.

Alema, S., Calissano, P., Rusca, G. & Giuditta, A. (1973). Identification of a calcium-binding protein in the axoplasm of squid giant axons. *Journal of Neurochemistry*, 20, 681–9.

Alnaes, E. & Rahamimoff, R. (1974). Dual action of praseodymium on transmitter release at the frog neuromuscular junction. *Nature, London*, 247, 478–9.

Alnaes, E. & Rahamimoff, R. (1975). On the role of mitochondria in transmitter release from motor nerve terminals. *Journal of Physiology, London*, 248, 285–306.

Altman, J. (1971). Coated vesicles and synaptogenesis. A developmental study in the cerebellar cortex of the rat. *Brain Research*, 30, 311–22.

Anderson, R.G.W., Brown, M.S. & Goldstein, J.L. (1977). Role of the coated endocytic vesicle in the uptake of receptor-bound low density lipoprotein in human fibroblasts. *Cell*, 10, 351–64.

Andres, K.H. (1964). Mikropinozytose im Zentralnervensystem. *Zeitschrift für Zellforschung und Mikroskopische Anatomie*, 64, 63–73.

Ashkong, Q.F., Fisher, D., Tampion, W. & Lucy, J.A. (1975). Mechanisms of cell fusion. *Nature, London*, 253, 194–5.

Atwood, H.L., Lang, F. & Morin, W.A. (1972). Synaptic vesicles: selective depletion in crayfish excitatory and inhibitory axons. *Science*, 176, 1353.

Baker, P.F. & Crawford, A.C. (1972). Mobility and transport of magnesium in squid giant axons. *Journal of Physiology, London*, 227, 855–74.

Baker, P.F. & McNaughton, P.A. (1978). The influence of extracellular calcium binding on the calcium efflux from squid axons. *Journal of Physiology, London*, 276, 127–50.

Baker, P.F. & Schlaepfer, W. (1975). Calcium uptake by axoplasm extruded from giant axons of *Loligo*. *Journal of Physiology, London*, 249, 37P–38P.

Baker, P.F. & Schlaepfer, W.W. (1978). Uptake and binding of calcium by axoplasm isolated from giant axons of *Loligo* and *Myxicola*. *Journal of Physiology, London*, 276, 103–25.

Banker, G., Churchill, L. & Cotman, C.W. (1974). Proteins of the postsynaptic density. *Journal of Cell Biology*, 63, 456–65.

Banks, P. (1965). The ATPase activity of adrenal chromaffin granules. *Biochemical Journal*, 95, 490–6.

Becker, N.H., Hirano, A. & Zimmerman, H. (1968). Observations of the distribution of exogenous peroxidase in the rat cerebrum. *Journal of Neuropathology and Experimental Neurology*, 27, 439–52.

Becker, N.H., Novikoff, A.B. & Zimmerman, H.M. (1967). Fine structure observations of the uptake of intravenously injected peroxidase by the rat choroid plexus. *Journal of Histochemistry and Cytochemistry*, 15, 160–5.

Benedeczky, I. & Smith, A.D. (1972). Ultrastructural studies on the adrenal medulla of the golden hamster: origin and fate of secretory granules. *Zeitschrift für Zellforschung und Mikroskopische Anatomie*, 124, 367–86.

Bennet, M.V.L., Highstein, S.M. & Model, P.G. (1976). Depletion of vesicles and fatigue at a vertebrate central synapse. In *Electrobiology of Nerve, Muscle, and Synapse*, ed. J.P. Reuben, D.P. Purpura & M.V.L. Bennet, pp. 105–22. New York: Raven Press.

Bennet, M.V.L., Model, P.G. & Highstein, S.M. (1975). Stimulation-induced depletion of vesicles, fatigue of transmission and recovery processes at a vertebrate central synapse. *Cold Spring Harbor Symposia on Quantitative Biology*, 40, 25–35.

Birks, R.I., Huxley, H.E. & Katz, B. (1960). The fine structure of the neuromuscular junction of the frog. *Journal of Physiology, London*, 150, 134–44.

Birks, R.I., Mackey, M.C. & Weldon, P.R. (1972). Organelle formation from pinocytotic elements in neurites of cultured sympathetic ganglia. *Journal of Neurocytology*, 1, 311–40.

Bittner, G.D. & Kennedy, D. (1970). Quantitative aspects of transmitter release. *Journal of Cell Biology*, 47, 585–92.

Blaustein, M.P. & Ector, A.C. (1976). Carrier-mediated sodium-dependent and calcium-dependent calcium efflux from pinched off presynaptic nerve terminals (synaptosomes) *in vitro*. *Biochimica et Biophysica Acta*, 419, 295–308.

Blaustein, M.P. & Hodgkin, A.L. (1969). The effect of cyanide on the efflux of calcium from squid axons. *Journal of Physiology, London*, 200, 497–527.

Blaustein, M.P. & Oborn, C.J. (1975). The influence of sodium on calcium fluxes in pinched-off nerve terminals *in vitro*. *Journal of Physiology, London*, 247, 657–86.

Blitz, A.L., Fine, R.E. & Toselli, P.A. (1977). Evidence that coated vesicles isolated from brain are calcium sequestering organelles resembling sarcoplasmic reticulum. *Journal of Cell Biology*, 75, 735–47.

Bohan, T.P., Boyne, A.F., Guth, P.S., Narayanan, Y. & Williams, T.H. (1973). Electron-dense particle in cholinergic synaptic vesicles. *Nature, London*, 244, 32–4.

Borowitz, J.L. (1967). Calcium binding by subcellular fractions of bovine adrenal medulla. *Journal of Cellular and Comparative Physiology*, 69, 305–10.

Borowitz, J.L. (1969). Effect of acetylcholine on subcellular distribution of ^{45}Ca in bovine adrenal medulla. *Biochemical Pharmacology*, 18, 715–23.

Boyne, A.F., Bohan, T.P. & Williams, T.H. (1974). Effects of calcium-containing fixation solutions on cholinergic synaptic vesicles. *Journal of Cell Biology*, 63, 780–95.

Brightman, M.W. (1965). The distribution within the brain of ferritin injected into cerebrospinal fluid compartments. II. Parenchymal distribution. *American Journal of Anatomy*, 117, 193–220.

Bunge, M.B. (1973). Fine structure of nerve fibre and growth cones of isolated sympathetic neurons in culture. *Journal of Cell Biology*, 56, 713–35.

Bunge, M.B. (1977). Initial endocytosis of peroxidase or ferritin by growth cones of cultured nerve cells. *Journal of Neurocytology*, 6, 407–39.

Bunt, A.H. (1969). Formation of coated and 'synaptic' vesicles within neurosecretory axon terminals of the crustacean sinus gland. *Journal of Ultrastructure Research*, 28, 411–21.

Ceccarelli, B., Hurlbut, W.P. & Mauro, A. (1972). Depletion of vesicles from frog neuromuscular junctions by prolonged tetanic stimulation. *Journal of Cell Biology*, 54, 30–8.

Ceccarelli, B., Hurlbut, W.P. & Mauro, A. (1973). Turnover of transmitter and synaptic vesicles at the frog neuromuscular junction. *Journal of Cell Biology*, 57, 499–524.

Ceccarelli, B., Pensa, P., DeGiuli, C. & Pelosi, G. (1967). Sur les méthodes de fixation pour étudier les terminaisons synaptiques centrales du rat. *Journal de Microscopie*, 6, 42a–43a.

Chi, E.Y., Lagunoff, D. & Koehler, J.K. (1976). Freeze-fracture study of mast cell secretion. *Proceedings of the National Academy of Sciences, USA*, 73, 2823–7.

Clark, A.W., Hurlbut, W.P. & Mauro, A. (1972). Changes in the fine structure of the neuromuscular junction of the frog caused by black widow spider venom. *Journal of Cell Biology*, 52, 1–14.

Clark, A.W., Mauro, A., Longenecker, H.E., Jr & Hurlbut, W.P. (1970). Effects of black widow spider venom on the frog neuromuscular junction. *Nature, London*, 225, 701–5.

Couteaux, R. & Pécot-Dechavassine, M. (1970). Vésicules synaptique et poches au niveau des 'zones actives' de la jonction neuromusculaire. *Comptes Rendus des Séances de l'Académie des Sciences, Paris, Series D*, 271, 2346–9.

Crowther, R.A., Finch, J.T. & Pearse, B.M.F. (1976). On the structure of coated vesicles. *Journal of Molecular Biology*, 103, 785–98.

Dahl, G., Gratzl, M. & Ekerdt, R. (1976). *In vitro* fusion of secretory vesicles isolated from pancreatic B-cells and from the adrenal medulla. *Journal of Cell Biology*, 70, 180a.

Damassa, D.A., Davis, T.L., Shotten, D.M., Heuser, J.E., Pumplin, D. & Reese,

T.S. (1976). Structure of nerve terminals in frog muscle after prolonged treatment with brown widow spider venom. *Biological
Bulletin*, 151, 406–7.

Del Castillo, J. & Katz, B. (1956). Biophysical aspects of neuromuscular transmission. *Progress in Biophysics and Biophysical Chemistry*, 6, 121–70.

Del Castillo, J. & Stark, L. (1952). The effect of calcium ions on the motor
end-plate potentials. *Journal of Physiology, London*, 116, 507–15.

DeMeis, L., Rubin-Altschul, B.M. & Machado, R.D. (1970). Comparative
data of Ca^{2+} transport in brain and skeletal muscle microsomes.
Journal of Biological Chemistry, 245, 1883–9.

DeRobertis, E.D.P. (1958). Submicroscopic morphology and function of the
synapse. *Experimental Cell Research, Supplement* 5, 347–69.

DeRobertis, E.D.P. & Bennett, H.S. (1955). Some features of the submicroscopic morphology of synapses in frog and earthworm. *Journal of
Biophysical and Biochemical Cytology*, 1, 47–58.

Diner, O. (1967). L'expulsion des granules de la médullo-surrénale chez le
Hamster. *Comptes Rendus des Séances de l'Académie des Sciences,
Paris, Series D*, 265, 616–19.

Dipolo, R., Requena, J., Brinley, F.J., Jr, Mullins, L.J., Scarpa, A. & Tiffert,
T. (1976). Ionized calcium concentrations in squid axons. *Journal
of General Physiology*, 67, 433–67.

Douglas, W.W. (1973). How do neurons secrete peptides? Exocytosis and its
consequences, including 'synaptic vesicle' formation, in the
hypothalamo-neurohypophyseal system. *Progress in Brain Research*,
39, 21–39.

Douglas, W.W. & Nagasawa, J. (1971). Membrane vesiculation at sites of
exocytosis in the neurohypophysis, adenohypophysis and adrenal
medulla: a device for membrane conservation. *Journal of Physiology,
London*, 218, 94P–95P.

Douglas, W.W., Nagasawa, J. & Schultz, R.A. (1971). Coated microvesicles in
neurosecretory terminals of posterior pituitary glands shed their
coats to become 'smooth' synaptic vesicles. *Nature, London*, 232,
340–1.

Dreifuss, J.J., Akert, K., Sandri, C. & Moor, H. (1973). The fine structure of
freeze-fractured neurosecretory nerve endings in the neurohypophysis.
Brain Research, 62, 367–72.

Dreyer, F., Peper, K., Akert, K., Sandri, C. & Moor, H. (1973). Ultrastructure
of the 'active zone' in the frog neuromuscular junction. *Brain Research*,
62, 373–80.

Duggan, P.F. & Kelleher, D.N. (1975). Effects of univalent cations and neurotransmitters on the adenosine triphosphate-dependent uptake of
calcium ions by brain microsomal fraction. *Biochemical Society
Transactions*, 3, 1226–8.

Duncan, C.J. (1976). Properties of the Ca^{2+}-ATPase activity of mammalian
synaptic membrane preparation. *Journal of Neurochemistry*, 27,
1277–9.

Evans, E.E. (1966). On the ultrastructure of the synaptic region of visual receptors in certain vertebrates. *Zeitschrift für Zellforschung und Mikroskopische Anatomie,* 71, 499–516.

Fatt, P. & Katz, B. (1952). Spontaneous subthreshold activity at motor nerve endings. *Journal of Physiology, London,* 117, 109–28.

Gray, E.G. (1961). The granule cells, mossy synapses and Purkinje spine synapses of the cerebellum: light and electron microscope observation. *Journal of Anatomy,* 95, 345–56.

Gray, E.G. (1972). Are the coats of coated vesicles artefacts? *Journal of Neurocytology,* 1, 363–82.

Gray, E.G. (1975). Synaptic fine structure and nuclear, cytoplasmic, and extracellular networks. The stereo-framework concept. *Journal of Neurocytology,* 4, 315–39.

Gray, E.G. & Parsegian, V.A. (1977). Delivery of synaptic vesicles to the active zones. *Neuroscience Research Program Bulletin,* 15, 614–22.

Gray, E.G. & Pease, H.L. (1971). On understanding the organisation of the retinal receptor synapses. *Brain Research,* 35, 1–15.

Gray, E.G. & Willis, R.A. (1970). On synaptic vesicles, complex vesicles and dense projections. *Brain Research,* 24, 149–68.

Grynszpan-Winograd, O. (1971). Morphological aspects of exocytosis in the adrenal medulla. *Philosophical Transactions of the Royal Society of London, Series B,* 261, 291–2.

Hama, K. & Saito, K. (1977). Fine structure of the afferent synapse of the hair cells in the saccular macula of the goldfish, with special reference to the anastomosing tubules. *Journal of Neurocytology,* 6, 361–73.

Heuser, J.E. (1976). Morphology of synaptic vesicle discharge and reformation at the frog neuromuscular junction. In *Motor Innervation of Muscle,* ed. S. Thesleff, pp. 51–115. New York & London: Academic Press.

Heuser, J.E. (1977). Synaptic vesicle exocytosis revealed in quick frozen frog neuromuscular junctions treated with 4-aminopyridine and given a single electrical shock. In *Society for Neuroscience Symposia,* ed. W.M. Cowan & J.A. Ferrendelli, vol. 2, pp. 215–59. Bethesda: Society for Neuroscience.

Heuser, J.E. & Miledi, R. (1971). Effect of lanthanum ions on function and structure of frog neuromuscular junctions. *Proceedings of the Royal Society of London, Series B,* 179, 247–60.

Heuser, J.E. & Reese, T.S. (1972). Stimulation induced uptake and release of peroxidase from synaptic vesicles in frog neuromuscular junctions. *Anatomical Record,* 172, 329–30.

Heuser, J.E. & Reese, T.S. (1973). Evidence for recycling of synaptic vesicle membrane during transmitter release at the frog neuromuscular junction. *Journal of Cell Biology,* 57, 315–44.

Heuser, J.E. & Reese, T.S. (1975a). Redistribution of intramembranous particles from synaptic vesicles: direct evidence for vesicle recycling. *Anatomical Record,* 181, 374.

Heuser, J.E. & Reese, T.S. (1975b). Structural changes at the 'active zones' of synapses in stimulated *Torpedo* electrical organs. *Biological Bulletin,* 149, 429.

Heuser, J.E., Reese, T.S. & Landis, D.M.D. (1974). Functional changes in frog neuromuscular junctions studied with freeze fracture. *Journal of Neurocytology*, 3, 109–31.

Heuser, J.E., Reese, T.S. & Landis, D.M.D. (1975). Preservation of synaptic structure by rapid freezing. *Cold Spring Harbor Symposia on Quantitative Biology*, 40, 17–24.

Hillman, D.E. & Llinás, R. (1974). Calcium-containing electron-dense structures in the axons of the squid giant synapse. *Journal of Cell Biology*, 61, 146–55.

Hodgkin, A.L. & Keynes, R.D. (1957). Movements of labelled calcium in squid giant axons. *Journal of Physiology, London*, 138, 253–81.

Holtzman, E. (1971). Cytochemical studies of protein transport in the nervous system. *Philosophical Transactions of the Royal Society of London, Series B*, 261, 407–21.

Holtzman, E. & Dominitz, R. (1968). Cytochemical studies of lysosomes, Golgi apparatus and endoplasmic reticulum in secretion and protein uptake by adrenal cells of the rat. *Journal of Histochemistry and Cytochemistry*, 16, 320–36.

Holtzman, E., Freeman, A.R. & Kashner, L.A. (1970). A cytochemical and electron microscope study of channels in Schwann cells surrounding lobster giant axons. *Journal of Cell Biology*, 44, 438–45.

Holtzman, E., Freeman, A.R. & Kashner, L.A. (1971). Stimulation-dependent alterations in peroxidase uptake at lobster neuromuscular junctions. *Science*, 173, 733–6.

Holtzman, E. & Peterson, E.R. (1969). Uptake of protein by mammalian neurons. *Journal of Cell Biology*, 40, 863–9.

Holtzman, E., Schacher, S., Evans, J. & Teichberg, S. (1977). Origin and fate of the membranes of secretion granules and synaptic vesicles: membrane circulation in neurons, gland cells and retinal photoreceptors. In *The Synthesis, Assembly and Turnover of Cell Surface Components*, de. G. Poste & G.L. Nicolson, pp. 165–246. Amsterdam: Elsevier/North-Holland Biomedical Press.

Holtzman, E., Teichberg, S., Abrahams, S.J., Citkowitz, E., Crain, S.M., Kawai, N. & Peterson, E.R. (1973). Notes on synaptic vesicles and related structures, endoplasmic reticulum, lysosomes and peroxisomes in nervous tissue and the adrenal medulla. *Journal of Histochemistry and Cytochemistry*, 21, 349–85.

Ingolia, T.D. & Koshland, D.E., Jr (1978). The role of calcium in fusion of artificial vesicles. *Journal of Biological Chemistry*, 253, 3821–9.

Jones, S.F. & Kwanbunbumpen, S. (1970). The effects of nerve stimulation and hemicholinium on synaptic vesicles at the mammalian neuromuscular junction. *Journal of Physiology, London*, 207, 31–50.

Jorgensen, O.S. & Mellerup, E.T. (1974). Endocytotic formation of rat brain synaptic vesicles. *Nature, London*, 249, 770–1.

Kadota, K. & Kadota, T. (1973a). Isolation of coated vesicles, plain synaptic vesicles and fine particles from synaptosomes of guinea pig whole brain. *Journal of Electron Microscopy*, 22, 91–8.

Kadota, K. & Kadota, T. (1973*b*). Isolation of coated vesicles, plain synaptic vesicles, and flocculent material from a crude synaptosome fraction of guinea pig whole brain. *Journal of Cell Biology,* 58, 135–51.

Kadota, T., Kadota, K. & Gray, E.G. (1976). Coated-vesicle shells, particle/ chain material, and tubulin in brain synaptosomes. *Journal of Cell Biology,* 69, 608–21.

Kanaseki, T. & Kadota, K. (1969). The 'vesicle in a basket'. A morphological study of the coated vesicle isolated from the nerve endings of the guinea pig brain, with special reference to the mechanism of membrane movements. *Journal of Cell Biology,* 42, 202–20.

Katz, B. & Miledi, R. (1965). The effect of calcium on acetylcholine release from motor nerve terminals. *Proceedings of the Royal Society of London, Series B,* 161, 496–503.

Katz, B. & Miledi, R. (1969). Tetrodotoxin-resistant electrical activity in presynaptic terminals. *Journal of Physiology, London,* 203, 459–87.

Kelly, P.T. & Cotman, C.W. (1977). Identification of glycoproteins and proteins at synapses in the central nervous system. *Journal of Biological Chemistry,* 252, 786–93.

Kendrick, N.C., Blaustein, M.P., Fried, R.C. & Ratzlaff, R.W. (1977). ATP-dependent calcium storage in presynaptic terminals. *Nature, London,* 265, 246–8.

Koenig, H.L., Giamberardino, L.D. & Bennett, G. (1973). Renewal of proteins and glycoproteins of synaptic constituents by means of axonal transport. *Brain Research,* 62, 413–17.

Korneliussen, H. (1972). Ultrastructure of normal and stimulated motor endplates. *Zeitschrift für Zellforschung und Mikroskopische Anatomie,* 130, 28–57.

Landis, D.M.D. & Reese, T.S. (1974). Differences in membrane structure between excitatory and inhibitory synapses in the cerebellar cortex. *Journal of Comparative Neurology,* 155, 93–126.

LaVail, J.H. & LaVail, M.M. (1974). The retrograde intra-axonal transport of horseradish peroxidase in the chick visual system: a light and electron microscope study. *Journal of Comparative Neurology,* 157, 303–58.

Litchy, W.J. (1973). Uptake and retrograde transport of horseradish peroxidase in frog sartorius nerve *in vitro. Brain Research,* 56, 377–81.

Llinás, R., Blinks, J.R. & Nicholson, C. (1972). Calcium transient in pre-synaptic terminal of giant squid synapse: detection with aequorin. *Science,* 176, 1127–9.

Llinás, R. & Nicholson, C. (1975). Calcium role in depolarisation–secretion coupling: an aequorin study in squid giant synapse. *Proceedings of the National Academy of Sciences, USA,* 72, 187–90.

Llinás, R., Steinberg, I.Z. & Walton, K. (1976). Presynaptic calcium currents and their relation to synaptic transmission: voltage clamp study in squid giant synapse and theoretical model for the calcium gate. *Proceedings of the National Academy of Sciences, USA.* 73, 2918–22.

Llinás, R., Walton, K. & Hess, R. (1976). Voltage clamp study of presynaptic calcium current in squid giant synapses. *Federation Proceedings*, 35, 696.

MacLennan, D.H., Yip, C.C., Iles, G.H. & Seeman, P. (1972). Isolation of sarcoplasmic reticulum proteins. *Cold Spring Harbor Symposia on Quantitative Biology*, 37, 469–77.

Matus, A.I., Walters, B.B. & Jones, D.H. (1975). Junctional ultrastructure in isolated synaptic membranes. *Journal of Neurocytology*, 4, 357–67.

Matus, A.I., Walters, B.B. & Mughal, S. (1975). Immunohistochemical demonstration of tubulin associated with microtubules and synaptic junctions in mammalian brain. *Journal of Neurocytology*, 4, 733–44.

Miledi, R. (1973). Transmitter release induced by injection of calcium ions into nerve terminals. *Proceedings of the Royal Society of London, Series B*, 183, 421–5.

Miledi, R. & Slater, C.R. (1966). The action of calcium on neuronal synapses in the squid. *Journal of Physiology, London*, 184, 473–98.

Model, P.G., Highstein, S.M. & Bennet, M.V.L. (1975). Depletion of vesicles and fatigue of transmission at a vertebrate central synapse. *Brain Research*, 98, 209–28.

Mullins, L.J. (1976). Steady-state calcium fluxes: membrane versus mitochondrial control of ionized calcium in axoplasm. *Federation Proceedings*, 35, 2583–8.

Mullins, L.J. & Brinley, F.J., Jr (1975). Sensitivity of calcium efflux from squid axons to changes in membrane potential. *Journal of General Physiology*, 65, 135–52.

Nagasawa, J., Douglas, W.W. & Schulz, R.A. (1970). Ultrastructural evidence of secretion by exocytosis and of 'synaptic vesicle' formation in posterior pituitary glands. *Nature, London*, 227, 407–9.

Nagasawa, J., Douglas, W.W. & Schulz, R.A. (1971). Micropinocytotic origin of coated and smooth microvesicles ('synaptic vesicles') in neuro-secretory terminals of posterior pituitary glands demonstrated by incorporation of horseradish peroxidase. *Nature, London*, 232, 341–2.

Nickel, E., Vogel. A. & Waser, P.G. (1967). 'Coated Vesicles' in der Umgebung der neuro-muskularen Synapsen. *Zeitschrift für Zellforschung und Mikroskopische Anatomie*, 78, 261–6.

Normann, T.C. & Hall, T.A. (1978). Calcium and sulphur in neurosecretory granules and calcium in mitochondria as determined by electron microscope X-ray microanalysis. *Cell and Tissue Research*, 186, 453–63.

O'Driscoll, G. & Duggan, P.F. (1977). Distribution of the calcium-ion-transport system in subcellular fractions from rabbit brain. *Biochemical Society Transactions*, 5, 1708–10.

Ohtsuki, I. (1969). ATP-dependent Ca uptake by brain microsomes. *Journal of Biochemistry, Tokyo*, 66, 645–50.

Palay, S.L. (1963). Alveolate vesicles in Purkinje cells of the rat's cerebellum. *Journal of Cell Biology*, 19, 89A.

Pappas, G.D. & Rose, S. (1976). Localisation of calcium deposits in the frog neuromuscular junctions at rest and following stimulation. *Brain Research*, 103, 362–5.

Párducz, Á. & Fehér, O. (1970). Fine structural alterations of presynaptic endings in the superior cervical ganglion of the cat after exhausting preganglionic stimulation. *Experientia*, 26, 629–30.

Párducz, Á., Fehér, D. & Joó, F. (1971). Effects of stimulation and hemicholium (HC-3) on the fine structure of nerve endings in the superior cervical ganglion of the cat. *Brain Research*, 34, 61–72.

Paula-Barbosa, M. & Gray, E.G. (1974). The effects of various fixatives at different pH on synaptic coated vesicles, reticulosomes and cytonet. *Journal of Neurocytology*, 3, 471–86.

Pearse, B.M.F. (1975). Coated vesicles from pig brain: purification and bio-chemical characterisation. *Journal of Molecular Biology*, 97, 93–8.

Pearse, B.M.F. (1976). Clathrin: a unique protein associated with intracellular transfer of membrane by coated vesicles. *Proceedings of the National Academy of Sciences, USA*, 73, 1255–9.

Perri, V., Sacchi, O., Raviola, E. & Raviola, G. (1972). Evaluation of the number and distribution of synaptic vesicles at cholinergic nerve-endings after sustained stimulation. *Brain Research*, 39, 526–9.

Pfenninger, K., Akert, K., Moor, H. & Sandri, C. (1971). Freeze-fracturing of presynaptic membranes in the central nervous system. *Philosophical Transactions of the Royal Society of London, Series B*, 261, 387.

Pfenninger, K., Akert, K., Moor, H. & Sandri, C. (1972). The fine structure of freeze-fractured presynaptic membranes. *Journal of Neurocytology*, 1, 129–49.

Politoff, A.L., Rose, S. & Pappas, G.D. (1974). The calcium binding sites of synaptic vesicles of the frog neuromuscular junction. *Journal of Cell Biology*, 61, 818–23.

Pumplin, D.W. & Reese, T.S. (1977). Action of brown widow spider venom and botulinum toxin on frog neuromuscular junction examined with the freeze-fracturing technique. *Journal of Physiology, London*, 273, 443–57.

Puszkin, S., Schook, W., Puszkin, E., Roualt, C., Ores, C., Schlossberg, J., Kochwa, S. & Rosenfield, R.E. (1976). Contractile elements in brain tissue: evidence of α-actinin in synaptosomes. In *Contractile Systems in Non-Muscle Tissues*, ed. S.V. Perry, R. Margreth & R.S. Adelstein, pp. 67–80. Amsterdam: Elsevier/North-Holland Biomedical Press.

Pysh, J.J. & Wiley, R.G. (1972). Morphological alterations of synapses in electrically stimulated superior cervical ganglia of the cat. *Science*, 176, 191–3.

Pysh, J.J. & Wiley, R.G. (1974). Synaptic vesicle depletion and recovery in cat sympathetic ganglia electrically stimulated *in vivo*. Evidence for transmitter secretion by exocytosis. *Journal of Cell Biology*, 60, 365–74.

Rahamimoff, R. & Alnaes, E. (1973). Inhibitory action of ruthenium red in neuromuscular transmission. *Proceedings of the National Academy of Science, USA*, 70, 3613–16.

Rahamimoff, R., Erulkar, S.D., Alnaes, E., Meiri, H., Rothenker, H. & Rahamimoff, H. (1975). Modulation of transmitter release by calcium ions. *Cold Spring Harbor Symposia on Quantitative Biology,* **40,** 107–16.

Raviola, E. & Gilula, N.B. (1975). Intramembrane organisation of specialised contacts in the outer plexiform layer to the retina. A freeze-fracture study in monkeys and rabbits. *Journal of Cell Biology,* **65,** 192–222.

Rees, R.P., Bunge, M.B. & Bunge, R.P. (1976). Morphological changes in the neuritic growth cone and target neuron during synaptic junction development in culture. *Journal of Cell Biology,* **68,** 240–63.

Ripps, H., Shakib, M. & MacDonald, E.D. (1976). Peroxidase uptake by photoreceptor terminals of the skate retina. *Journal of Cell Biology,* **70,** 86–96.

Robertson, J.D. (1954). Electron microscopic study of an invertebrate synapse. *Federation Proceedings,* **13,** 119.

Robinson, J.D. & Lust, W.D. (1968). Adenosine triphosphate-dependent calcium accumulation by brain microsomes. *Archives of Biochemistry and Biophysics,* **125,** 286–94.

Rosenbluth, J. & Wissig, S.L. (1963). The uptake of ferritin by toad spinal ganglion cells. *Journal of Cell Biology,* **19,** 91a.

Rosenbluth, J. & Wissig, S.L. (1964). The distribution of exogenous ferritin in toad spinal ganglia and the mechanism of its uptake by neurons. *Journal of Cell Biology,* **23,** 307–25.

Roth, T.F. & Porter, K.R. (1964). Yolk protein uptake in the oocyte of the mosquito *Aedes aegypti* L. *Journal of Cell Biology,* **20,** 313–32.

Sack, D.H., Fine, R.E. & Blitz, A.L. (1978). Evidence for a specific brain coated vesicle–intermediate filament complex. *Federation Proceedings,* **37,** 278.

Sandri, C., Akert, K., Livingston, R.B. & Moor, H. (1972). Particle aggregations at specialised sites in freeze-etched postsynaptic membranes. *Brain Research,* **41,** 1–16.

Satir, B., Schooley, C. & Satir, P. (1973). Membrane fusion in a model system. Mucocyst secretion in *Tetrahymena. Journal of Cell Biology,* **56,** 153–76.

Scales, D.J. & Inesi, G. (1976). Localisation of ATPase protein in sarcoplasmic reticulum membrane. *Archives of Biochemistry and Biophysics,* **176,** 392–4.

Schacher, S., Holtzman, E. & Hood, D.C. (1976). Synaptic activity of frog retinal photoreceptors. A peroxidase uptake study. *Journal of Cell Biology,* **70,** 178–92.

Schaeffer, S.F. & Raviola, E. (1977). Membrane recycling in the cone cell endings of the turtle retina. *Journal of Cell Biology,* **75,** 106a.

Singer, S.J. & Nicolson, G.L. (1972). The mosaic model of the structure of cell membranes. *Science,* **175,** 720–31.

Smith, U. (1971). Uptake of ferritin into neurosecretory terminals. *Philosophical Transactions of the Royal Society of London, Series B,* **261,** 391–4.

Stelzner, D.J., Martin, A.H. & Scott, G.L. (1973). Early stages of synaptogenesis in the cervical spinal cord of the chick embryo. *Zeitschrift für Zellforschung und Mikroskopische Anatomie,* **138,** 475–88.

Tanaka, R., Takeda, M. & Jaimovich, M. (1976). Characterisation of ATPases of plain synaptic vesicle and coated vesicle fractions isolated from rat brains. *Journal of Biochemistry, Tokyo*, **80**, 831−7.

Teichberg, S., Holtzman, E., Crain, S.M. & Peterson, E.R. (1975). Circulation and turnover of synaptic vesicle membrane in cultured fetal mammalian spinal cord neurons. *Journal of Cell Biology*, **67**, 215−30.

Therien, H.M. & Mushynski, W.E. (1976). Isolation of synaptic junctional complexes of high structural integrity from rat brain. *Journal of Cell Biology*, **71**, 807−22.

Thorn, N.A., Russell, J.T. & Villardt, H. (1975). Hexosamine, calcium, and neurophysin in secretory granules and the role of calcium in hormone release. *Annals of the New York Academy of Sciences*, **248**, 202−17.

Trotta, E.E. & DeMeis, L. (1975). ATP-dependent calcium accumulation in brain microsomes. Enhancement by phosphate and oxalate. *Biochimica et Biophysica Acta*, **394**, 239−47.

Tulkens, P., Schneider, Y.-J. & Trouet, A. (1977). The fate of plasma membrane during endocytosis. *Biochemical Society Transactions*, **5**, 1809−15.

Turner, P.T. & Harris, A.B. (1973). Ultrastructure of synaptic vesicle formation in cerebral cortex. *Nature, London*, **242**, 57−9.

Turner, P.T. & Harris, A.B. (1974). Ultrastructure of exogenous peroxidase in cerebral cortex. *Brain Research*, **74**, 305−26.

Van Herreveld, A. & Trubatch, J. (1975). Synaptic changes in frog brain after stimulation with potassium chloride. *Journal of Neurocytology*, **4**, 33−46.

von Hungen, K., Mahler, H.R. & Moore, W.J. (1968). Turnover of protein and ribonucleic acid in synaptic subcellular fractions of the rat. *Journal of Biological Chemistry*, **243**, 1415−23.

Wagner, J.A., Carlson, S.S. & Kelly, R.B. (1978). Chemical and physical characterisation of cholinergic synaptic vesicles. *Biochemistry*, **17**, 1199−207.

Walters, B.B. & Matus, A.I. (1975a). Proteins of the synaptic junction. *Biochemical Society Transactions*, **3**, 109−12.

Walters, B.B. & Matus, A.I. (1975b). Tubulin in postsynaptic junctional lattice. *Nature, London*, **257**, 496−8.

Waxman, S.G. & Pappas, G.D. (1969). Pinocytosis at postsynaptic membranes: electron microscopic evidence. *Brain Research*, **14**, 240−4.

Weldon, P.R. (1975). Pinocytotic uptake and intracellular distribution of colloidal thorium dioxide by cultured sensory neurites. *Journal of Neurocytology*, **4**, 341−56.

Wessells, N.K., Ludueña, M.A., Letourneau, P.C., Wrenn, J.T. & Spooner, B.S. (1974). Thorotrast uptake and transit in embryonic glia, heart fibroblasts and neurons *in vitro*. *Tissue and Cell Research*, **6**, 757−76.

Westrum, L.E. (1965). On the origin of synaptic vesicles in the cerebral cortex. *Journal of Physiology, London*, **179**, 4P−6P.

Westrum, L.E. & Gray, E.G. (1977). Microtubules associated with postsynaptic 'thickenings'. *Journal of Neurocytology*, **6**, 505−18.

Woods, J.W., Woodward, M.P. & Roth, T.F. (1978). Common features of coated vesicles from dissimilar tissues: composition and structure. *Journal of Cell Science,* **30,** 87–97.

Zacks, S.I. & Saito, A. (1969). Uptake of exogenous horseradish peroxidase by coated vesicles in mouse neuromuscular junctions. *Journal of Histochemistry and Cytochemistry,* **17,** 161–70.

Zakai, N., Kulka, R.G. & Loyter, A. (1976). Fusion of human erythrocyte ghosts promoted by the combined action of calcium and phosphate ions. *Nature, London,* **263,** 696–9.

Plate

Plate 1. Electron micrograph of coated vesicle/neurofilament complex formed *in vitro*. A thin section through a pellet produced by low-speed centrifugation of a suspension of brain CVs incubated at 4 °C overnight, stained with uranyl acetate and lead citrate, is shown. The bar represents 200 nm.

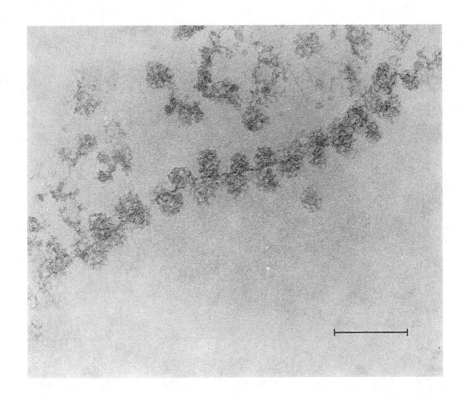

6

Coated vesicles in the oocyte

FRANCO GIORGI

One of the most striking aspects of oogenesis is the remarkable growth that is undertaken by the oocyte. This is normally taken to be an adaptation for providing the embryo with the substances needed to sustain full development. In fact, with the exception of mammals, oocytes of most species achieve this by accumulating large quantities of reserve material generally referred to as yolk. The latter may either be derived from the maternal blood stream, in which case it is called heterosynthetic, or, alternatively, synthesised by the oocyte itself, and is then termed as autosynthetic (Schechtman, 1955).

Although coated vesicles may form in oocytes of numerous species, irrespective of their reproductive biology, it is quite apparent that micropinocytosis in oocytes plays a major role only when yolk is formed by heterosynthetic processes. In fact, as in other cell types, micropinocytosis in these instances provides a method of engulfing large molecules and sequestering them within an intracellular compartment which remains separated from the rest of the cytoplasm by a limiting membrane. However, substances ingested into the oocyte normally do not undergo intracellular digestion soon after uptake but are stored for a definite period of time prior to their final utilisation.

For the reasons outlined above, it is clear that certain species are better suited than others for studying the processes of micropinocytosis in oocytes (by biochemical as well as ultrastructural methods). However in spite of the fragmentary knowledge presently available, the general picture that emerges allows a comparison to be made between oocytes and other cell types in which micropinocytosis has been more thoroughly investigated. This chapter will mainly review the evidence concerning oocytes of those species more widely employed for studying micropinocytosis, but comparative information will also be given whenever it is helpful.

Occurrence of coated vesicles in developing oocytes

Oogenesis in various animal species has been subdivided into a number of typical developmental stages which are assumed to follow one another in a temporal sequence (Cummings & King, 1969; Dumont, 1972; Matsuzaki, 1975). Although ovarian development may have characteristics that differ from species to species, it appears that oocytes undergoing yolk deposition by heterosynthetic processes exhibit a number of ultrastructural features common to numerous species. The description which follows pertains mainly to those aspects of micropinocytosis which are of more frequent occurrence.

In general, the onset of vitellogenesis is marked by the appearance of the first yolk spheres in the oocyte. This event is preceded by or is coincident with the formation of coated vesicles in the cortical ooplasm of the developing oocyte. At this point in the growth of the oocyte, the oolemma is thrown into a series of infoldings or microvilli which extend into the adjacent extra-cellular space. At the bases of the microvilli there appear invaginations which are structured elements, generally referred to as pits, that resemble quite closely the coated vesicles of the cortical ooplasm (Plate 1). Both of these elements are three-layered structures: the outermost layer consists of a lattice (previously called a bristle-coat) with a thickness of about 20 nm, immediately inside which is a unit membrane and then a so-called fuzzy or fibrous coat of about 25–40 nm. The middle layer appears to be continuous with the rest of the oolemma. The inner layer consists of a fibrous material which is structurally comparable to that present in the follicle oocyte border (Roth & Porter, 1964). Such material has been considered equivalent to the cell coat known to be present in almost every cell type (Stockem, 1977) and known to contain much carbohydrate (Luft, 1976). However, vitellogenic oocytes differ from other cell types in that the location of this material appears to be restricted to discrete patches along the external side of the oolemma (Anderson, 1969). This spatial distribution would presumably reflect a preferential deposition in sites corresponding to the pits. This is supported by observations in hen ovarian follicles where the fibrous material appears restricted to the non-disc region of the oolemma that contains most of the coated vesicles (Perry, Gilbert & Evans, 1978b).

According to another view the inner lining of the pits and coated vesicles may be blood proteins bound to the oolemma in the process of being incorporated into the oocyte (Beams & Kessel, 1969; Huebner, Tobe & Davey, 1975). Both suggestions are compatible with the known ultrastructure of the cortical ooplasm. At the present, however, cytochemical methods cannot resolve the alternatives because both blood proteins and cell coat constituents exhibit carbohydrate-like reactivity (Anderson & Spielman, 1971).

The outer layer of the coated vesicles is morphologically characterised in thin-sectioned material by the presence of subunits which radiate outwards from the middle layer. It has been suggested that the role of this layer may be related to the mechanical process of invagination of proteins of the oolemma to which blood proteins would be bound, and/or to prevent refusion of the internalised vesicles with the cell surface (Huebner & Anderson, 1972). Alternatively, such a layer could participate in the process of binding proteins (Roth & Porter, 1962), either directly by providing the receptor sites or indirectly by holding them in place on the oolemma.

The observation that pits and coated vesicles are structurally similar in all oocytes so far examined, suggests that they may also be functionally related. The most accredited view is that the pits constitute a stage in the formation of coated vesicles in which they appear still connected to the oolemma by a neck. This interpretation obviously implies that coated vesicles in oocytes are moving away from the oolemma, rather than moving toward it. Although a purely structural analysis would not allow one to distinguish between these alternatives, the cytochemical experiments to be described later are consistent with this interpretation.

While most oocytes have been described as possessing only one type of coated vesicle, generally restricted to the vitellogenic period, oocytes in *Lebistes* (Droller & Roth, 1966) and in *Periplaneta americana* (Anderson, 1964) appear to form two types of coated vesicles. These vesicles, although structurally similar, differ in size, fate of the ingested material, and in time of appearance. The ones formed in previtellogenic stages are smaller than those formed during vitellogenesis and serve to convey the adsorbed material to tubules or multivesicular bodies rather than to yolk spheres. The presence of two distinct types of coated vesicles has been reported also in sea urchin oocytes (Tsukahara & Sugiyama, 1969; Tsukahara, 1970). They have been termed alpha- and beta-pinosomes and differ in size and in the nature of the enclosed material. In vitellogenic oocytes of the newt *Triturus vulgaris* there appear to be two types of vesicles only one of which is coated, the other being devoid of a coat on either side of its limiting membrane. The latter type of vesicle is present only underneath the microvilli and is assumed to be formed by invagination of the microvillous plasma membrane (Spornitz & Kress, 1973) (Plate 2).

The differential appearance of coated vesicles may indicate, as suggested by Anderson (1964) for *P. americana*, that the oolemma is undergoing changes in its capacity to bind blood proteins. This hypothesis, if true, may apply also to species other than those cited above. It may be that coated vesicles in oocytes of other species, although morphologically indistinguishable, do also undergo changes in their binding capacity. Cutting & Roth (1970, 1973) have, in fact, demonstrated this in laying chickens.

It had been thought that the time of appearance of specific coated vesicles in the oocyte could have been determined by the type of extracellular material present on the oolemma. This hypothesis, however, was contraindicated by the observation that blood proteins have access to the oocyte surface at all developmental stages. It would thus seem more likely that the process is under cellular control (Droller & Roth, 1966).

Certain interesting adaptations of the chicken oocyte, probably resulting from the sequestering of large quantities of blood proteins, have recently been reported. Perry, Gilbert & Evans (1978a) have observed that the coated vesicles present in large-size follicles may reach dimensions two to three times greater than those of the vesicles present in younger follicles (Plate 3). In addition, the coat material which in other oogenic periods would have lined the ooplasmic side of the pits, during the period of active vitellogenesis comes instead to form a continuous array of projections toward the ooplasm (Roth, Cutting & Atlas, 1976). This structural modification is interpreted by the authors as an indication of the high rate at which coated vesicles move away from the oolemma.

In the early stages of chicken oogenesis, the cortical ooplasm contains numerous projections which emerge from the overlying follicle cells (Bellairs, 1965). These structural elements, which have been termed lining bodies, are assumed to play a major role in the initial development of the yolk spheres (Paulson & Rosenberg, 1972). Similar structures have also been described in growing oocytes of the turtle *Pseudemy scripta* (Rahil & Narbaitz, 1973).

In numerous animal species, the cortical ooplasm of vitellogenic oocytes contains several vesicles which appear either to be derived from the coated vesicles or are in some other way related to the process of pinocytic uptake. Although a strict morphological analysis alone does not provide information as to the direction of the flow involved, if any, the current view is that the coated vesicles, by moving inward, fuse with each other and in so doing give rise to vesicles of larger size (Matsuzaki, 1971; Mahowald, 1972a). During this inward movement and subsequent fusion the coated vesicles lose their coats on both inner and outer sides and acquire an electron-dense content which fills them to varying degrees (Follett & Redshaw, 1974). Several authors have also reported the presence in the cortical ooplasm of tubular profiles of variable length, which are occasionally observed in actual continuity with the limiting membrane of the forming yolk spheres (Ulrich, 1969; De Loof & Lagasse, 1970; Herbaut, 1972; Giorgi, 1975). These tubules have been variously interpreted by investigators working on vitellogenesis. Some have assumed that they are derived from the endoplasmic reticulum (Favard-Sereno, 1964; Cummings & King, 1970), while others are of the opinion that they form through a process of progressive fusion of coated vesicles (Anderson, 1969; Mahowald, 1972b; Giorgi & Jacob, 1977a). Yet another hypothesis predicts that they represent excess membrane being withdrawn from the forming yolk spheres that have completed their yolk intake (Roth & Porter, 1964; Wallace & Dumont, 1968).

It has been noted that the tubules present in the cortical ooplasm exhibit, like the coated vesicles, an inner lining of fibrous material. This obviously favours the view of them being a residue of the vesicles. The use of extracellular tracers has helped to substantiate this view. On the other hand, the question of whether the tubules seen in continuity with the forming yolk spheres are in the process of fusing with the spheres or being withdrawn from them has still not been definitely answered.

Microtubules have frequently been observed in the cortical ooplasm of a number of species. Although they bear no structural relationship to the coated vesicles, their presence in oocytes may be necessary for protein uptake. This is suggested by the inhibitory effect on protein incorporation resulting in *Xenopus* oocytes from treatment with vinblastine (Dumont & Wallace, 1972). An altered cytology has also been observed in *Rhodnius* ovaries following in-vivo exposure to vinblastine (Huebner & Anderson, 1970).

Role of coated vesicles in protein incorporation
Cytochemical evidence

The limitations of structural analysis alone constitute a very well-known problem to many morphologists. Fortunately, in the study of vitellogenesis, the use of extracellular tracers has helped to clarify the dynamics of pinocytic uptake as well as the ultimate location within the cell of the ingested material. A wide variety of molecules has been employed to achieve this end.

In the pioneering work of Telfer (1961), vitellogenic oocytes of the saturniid moth were exposed to fluorescein-labelled antibodies prepared against yolk proteins. Using this method, the author was able to show that blood proteins reach the oocyte surface by passing through the intercellular spaces present in the follicular epithelium overlying the oocyte; and that once internalised into the ooplasm by micropinocytosis they are sequestered into yolk spheres. Subsequent cytochemical studies have substantiated the general applicability of Telfer's observations to other insect species (Stay, 1965; Ramamurty & Majumdar, 1967; Ramamurty, 1968; Yonge & Hagedorn, 1977).

Anderson & Spielman (1971) have used several extracellular tracers to follow the vitellogenic uptake in mosquito oocytes. They observed that among the various tracers injected into blood-fed females, only those having molecular sizes with a long axis less then 11 nm could reach the oocyte surface, larger molecules being blocked by the external basement lamina.

Essentially similar results were obtained in oocytes of *Drosophila melanogaster* by Mahowald (1972*b*) and in those of lizard by Neaves (1972). Recently, Giorgi & Jacob (1977*b*) have observed that when vitellogenic oocytes of *Drosophila* are exposed *in vivo* to peroxidase, several structural elements in the cortical ooplasm become intensely labelled. These include the coated vesicles, the tubules and the forming yolk spheres (Plate 4). These observations add further support to the hypothesis that the tubules present in this region of the oocyte are structurally derived from coated vesicles, presumably through repetitive fusion.

In amphibian oocytes exposed to iron-dextran, Wartenberg (1964) observed that the injected marker appears first in the pinocytic vesicles and is subsequently transferred to the forming yolk spheres. In the fowl, Knight & Schechtman (1954) showed that heterologous serum proteins such as bovine serum albumin (BSA), bovine γ-globulin or lobster serum are transferred from the blood circulation to the ovarian egg in an unmodified form. Similar observations have been carried out in the quail, in which radioactive mercury has been shown to gain access to the yolk spheres in the oocyte by passing through intercellular spaces between the cells of the granulosa and the theca interna (Nishimura, Urakawa & Iwata, 1976).

In vitellogenic oocytes of the annelid *Enchytraeus albidus,* micropinocytic incorporation of exogenously synthesised components occurs along with endogenous synthesis of core body substances. The use of peroxidase as an extracellular tracer has shown that material transferred to the oocyte by way of the coated vesicles is eventually stored in the matrix of the yolk spheres (Dumont, 1969).

The general picture emerging from these tracer studies indicates that any molecule below a certain size can gain access to the oocyte surface from the extracellular compartment. Indications are that molecules having dissimilar chemical properties enter the oocyte equally well, without any apparent selection having taken place (Anderson & Spielman, 1971). All tracers used so far, however, when detected in the oocyte, appear restricted to the superficial layer of the yolk spheres and only rarely become detectable in the crystalline main body (Wartenberg, 1964; Dumont, 1969).

It needs to be said that almost every tracer employed for studying vitellogenic uptake is incorporated adventitiously by the oocyte; that is to say, the tracers are not adsorbed onto the oolemma as the yolk precursors presumably are. For this reason such studies can have little bearing on the question

of selectivity of uptake. They simply indicate the route of entry into vitellogenic oocytes that is presumably the one also followed by yolk precursors
in in-vivo conditions.

Exceptions can be made to the conclusions drawn above when (*a*) coated
vesicles are formed during previtellogenic stages, (*b*) yolk material is autosynthetic, or (*c*) yolk reserves never form, as in mammals. An example of the
first situation can be found in oocytes of *P. americana,* where Anderson (1969)
has observed that uptake of peroxidase by previtellogenic oocytes results in
the tracer being sequestered into multivesicular bodies rather than into the
forming yolk spheres. In vitellogenic oocytes of the crayfish, yolk precursors
are synthesised by the oocyte itself. Coated vesicles in this species play a
minor role in yolk formation, since they convey extracellular material, including peroxidase, to multivesicular bodies and not to forming yolk spheres
(Ganion & Kessel, 1972).

Mammalian oocytes do not appear to store any type of reserve material
(Anderson & Beams, 1960; Adams & Hertig, 1964; Zamboni, 1970). They
nevertheless incorporate foreign protein antigens, as shown by the study of
Glass & Cons (1968) on mouse ovarian follicles. This process appears to occur
through formation of coated vesicles which serve to convey the ingested
material to multivesicular bodies either by fusing with them or via formation
of tubules (Anderson, 1972).

Autoradiographic evidence

The observation that exogenously applied molecules are taken into
the oocyte and sequestered in the yolk spheres suggests a route for the entry
of blood proteins. To ascertain whether this is really the case, a number of
investigators have studied the uptake of labelled precursors into vitellogenic
oocytes, mainly by using autoradiography.

If vitellogenic oocytes are exposed to a radioactively labelled amino acid,
every protein synthesised during the period of exposure will obviously be
radioactively marked. Despite this lack of specificity inherent in the autoradiographic technique, Bier (1962, 1963) was able to show that in-vivo
labelling of vitellogenic oocytes in *Musca domestica* occurs in two successive
periods. The first period occurs immediately after amino acid administration,
while the second appearance of the label is delayed in time. During the first
period of incorporation the label appears randomly distributed within the
ooplasm, while following the second one yolk spheres become preferentially
labelled.

Essentially similar observations have since been seen in vitellogenic oocytes
of other insect species (Melius & Telfer, 1969; Kunz & Petzelt, 1970; Chia &

Morrison, 1972). In *Drosophila melanogaster* the second wave of incorporation coincides with a distinctive labelling of yolk spheres and occurs only six hours after injection of tritiated lysine (Giorgi & Jacob, 1977*b*) (Plate 5). Two successive waves of incorporation have also been detected by autoradiography in vitellogenic oocytes of the zebrafish (Korfsmeier, 1966).

All available data are consistent with the interpretation that vitellogenic oocytes in many animal species, when exposed *in vivo* to radioactive amino acids, undergo first a period of incorporation which is primarily due to endogenous synthesis. This is also supported by the observation that vitellogenic oocytes exposed *in vitro* to radioactively labelled amino acids mimic only the first wave of incorporation (Giorgi & Jacob, 1977*b*). It is believed that the time lapse which characterises the appearance of the second wave of incorporation depends on the time required for synthesis in an extraovarian tissue followed by incorporation into the ovary. As such, the second period of incorporation is believed to reflect uptake of exogenously synthesised material and not endogenous synthesis. This assumption is corroborated by the observation that ovaries transplanted into female flies previously injected with amino acids become labelled according to a pattern comparable to the second wave of incorporation (Bier, 1963).

When ovarian follicles of cecropia moths are analysed autoradiographically during their terminal growth period, it is found that yolk spheres are still being formed in their cortical ooplasm. Because labelling of the follicles is observed in in-vitro conditions and at a time when micropinocytic activity has been completed, these yolk spheres probably result from endogenous synthesis rather than protein incorporation (Telfer & Anderson, 1968).

Brummett & Dumont (1977) have followed the process of protein incorporation into stage IV vitellogenic oocytes of *Xenopus laevis* using autoradiography. Using [^{3}H] vitellogenin as a labelled precursor they showed that the first organelles to become labelled in the cortical ooplasm are the coated vesicles. The label is then transferred to the so-called primordial yolk platelets and ultimately stored in the mature yolk platelets.

When autoradiographical methods are employed to study vitellogenesis in those species in which yolk has an autosynthetic origin, it is found that the labelling of the oocyte follows a pattern different from the one reported for heterosynthetic systems. The most extensive autoradiographical study to date is that carried out on the crayfish oocyte by Ganion & Kessel (1972). The results of this study indicate that yolk is synthesised by the endoplasmic reticulum and that no involvement of coated vesicles is required for this to occur.

Immunological and electrophoretic evidence

Telfer (1954) was apparently the first to demonstrate, in the cecropia silkworm, the antigenic identity of several blood proteins with those in oocytes. He observed that in the female the onset of vitellogenesis is accompanied by a sharp decrease in the blood concentration of antigen seven and that following transplantation of the ovary into male abdomens, the oocyte failed to undergo yolk deposition. Furthermore, the observation that in ovariectomised females antigen seven increases its concentration over the level present in control females, constitutes convincing evidence in support of the view that this blood protein is a yolk precursor. Telfer (1954) interpreted these observations as indicating that a tissue other than the ovary synthesises the yolk precursor and that while the ovary is capable of absorbing this precursor from the blood it cannot synthesise it.

Since these early studies on cecropia silkworm, immunological techniques have been widely employed to identify yolk precursors in numerous insect species. Sex-specific antigens have been detected in the haemolymph, fat body and ovary extracts of the house fly (Bodnaryk & Morrison, 1968), of *Leucophaea maderae* (Engelmann & Penney, 1966; Scheuer, 1969), of *Schistocerca gregaria* (Dufour, Taskar & Perron, 1970), of *Periplaneta americana* (Bell, 1970; Clore, Petrovitch, Koeppe & Mills, 1978), of *Blattella germanica* (Tanaka & Ishizaki, 1974) and of *Aedes aegypti* (Hagedorn & Judson, 1972).

In the blue crab *Callinectes sapidus,* a lipoprotein in the haemolymph has been found to be serologically identical to a lipovitellin isolated from the oocyte, a finding which reflects the lipoprotein's yolk-like nature (Kerr, 1969). Two lipoproteins isolated from eggs of *Dermacentor andersoni* have been found to be immunologically indistinguishable but physically dissimilar from their common haemolymph precursor (Boctor & Kamel, 1976).

In an attempt to correlate protein contents of the ovary with those of the haemolymph and fat body, several authors have made extensive use of various types of electrophoretic analysis. With these techniques, yolk precursors have been identified in several species of Lepidoptera (Whitmore & Gilbert, 1974) in *Bombyx mori* (Doira & Kawaguchi, 1972), in *Apis mellifica* (Engels, 1972), in *Drosophila melanogaster* (Gelti-Douka, Gingeras & Kambysellis, 1974; Gavin & Williamson, 1976a; Bownes & Hames, 1977), and in the American cockroach (Nielsen & Mills, 1968; Krolak, Clore, Petrovitch & Mills, 1977).

Two endogenously synthesised protein fractions have also been identified by polyacrylamide gel electrophoresis in the ovary of the zebrafish *Brachydanio rerio.* A third ovarian protein fraction in this species is electrophoretically comparable to a certain serum protein and for this reason is believed to represent exogenous yolk (Heesen te & Engels, 1973).

The similarities in the protein contents of blood and oocytes as revealed by immunological and electrophoretic methods clearly indicate the existence of a mechanism by which the oocyte can sequester proteins from the blood. The alternative interpretation – that ovarian proteins could be secreted into the blood stream – is incompatible with the known ultrastructure of the ovary. The above conclusion is further demonstrated by analysis of mutants with defective micropinocytic function, in which yolk precursor proteins accumulate in the serum (Kawaguchi & Doira, 1973; Schjeide, Briles, Holshouser & Jones, 1976).

In recent years biochemical methods have been devised to isolate coated vesicles from cells of various animal tissues (Kanaseki & Kadota, 1969; Schjeide *et al.,* 1969; Pearse, 1975). Coated vesicles from various tissues, including those from oocytes, all seem to have three protein fractions in common (Woods, Woodward & Roth, 1978). Of these fractions, the one exhibiting the highest molecular weight, termed clathrin by Pearse (1975), is assumed to constitute an essential component of the cytoplasmic coat of the coated vesicles (Crowther, Finch & Pearse, 1976).

Selectivity of protein incorporation
In his original experiments on cecropia silkworm, Telfer (1954) noticed that antigen seven is 20 times more concentrated in the ovary than in the blood. Other blood proteins, however, exhibit a similar concentration in the blood and in the ovary. These observations suggest that vitellogenic oocytes of saturniid moths possess a mechanism for selectively removing certain proteins, but not others, from the circulation (Telfer, 1960). A difference in the relative concentration of yolk proteins in the oocyte and in the haemolymph has also been detected immunologically in *Periplaneta americana* (Bell, 1970). In *Xenopus laevis,* incorporation of vitellogenin into oocytes stimulated with human chorionic gonadotrophin (HCG) has been estimated to occur 25 times more rapidly than incorporation of other serum proteins (Wallace & Dumont, 1968). During the period of HCG stimulation, the amount of vitellogenin entering the oocyte by micropinocytosis has been found to be balanced by the amount of vitellogenin released into the circulation following its synthesis in the liver (Wallace & Jared, 1969) (Fig. 1).

Recently, selectivity of vitellogenin uptake in *Xenopus* oocytes has been tested in an in-vitro system and found to be comparable to that previously estimated in intact females. In addition, when other proteins such as BSA or ferritin are included in the culture medium along with vitellogenin, they appear not to compete with vitellogenin incorporation (Wallace & Jared, 1976). Selectivity of uptake has also been tested in cecropia, by incubating ovarian follicles in the presence of male blood proteins or purified vitellogenin. The cortical ooplasm of the incubated follicles exhibited labelled yolk spheres only when the radioactive component in the incubation mixture was vitellogenin (Hausman, Anderson & Telfer, 1971). When male blood protein was the radioactive precursor much less labelling occurred on the yolk spheres. This is indicative of a low rate of uptake.

In white leghorn chickens, Cutting & Roth (1973) have noted that phosvitin in large-sized oocytes is six times more concentrated than in the serum, while IgG is only 0.6 times as concentrated. When the same proteins are examined in younger oocytes, it is found that an equal rate of phosvitin uptake corresponds to a lowered rate of IgG accumulation. These observations are taken to indicate that during oocyte development there is a change in the sequestration capacity with regard to specific proteins (Cutting & Roth, 1970, 1973).

The general conclusion which can be drawn from the available evidence points to the existence in oocytes of a selective mechanism preferentially adapted for uptake of one or a few specific proteins. The selectivity, however, is not thought of as being absolute (Telfer, 1965; Wallace, 1978) because other serum proteins (Glass, 1959) or even heterologous proteins may ultimately be pinocytosed and sequestered into the yolk spheres.

Fig. 1. Loss of labelled protein from the serum (*a*) and its incorporation into the ovary (*b*) following injection of 5 mg of [^{3}H]vitellogenin (open circles) or serum protein fractions [^{3}H]S1 (solid circles) and [^{3}H]S2 (triangles) into vitellogenic females of *Xenopus laevis* (given 1000 units HCG per animal seven days previously). (From Wallace & Jared, 1969.)

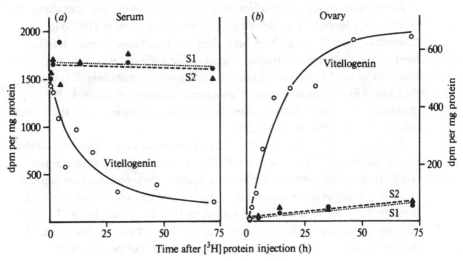

In his review article on endocytosis, Jacques (1969) stated that selectivity is only a characteristic of adsorptive types of endocytosis: liquid endocytosis would, in fact, engulf all substances present in the extracellular compartment at a rate which would depend on their concentration therein. The fact that vitellogenic oocytes may undergo adsorptive pinocytosis provides a clue as to how female-specific proteins are selectively sequestered from the blood (Telfer, 1965). This could be achieved by adsorption of molecules of vitellogenin onto the portions of the oolemma which give rise to the coated vesicles. By the same token, the non-selective components and/or the heterologous molecules experimentally given to the oocyte could be adventitiously pinocytosed by being enclosed within the small volume of the forming coated vesicles. For instance, in in-vitro cultured cells, peroxidase is incorporated in amounts proportional to the actual concentration in the medium and is believed to be taken up in an unbound form (Steinman, Silver & Cohn, 1974). Similarly, in mouse macrophages, uptake of ^{198}Au increases with increasing concentration in the culture medium (Davis, Allison & Haswell, 1973). On the other hand, when uptake of vitellogenin into oocytes is considered, it is found that its rate of uptake depends on factors other than its concentration in the extracellular compartment (see section on 'Control of protein incorporation').

This type of reasoning brings me to the conclusion that adsorption of vitellogenin could be attained by attachment onto specific binding sites or receptors of the oocyte plasma membrane. This is a subject which will be fully explored in a later chapter of this book. In this article, however, I will refer to the recent evidence that is clearly indicative of receptor sites in vitellogenic oocytes.

An interesting example of differential distribution of binding sites along the oolemma has recently been reported by Brummett & Dumont (1976) in *Xenopus* oocytes. These authors have observed that, during ovarian development, anionic sites are preferentially located along the distal parts of the microvilli and not in the pit regions. The differential distribution of these sites during ovarian development is interpreted as being a mechanism for channelling negatively charged molecules, such as vitellogenin, onto the coated vesicles for their subsequent uptake.

Receptor sites in oocytes exhibit a high binding capacity for specific vitellogenin molecules (Yusoko & Roth, 1976). As measured by the differential rate of vitellogenin uptake in vitellogenic oocytes, these receptors are capable of distinguishing between molecules of related species, even though these may share similar physical and chemical properties (Kunkel & Pan, 1976). Binding capacity for specific vitellogenin molecules by the host oocyte appears to be relatively stable at least within a taxonomical family, but declines with increasing phyletic distance between the species tested (Kunkel, Johnson, Haggerty & Sargent, 1976).

Such conclusions have also been supported by transplantation experiments between related species of cockroaches. When oocytes are transplanted into a female cockroach of another species, their capacity to sustain vitellogenesis decreases proportionally with the phyletic distance between the donor and the host species (Bell, 1972).

These observations are in line with recent models of the plasma membrane, whereby globular proteins, including specific receptors, are believed to be capable of translation and of dynamic insertion into the lipid bilayer (Singer & Nicolson, 1972; Nicolson & Yanagimachi, 1974; Jaffe, 1977). Structural modifications of the plasma membrane may conceivably be related to a selective internalisation process leading perhaps to formation of coated vesicles (Orci & Perrelet, 1973; Humbert *et al.*, 1977).

Selective pinocytic uptake in oocytes may be aided by the overlying follicular epithelium. Isolated cecropia follicles are capable of incorporating amino acids into their follicle cells and of transferring the synthesised product to the forming yolk spheres (Anderson & Telfer, 1969). Following inhibition of pinocytic uptake by treatment with trypan blue, the follicle cell product is found to accumulate, along with vitellogenin and a carotenoid pigment, in the follicle oocyte border (Ramamurty, 1964; Anderson & Telfer, 1970*a, b*). These observations, together with the evidence that radioactively labelled follicle cell product is electrophoretically detectable in the yolk spheres (Bast & Telfer, 1976), indicate quite clearly that follicle cells are capable of synthesising and secreting into the adjacent extracellular space a protein which then enters the oocyte via coated vesicles. Anderson & Telfer (1970*a*) have suggested that an interaction with the follicle cell product would allow acid blood-proteins to be bound onto the oocyte surface, thus facilitating their uptake.

On the basis of ultrastructural and autoradiographical analyses, Chia & Morrison (1972) have suggested that a similar transfer of material derived from the follicle cells could occur in the vitellogenic oocytes of *Musca domestica*. No such evidence has yet been obtained in amphibian oocytes, where, unlike cecropia oocytes (Anderson, 1971), a follicle cell product is not required, at least in-vitro, for vitellogenin uptake to occur (Wallace, Ho, Salter & Jared, 1973).

Fate of coated vesicles
Ultrastructural and cytochemical analyses of yolk deposition
Ultrastructural studies of vitellogenic oocytes have aimed to resolve yolk-sphere structure into parts formed by different constituent organelles,

as well as to account for the differing roles of these constituent organelles in relation to yolk deposition. In the last two decades this kind of approach has been pursued by examination of the ultrastructure and cytochemistry of the earliest yolk spheres formed during oogenesis.

In their study on previtellogenic oocytes of the salamander *Triturus viridescens,* Hope, Humphreys & Bourne (1964) noticed the presence of circular or semicircular membranous structures which enclosed variable numbers of vesicles and granules. These structures were interpreted by the authors as precursor bodies of the yolk platelets. Similar structures, although originally described as constituting a later transformation of mitochondria, have also been reported by Balinski & Devis (1963) in oocytes of *Xenopus laevis.*

A number of distinctive structural features have been described by Massover (1971) in a class of yolk platelets that have been seen in the early stages of vitellogenesis in amphibian oocytes. In addition to the three basic components, i.e. the limiting membrane, the superficial layer and the main body, this type of organelle, defined as a nascent yolk platelet, is characterised by the presence of a variety of inclusions and materials within the irregularly shaped superficial layer. By inference, bodies formed previous to the nascent yolk platelets are thought to be structurally similar but devoid of the crystalline main body (Massover, 1971).

In the newt *Triturus vulgaris,* Spornitz & Kress (1973) have recently described two types of yolk platelet formation, the first of which, defined as primary yolk, is apparently equivalent to a later transformation of multivesicular bodies. When pinocytic uptake begins in the oocyte, yolk precursors are conveyed to the multivesicular bodies through transfer of coated vesicles. Continuous addition of exogenously derived material to these organelles finally results in the formation of secondary yolk. A similar description seems to apply to vitellogenic oocytes of *Rana esculenta* and *R. temporaria* (Kress & Spornitz, 1972) (Plate 6). In *Periplaneta americana,* multivesicular bodies have been shown to acquire exogenous material through inclusion of small-sized coated vesicles which form during previtellogenesis (Anderson, 1969).

Multivesicular bodies may also occur in autosynthetic types of vitellogenesis (Ganion & Kessel, 1972), as well as in oocytes of those organisms like fishes (Shacley & King, 1977) and annelids (Dumont, 1969) in which yolk is made by both hetero- and autosynthesis. A similar situation has been reported by Dumont & Anderson (1967) for the horseshoe crab. The presence of multivesicular bodies has also been described in mammalian oocytes, but unlike in other organisms they never accumulate material of either exogenous or endogenous origin (Anderson, 1972).

During vitellogenesis, mature yolk spheres acquire a definite structural

appearance characterised by a limiting membrane which encloses both a trans-
lucent superficial layer and an electron-dense main body (Karasaki, 1963,
1967; Yamamoto & Oota, 1967). Of these components, the main body is
thought to contain pinocytosed blood proteins — vitellogenin or its subunits —
in a crystallised form (Redshaw & Follett, 1971; Spornitz, 1972). How
materials initially bound to the luminal surface of the coated vesicles are
ultimately taken into the forming yolk spheres is still a matter of controversy.
One hypothesis predicts that the coated vesicles may enter the yolk spheres
through interruptions present along the limiting membrane (Hope *et al.*, 1964;
Kress & Spornitz, 1972). As a result of this process, intact vesicles could gain
access to the yolk spheres. Subsequent dissolution of the limiting membrane
of the vesicles would then cause their contents to be released within the yolk
spheres. Another view holds that the vesicle and yolk sphere membranes fuse,
thus allowing the direct release of the vesicle contents into the yolk spheres
(Massover, 1971).

Various models have been proposed to account for the type of periodicity
observed in the crystal of the main body (Wallace, 1963; Ohlendorf *et al.*,
1975). That such a structure does not result from artifactual fixation of the
oocyte is confirmed by freeze-fracture observations in both amphibian
(Leonard, Deamer & Armstrong, 1972) and insect (Liu, 1973) oocytes. The
superficial layer occupies the yolk sphere periphery with a variable thickness
that depends on the stage of maturation reached. In *Drosophila* oocytes the
superficial layer of the yolk spheres contains a roundish dense body, termed
the associated body, which renders the layer itself asymmetrical (Giorgi &
Jacob, 1977*a, c*) (Plate 7).

In oocytes of various organisms, the presence of carbohydrates in the
superficial layer of the yolk spheres has been revealed by the use of several
cytochemical techniques (Ohno, Karasaki & Takata, 1964; Dumont &
Anderson, 1967; Tandler & La Torre, 1967; Dumont, 1969; Favard & Favard-
Sereno, 1969; Takashima, 1971). As to the origin of such a component of the
yolk spheres, several lines of evidence point to the oolemma as one of the
major sources (Wallace & Jared, 1969). The carbohydrate component present
in the superficial layer could, in fact, be equivalent to the fuzzy or fibrous
coat of the pits of the oolemma, which is brought to the yolk spheres following internalisation of the coated vesicles (Favard-Sereno, 1969; Jollie & Triche,
1971; Giorgi & Jacob, 1977*c*).

On the other hand, the possibility that carbohydrates present in the superficial layer are transferred to the yolk spheres by endogenously derived vesicles
may not be excluded (Dumont & Anderson, 1967). In fact, the occurrence of
Golgi complexes in cytoplasmic regions close to the forming yolk spheres may

substantiate this view (Spornitz & Kress, 1973). This is further corroborated
by the finding that in *Drosophila* oocytes the superficial layer of the forming
yolk spheres shares a common reactivity with the Golgi complex (Giorgi &
Jacob, 1977c; Giorgi, Bucci Innocenti & Ragghianti, 1976) (Plate 8).

DNA is another component of the yolk spheres in several organisms (Hanocq-
Quertier, Baltus, Ficq & Brachet, 1968; Emanuelsson, 1971; Bruce &
Emanuelsson, 1975). [^{14}C] actinomycin binding to yolk spheres in vitellogenic
oocytes has confirmed the non-artifactual nature of the biochemical detection
of DNA (Brachet & Ficq, 1965; Emanuelsson, 1969). Over the years, a variety
of functions has been attributed to yolk DNA, but there is still no more than
speculative evidence regarding its function. Recently, Opresko, Wiley & Wallace
(1978) have shown that yolk DNA in *Xenopus* oocytes is taken up from the
maternal blood circulation with no apparent selectivity. Because of this
adventitious incorporation into the oocyte, no informational role for yolk
DNA has been suggested.

Biochemical analysis of yolk deposition

During protein incorporation into vitellogenic oocytes and subsequent
deposition of yolk, the haematic yolk precursor may undergo structural changes
in its protein moiety. This process has been extensively studied in *Xenopus*
oocytes by Wallace and co-workers.

When serum of HCG-treated *Xenopus* vitellogenic females is analysed by
chromatography on TEAE-cellulose, a major component, referred to as
vitellogenin, becomes evident at an elution position of 0.64 (Wallace, 1965).
This is a lipoglycophosphoprotein, which can also be detected in normal
females depending on their nutritional state (Wallace, 1970). Its synthesis can
be induced in the liver of both sexes following treatment with oestrogens
either in *in vivo* (Wallace & Jared, 1969) or *in vitro* (Wangh & Knowland,
1975; Green & Tata, 1976). While oestrogen induces the hepatic synthesis of
vitellogenin but not its turnover, it appears that HCG is necessary for vitellogenin to undergo a steady-state turnover in the blood circulation (Wallace &
Jared, 1969). A similar hormonal dependence of vitellogenin turnover in the
blood has been reported in the species *Bufo bufo* (Emmerson & Kjaer, 1974).
Hepatic synthesis of vitellogenin as a response to oestrogen induction has also
been thoroughly investigated in roosters (Tata, 1976).

The relationship between vitellogenin and yolk proteins in *Xenopus laevis*
has been investigated by injecting females with doubly labelled vitellogenin.
With progressively longer periods of exposure to labelled vitellogenin, the
radioactivity recovered from the ovarian extracts becomes increasingly high
in the fractions of the two yolk proteins, lipovitellin and phosvitin (Fig. 2).

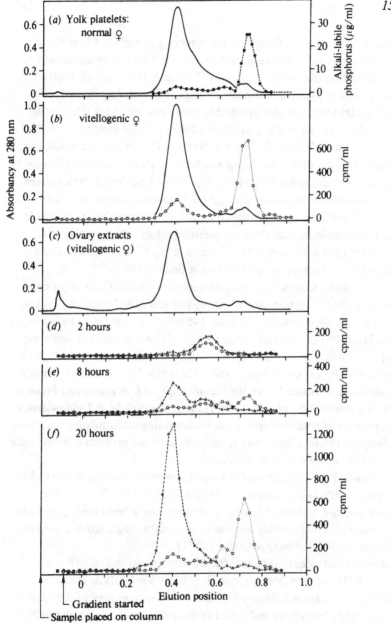

Fig. 2. Chromatography and labelling of ovarian extracts following injection of [^{3}H, ^{32}P] vitellogenin into vitellogenic females of *Xenopus laevis*. (a) and (b) Yolk platelet proteins extracted from the ovary. (c)–(e) Protein extracted with 1.5 M sodium chloride from the ovaries of HCG – treated females. Ovarian extracts were made 2 hours (d), 8 hours (e) and 20 hours (f) after the injection of [^{3}H, ^{32}P] vitellogenin. The absorbancy (a, b, c) is indicated by a solid line; [^{32}P] protein labelling and [^{3}H] protein labelling are indicated with open circles and open triangles respectively. (From Wallace & Jared, 1969.)

The data indicate that vitellogenin is the precursor molecule of lipovitellin and phosvitin and that its uptake by the oocyte is followed by conversion of the precursor into the two yolk proteins (Wallace & Jared, 1969). Such a conversion appears to take place through cleavage of the vitellogenin molecule at specific sites which are also susceptible to proteolytic attack (Bergink & Wallace, 1974). Partial proteolysis of the vitellogenin molecule has even been suggested by Chen, Couble, De Lucca & Wyatt (1976) in oocytes of *Locusta migratoria*, to account for the change in subunit pattern noticed in this species during vitellogenin uptake. Whatever change is involved during vitellogenesis, however, it seems to occur only through rearrangement of the molecule, rather than hydrolysis and resynthesis. In fact, the oocyte vitellin, at least in amphibians, remains immunologically indistinguishable from its haematic precursor (Wittliff & Kenney, 1972; Gelissen *et al.*, 1976) and maintains a similar amino acid composition (Redshaw & Follett, 1971).

In *Leucophaea maderae*, ovarian yolk proteins consist of two major components with sedimentation coefficients of 14S and 28S (Dejmal & Brookes, 1972). During yolk deposition the small 14S component is gradually assembled into the larger 28S component and as a result of this process the relative proportions of the two components appear to vary with the developmental stage of the oocyte (Brookes & Dejmal, 1968). The native form of the vitellogenin molecule has been isolated from the haemolymph of *L. maderae* and found to have a molecular weight of 5.25×10^5 (Engelmann, Friedel & Ladduwahetty, 1976). Because of possible degradation arising during electrophoretic analysis, no definite molecular relationship could be ascertained between this molecular form and the two ovarian yolk proteins.

A relatively pure and undegraded preparation of vitellogenin from chicken serum has recently been obtained by Bergink *et al.* (1974). Previous attempts to isolate a unique molecular form of yolk precursor in birds have apparently been unsuccessful, presumably because of proteolytic degradation during the isolation procedure (Beuving & Gruber, 1971).

Unlike the instances so far cited, there is evidence in cecropia (Pan & Wallace, 1974) and in *Philosomia cynthia* (Chino, Yamagata & Takahashi, 1976; Chino, Yamagata & Sato, 1977) that the protein moiety of the vitellogenin is simply pinocytosed and stored in the yolk platelets in an unmodified form.

Many of the data obtained from in-vivo experiments have later been confirmed by employing in-vitro systems designed for short-term culture of vitellogenic oocytes (Wallace, Jared & Nelson, 1970). Using these systems, Jared, Dumont & Wallace (1973) have studied the incorporation of radioactively labelled vitellogenin into vitellogenic oocytes isolated from previously

HCG-stimulated *Xenopus* females. When isolated oocytes are incubated under these conditions for a short time period and subsequently analysed on a PVP–sucrose gradient, it is found that most of the radioactivity is restricted to the coated vesicle region. If such a short incubation period is then followed by a long chase in unlabelled medium, the radioactivity is transferred from this region to that where yolk platelets sediment (Fig. 3). The observations are thus

Fig. 3. Distribution of radioactivity in centrifuged gradients derived from 15 oocytes which were incubated in the presence of [^{3}H, ^{32}P]-vitellogenin for 10 minutes and homogenised immediately (*a*) or after further incubation in unlabelled medium for either 45 minutes (*b*) or 18 hours (*c*). (From Jared, Dumont & Wallace, 1973.)

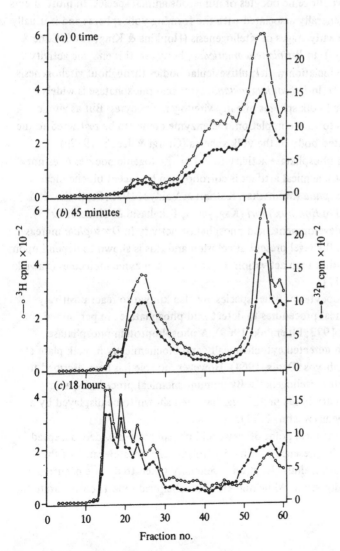

consistent with the generally accepted view that yolk precursors are taken into the oocyte and transferred to the yolk platelets via formation and fusion of the coated vesicles. No such pattern of labelling is observed when isolated oocytes are exposed to tritiated leucine or to ^{32}P rather than to labelled vitellogenin. Available evidence indicates that in *Xenopus* oocytes at least 99 % of the proteins stored in the yolk platelets are heterosynthetic in origin, thus precluding any role for autosynthetic processes in yolk formation. (Wallace, Nickol & Jared, 1972).

Yolk in relation to lysosomes
A number of authors have demonstrated the occurrence of acid phosphatase in vitellogenic oocytes of numerous animal species. In most insects this enzyme is generally associated with the forming yolk spheres and is usually restricted to the early stages of vitellogenesis (Hopkins & King, 1966; Cone & Eschenberg, 1966). In *Periplaneta americana,* however, this enzyme activity occurs only in association with multivesicular bodies throughout vitellogenesis (Anderson, 1969). In *Drosophila melanogaster* acid phosphatase is widely distributed among yolk spheres of early vitellogenic oocytes. But as vitellogenesis proceeds toward completion, the enzyme comes to be restricted to the so-called associated body of the yolk spheres (Giorgi & Jacob, 1977*c*).

That the acid phosphatase activity found in vitellogenic oocytes does not result from a cytochemical artifact is corroborated by recent biochemical evidence in *Drosophila* (Mulherkar, Kothari & Vaidya, 1972; Postlethwait & Gray, 1975) and in *Bombyx mori* (Kageyama, Takahashi & Ohnishi, 1973). During ovarian development, acid phosphatase activity in *Drosophila* increases to over 30 times the level present at eclosion and this is shown to depend upon new synthesis rather than activation of pre-existing enzyme molecules (Sawicki & McIntyre, 1977).

Vitellogenic oocytes of other species are also known to react positively when cytochemical procedures to detect acid phosphatase are performed (Pasteels, 1969, 1973; Bluemink, 1969). A phosphoprotein phosphatase activity has been detected cytochemically and biochemically in yolk platelets of amphibian embryos (Denis, 1964). However, no role has been ascribed to this enzyme during vitellogenesis. By immunochemical procedures, acid phosphatase activity in sea urchin eggs has been shown to be displayed by a number of antigens (Westin, 1975).

From a cytochemical point of view, acid phosphatase is widely accepted as a marker for the presence of other hydrolytic enzymes. Because of this assumption, its intracellular location is generally taken to denote occurrence of intracellular digestion. While numerous biochemical reports corroborate the above cytochemical assumption (De Duve & Wattiaux, 1966) less information is presently available concerning vitellogenic oocytes. Schuel, Wilson, Wilson & Bressler (1975) have offered evidence of the presence of several typical lysosomal enzymes in association with yolk platelets of unfertilised sea urchin eggs. Three types of lysosomal activity have also been detected biochemically in the yolk platelets of *Salmo gairdneri* during ovarian or embryonic development (Vernier & Sire, 1977).

There are thus data supporting the cytochemically based hypothesis that yolk platelets may be functionally related to storage granules of a lysosomal nature (Bluemink, 1969). More biochemical data are obviously needed to verify the hypothesis in other species, especially those in which yolk is heterosynthetic and as such more readily comparable to somatic systems undergoing endocytosis. The whole matter of lysosomal involvement in relation to yolk formation in vitellogenic oocytes may be viewed according to one of the following hypotheses.

The first hypothesis is that yolk material is sequestered into a cell compartment which may either never come into contact with lysosomal-like granules or alternatively may in fact become accessible to lysosomal enzymes. The observation that radioactively labelled vitellogenin undergoes degradation when injected into vitellogenic *Xenopus* oocytes shows that oocytes are capable of performing a proteolytic activity, presumably as a result of their enzymatic make-up (Dehn & Wallace, 1973). However, the fact that in in-vivo conditions vitellogenin is converted into stable yolk proteins suggests that few hydrolytic enzymes, if any, ever come into contact with coated vesicles or forming yolk platelets. Proteolytic degradation of proteins microinjected into vitellogenic oocytes could presumably occur through autophagy as suggested by Stacey & Allfrey (1977) for HeLa cells. Even so, the evidence would suggest that, unlike other systems, autophagic vacuoles and heterophagosomes in amphibian oocytes form two distinct types of cell compartment. This conclusion, however, does not rule out the possibility that certain proteolytic enzymes may become enclosed within the yolk platelets, although inactivated by the simultaneous presence of specific inhibitors (Slaughter & Triplett, 1975, 1976).

The second hypothesis could presumably be more consistent with the physiology of vitellogenesis in insects. A current speculation among cytochemists studying insect vitellogenesis (Hopkins & King, 1966; Cone & Eschenberg, 1966; Giorgi & Jacob, 1977c) is that lysosomal involvement in oocytes could be functionally related to the occurrence of a selective hydrolysis, thereby degrading all non-vitellogenic blood proteins which are adventitiously engulfed into the yolk spheres. This view, however, would also make it necessary to postulate the existence of a mechanism whereby yolk material is somehow 'protected' from proteolytic attack by the lysosomal enzymes (Giorgi & Jacob, 1977c). An interesting parallel has been suggested for the human placenta concerning immunity transmission from the mother to the foetus (Brambell, Hemmings & Morris, 1964; Brambell, 1966).

Membrane recycling

Formation of coated vesicles in vitellogenic oocytes results, as in other cell systems undergoing pinocytosis, in the continuous withdrawal of portions of the plasma membrane (Telfer, 1965). In order for the oocyte to keep the proper size or even to enlarge, this inward flow of membrane fragments has to be balanced by an equal or more extensive movement of exocytic vesicles toward the oolemma (Giorgi et al., 1976). On the basis of these considerations, it would be predictable that vitellogenic oocytes should form as much exocytic vesicle membrane as is required to replace plasma membrane endocytic losses. Ultrastructural observations carried out on vitellogenic oocytes have revealed the presence of vesicles of various sizes and, as already mentioned, some of them are assumed to represent a later transformation of the coated vesicles. Other vesicles in the cortical ooplasm could be exocytic in nature, but such a distinction may not be possible on the basis of a strict morphological study (Tesoriero, 1977).

As in other cell types, the Golgi apparatus in vitellogenic oocytes may very well function as a site for assembly of membrane fragments into exocytic vesicles (Herzog & Farquhar, 1977). A shift in the distribution of vesicles from the Golgi apparatus to the apical cytoplasm has been frequently observed in several cell types (Friend & Farquhar, 1967; Cole & Wynne, 1973; Franke & Herth, 1974) and it is not unlikely that a similar phenomenon could take place in vitellogenic oocytes also.

The forming yolk spheres themselves may also be involved in the process of membrane turnover since they constitute an obvious site for membrane convergence during vitellogenesis. If the occurrence of lysosomal enzymes in yolk spheres could be documented more extensively, the idea of membrane turnover in association with yolk deposition might certainly receive more credit.

Ultrastructural visualisation of vesicles in the cortical ooplasm may perhaps not be an adequate tool upon which to base our understanding of membrane balance in vitellogenic oocytes. Present knowledge of the molecular turnover of membranes makes it likely that large-scale movements of membrane components could occur at a level which is not detectable microscopically (Graham, Sumner, Curtis & Pasternak, 1973). However, whatever mechanism is involved, it appears that recycling rather than synthesis may prove a more likely mechanism for membrane balance during extensive micropinocytosis in several cell systems (Silverstein, Steinman & Cohn, 1977). This is supported by consideration of how rapid the process of membrane replacement needs to be to keep pace with membrane insertion (Silverstein *et al.*, 1977). Further support for this conclusion is derived from recent observations that plasma membrane immunoglobulin G molecules can be detected immunologically on the cell surface, following their internalisation into the lysosome fraction (Schneider, Tulkens & Trouet, 1977). Implicit in the concept of membrane recycling is the existence of a membrane precursor pool. Evidence obtained from isolated amphibian oocytes is consistent with this assumption in that sequestering activity in these cells is only moderately affected by inhibition of protein synthesis (Wallace & Ho, 1972).

Control of protein incorporation

Available evidence indicates that protein uptake in somatic cells is proportional, for a given period of exposure, to the actual concentration of proteins in the medium (Cohn & Benson, 1965; Contractor & Krakauer, 1976). By contrast, micropinocytosis in oocytes occurs at a rate which is characteristic of the developmental stage reached and is found only during a well-defined period of the oocyte's development, namely vitellogenesis. This difference in endocytosis between oocytes and somatic cells brings into question the possible control mechanisms. There seems to be little doubt that availability of serum proteins *per se* may constitute a condition sufficient for inducing micropinocytic activity in in-vitro cultured cells (Schellens, Brunk & Lindgren, 1976). On the other hand, presence of yolk protein precursors in the blood constitutes a condition which, though necessary (Hausman, Anderson & Telfer, 1971), is not sufficient for inducing micropinocytic uptake in vitellogenic oocytes. For instance, injection of vitellogenin into allatectomised females of *Periplaneta americana* fails to promote yolk deposition in vitellogenic oocytes (Bell, 1969). A similar inability to form yolk spheres is exhibited by vitellogenic oocytes transplanted into male cockroaches injected with yolk fluid (Bell & Barth, 1971). In all experimentally induced conditions, including allatectomy, decapitation or even starvation, topical application of juvenile hormone analogues to operated insects has been shown to restore yolk deposition in the oocyte, regardless of the titre of vitellogenin in the blood (Adams, 1974; Wilhelm & Luscher, 1974; Gillott & Elliott, 1976; Elliott & Gillott, 1976; Buhlmann, 1976; Kelly & Davenport, 1976; Lazarovici & Pener, 1977; Sakurai, 1977; Hagedorn *et al.*, 1977). Similar observations have been made on diapausing females of the species *Drosophila grisea* where the juvenile hormone titre is regulated by photoperiodicity (Kambysellis & Heed, 1974), and also in the *Drosophila melanogaster* mutant apterous-four (Postlethwait & Weiser, 1973). The data are consistent with the hypothesis that juvenile hormone acts directly on the ovary so as to render it competent to incorporate vitellogenin from the haemolymph (Postlethwait & Handler, 1977; Handler & Postlethwait, 1977). Promotion of vitellogenin uptake by the hormone does not result simply from induction of vitellogenin synthesis in the fat body cells (Pan, Bell & Telfer, 1969; Engelmann, 1969; Engelmann, Hill & Wilkens, 1971; Prabhu & Nayar, 1972; Koeppe & Ofengand, 1976; Brookes, 1976; Pan & Wyatt, 1976). If this were the case isolated abdomens of *Drosophila* females (Postlethwait, Handler & Gray, 1976) as well as adult females of the mutant apterous-four, both of which contain vitellogenin in their haemolymph (Gavin & Williamson, 1976b), would not need juvenile hormone application to restore vitellogenesis. That juvenile hormone alone constitutes the major requirement for induction of micropinocytic activity in oocytes is demonstrated by in-vitro incubation experiments in which protein incorporation has been shown to occur only in the presence of hormone analogues or of the corpus allatum itself (Lender & Laverdure, 1967; Ittycheriah & Stephanos, 1969; Adams & Eide, 1972; Laverdure, 1975). When juvenile hormone is absent from the incubation medium, structures such as coated pits and vesicles are absent from the cortical ooplasm (Gwadz & Spielman, 1973; Giorgi, unpublished observations).

Evidence seems to indicate that juvenile hormone does not act as a triggering factor, for if this were the case its action would be restricted to the initiation of vitellogenesis; its presence is instead continuously required for persistence of vitellogenin uptake (Nijhout & Riddiford, 1974; Mjeni & Morrison, 1975, 1976).

An interesting analogy exists in the physiology of vitellogenesis between insects and amphibians. Injection of HCG into adult *Xenopus* females results in the formation of elaborated microvilli and numerous coated vesicles by the oolemma of vitellogenic oocytes (Holland & Dumont, 1975). On the other hand, when a second oestrogen injection follows the primary HCG treatment, the fine structure of the oolemma of vitellogenic oocytes undergoes drastic changes resulting in the cessation of endocytic activity. Treatment of isolated amphibian oocytes with steroids produces a similar alteration of the micropinocytic activity (Schuetz, Wallace & Dumont, 1974).

Hitherto, attempts to stimulate pinocytic activity in isolated amphibian oocytes have apparently been unsuccessful. Recent evidence, however, tends to indicate that such a failure is likely to result from a lack of the enveloping cell layers in isolated oocytes. In fact, when ovarian fragments, instead of disvested oocytes, are employed for in-vitro treatment with HCG, both endocytic activity and protein incorporation are reported to increase significantly (Wiley & Dumont, 1978). These results suggest that the cell layers enveloping the oocyte are the most likely candidates for the target tissue of gonadotrophin. This suggestion is strengthened by the finding that these cells in amphibians, as well as in fishes, are known to be the sites of oestrogen synthesis and secretion (Redshaw & Nicholls, 1971; Nicholls & Maple, 1972; Nagahama, Chan & Hoar, 1976).

Ovarian follicle cells in insects apparently have a similar role in relation to vitellogenesis. For instance, formation of intercellular spaces does not take place in the ovarian follicular epithelium of allatectomised females (Masner, 1968; Abu-Hakima & Davey, 1975). In-vitro cultured oocytes are capable of incorporating peroxidase only when, as a response to juvenile hormone, large intercellular spaces form in the follicular epithelium (Davey & Huebner, 1974; Abu-Hakima & Davey, 1977). Bell & Sams (1974) are of the opinion that juvenile hormone acts on follicle cells by inducing them to synthesise a follicle cell product which, as shown by Anderson & Telfer (1970a) in cecropia, is needed for vitellogenin incorporation.

The observation that juvenile hormone has a role in inducing vitellogenesis does not entirely explain why oocytes undergo micropinocytic activity only after the attainment of a certain developmental stage. Two points of view are compatible with current literature on insect vitellogenesis. Juvenile hormone titre varies during the life cycle (Rankin & Riddiford, 1978), and as a result of this variation oocytes may initiate micropinocytic activity only above a certain concentration threshold (Lanzrein, 1974). This seems to be the case, for instance, in honey bees (Fluri, Wille, Gerig & Luscher, 1977) and in bumblebee workers (Roseler, 1977). Alternatively, if oocytes have no induction threshold, they become competent to respond to whatever titre of juvenile hormone is present by initiating vitellogenesis only once a certain developmental stage is attained. Recent findings on the presence of ecdysone in ovaries of several insect species (Hagedorn *et al.*, 1975, 1977; Lagueux, Hirn & Hoffmann, 1977), and particularly the observation that ecdysone concentration increases as the oocyte develops (Legay, Calvez, Hirn & De Reggi, 1976), favour the view of this hormone being a metabolic trigger of oocyte competence toward juvenile hormone. This hypothesis is further strengthened by the observation that juvenile hormone and ecdysone may act as antagonists even during adult reproductive life (Herman & Barker, 1976; Pappas & Fraenkel, 1978; Went, 1978).

The evidence provided so far underscores the notion that in vitellogenic oocytes, blood proteins have no inductive role in relation to micropinocytosis, but function simply to fill the internalised vesicles formed by virtue of the hormone-modulated activity of the oolemma.

References

Abu-Hakima, R. & Davey, K.G. (1975). Two actions of juvenile hormone on follicle cells of *Rhodnius prolixus*. *Journal of Insect Physiology*, 53, 1187–8.

Abu-Hakima, R. & Davey, K.G. (1977). The action of juvenile hormone on follicle cells of *Rhodnius prolixus in vitro*: the effect of colchicine and cytochalasin B. *General and Comparative Endocrinology*, 32, 360–70.

Adams, E.A. & Hertig, A.T. (1964). Studies on guinea pig oocytes. I. Electron microscopic observations on the development of cytoplasmic organelles in oocytes of primordial and primary follicles. *Journal of Cell Biology*, 21, 397–427.

Adams, T.S. (1974). The role of juvenile hormone in housefly ovarian follicle morphogenesis. *Journal of Insect Physiology*, 20, 263–76.

Adams, T.S. & Eide, P.E. (1972). A method for the *in vitro* estimation of house fly egg development with a juvenile hormone analog. *General and Comparative Endocrinology*, 18, 12–21.

Anderson, E. (1964). Oocyte differentiation and vitellogenesis in the roach *Periplaneta americana*. *Journal of Cell Biology*, 20, 131–52.

Anderson, E. (1969). Oogenesis in the cockroach *Periplaneta americana*, with special reference to the specialisation of the oolemma and the fate of the coated vesicles. *Journal de Microscopie*, 8, 721–38.

Anderson, E. (1972). The localisation of acid phosphatase and the uptake of horseradish peroxidase in the oocyte and follicle cells of mammals. In *Oogenesis*, ed. J.D. Biggers & A.W. Schuetz, pp. 87–115. Baltimore: University Park Press.

Anderson, E. & Beams, H.W. (1960). Cytological observations on the fine structure of the guinea pig ovary with special reference to the oogonium, primary oocyte and associated follicle cells. *Journal of Ultrastructure Research*, 3, 432–46.

Anderson, L.M. (1971). Protein synthesis and uptake in isolated cecropia oocytes. *Journal of Cell Science*, 8, 735–50.

Anderson, L.M. & Telfer, W.H. (1969). A follicle cell contribution to the yolk spheres of moth oocytes. *Tissue and Cell*, 1, 633–44.

Anderson, L.M. & Telfer, W.H. (1970a). Trypan blue inhibition of yolk deposition – a clue to follicle cell function in the cecropia moth. *Journal of Embryology and Experimental Morphology*, 23, 35–52.

Anderson, L.M. & Telfer, W.H. (1970b). Extracellular concentration of proteins in the cecropia moth follicle. *Journal of Cell Physiology*, 76, 37–54.

Anderson, W.A. & Spielman, A. (1971). Permeability of the ovarian follicle of *Aedes aegypti* mosquitoes. *Journal of Cell Biology*, 50, 201–21.

Balinski, B.I. & Devis, R.J. (1963). Origin and differentiation of cytoplasmic structures in the oocytes of *Xenopus laevis*. *Acta Embryologiae et Morphologiae Experimentalis*, 6, 55–108.

Bast, R.E. & Telfer, W.H. (1976). Follicle cell protein synthesis and its contribution to the yolk of the cecropia moth oocyte. *Developmental Biology*, 52, 83–97.

Beams, H.W. & Kessel, R.G. (1969). Synthesis and deposition of oocyte envelopes (vitelline membrane, chorion) and the uptake of yolk in the dragon fly (*Odonata: Aeschnidae*). *Journal of Cell Science*, 4, 241–64.

Bell, W.J. (1969). Dual role of juvenile hormone in the control of yolk formation in *Periplaneta americana*. *Journal of Insect Physiology*, 15, 1279–90.

Bell, W.J. (1970). Demonstration and characterisation of two vitellogenic blood proteins in *Periplaneta americana*: an immunochemical analysis. *Journal of Insect Physiology*, 16, 291–9.

Bell, W.J. (1972). Yolk formation by transplanted cockroach oocytes. *Journal of Experimental Zoology*, 181, 41–8.

Bell, W.J. & Barth, R.H. Jr (1971). Initiation of yolk deposition by juvenile hormone. *Nature New Biology*, 230, 220–1.

Bell, W.J. & Sams, G.R. (1974). Factors promoting vitellogenic competence and yolk deposition in the cockroach ovary: the post-ecdysis female. *Journal of Insect Physiology*, 20, 2475–85.

Bellairs, R. (1965). The relationship between oocyte and follicle in the hen's ovary as shown by electron microscopy. *Journal of Embryology and Experimental Morphology*, 13, 215–33.

Bergink, E.W. & Wallace, R.A. (1974). Precursor–product relationship between amphibian vitellogenin and the yolk proteins lipovitellin and phosvitin. *Journal of Biological Chemistry*, 249, 2897–903.

Bergink, E.W., Wallace, R.A., Van De Berg, J.A., Bos, E.S., Gruber, M. & Ab, G. (1974). Estrogen-induced synthesis of yolk proteins in roosters. *American Zoologist*, 14, 1177–93.

Beuving, G. & Gruber, M. (1971). Isolation of phosvitin from the plasma of estrogenized roosters. *Biochimica et Biophysica Acta*, 232, 524–8.

Bier, K. (1962). Autoradiographische Untersuchungen zur Dotterbildung. *Naturwissenschaften*, 14, 332–63.

Bier, K. (1963). Autoradiographische Untersuchungen über die Leistungen des Follikelepithels und der Nahrzellen bei der Dotterbildung und Eiweisssynthese im Fliegenovar. *Archiv für Entwicklungsmechanik*, 154, 552–75.

Bluemink, J.G. (1969). Are yolk granules related to lysosomes? *Zeiss Information*, 73, 95–9.

Boctor, F.N. & Kamel, M.Y. (1976). Purification and characterisation of two lipovitellins from eggs of the tick *Dermacentor andersoni*. *Insect Biochemistry*, 6, 233–40.

Bodnaryk, R.P. & Morrison, P.E. (1968). Immunochemical analysis of the origin of a sex-specific blood protein in female house flies. *Journal of Insect Physiology*, 14, 1141–6.

Bownes, M. & Hames, B.D. (1977). Accumulation and degradation of three major yolk proteins in *Drosophila melanogaster*. *Journal of Experimental Zoology*, 200, 149–56.

Brachet, J. & Ficq, A. (1965). Binding sites of [^{14}C]actinomycin in amphibian oocytes and autoradiographic technique for the detection of cytoplasmic DNA. *Experimental Cell Research*, 32, 153–9.

Brambell, F.W.R. (1966). The transmission of immunity from mother to young and the catabolism of immunoglobulins. *Lancet*, ii, 1088–93.

Brambell, F.W.R., Hemmings, W.A. & Morris, I.G. (1964). A theoretical model of γ-globulin catabolism. *Nature, London,* 203, 1352–5.

Brookes, V.J. (1976). Protein synthesis in the fat body of *Leucophaea maderae* during vitellogenesis. *Journal of Insect Physiology,* 22, 1649–57.

Brookes, V.J. & Dejmal, R.K. (1968). Yolk protein: structural changes during vitellogenesis in the cockroach *Leucophaea maderae. Science,* 160, 999–1001.

Bruce, L. & Emanuelsson, H. (1975). Analysis of DNA isolated from chick blastoderm. *Experimental Cell Research,* 92, 462–6.

Brummett, A.R. & Dumont, J.N. (1976). Oogenesis in *Xenopus laevis* (Daudin). III. Localisation of negative charges on the surface of developing oocytes. *Journal of Ultrastructure Research,* 55, 4–16.

Brummett, A.R. & Dumont, J.N. (1977). Intracellular transport of vitellogenin in *Xenopus* oocytes. An autoradiographic study. *Developmental Biology,* 60, 482–6.

Buhlmann, G. (1976). Haemolymph, vitellogenin, juvenile hormone and oocyte growth in the adult cockroach *Nauphoeta cinerea* during first pre-oviposition period. *Journal of Insect Physiology,* 22, 1101–10.

Chen, T., Couble, P., De Lucca, F. & Wyatt, G.R. (1976). Juvenile hormone control of vitellogenin synthesis in *Locusta migratoria.* In *The Juvenile Hormone,* ed. L.J. Gilbert, pp. 505–29. New York: Plenum Press.

Chia, W.K. & Morrison, P.E. (1972). Autoradiographic and ultrastructural studies on the origin of yolk protein in the house fly, *Musca domestica. Canadian Journal of Zoology,* 50, 1569–76.

Chino, H., Yamagata, M. & Sato, S. (1977). Further characterisation of lepidopteran vitellogenin from haemolymph and mature eggs. *Insect Biochemistry,* 7, 125–31.

Chino, H., Yamagata, M. & Takahashi, K. (1976). Isolation and characterisation of insect vitellogenin. Its identity with haemolymph lipoprotein II. *Biochimica et Biophysica Acta,* 441, 349–53.

Clore, J.H., Petrovitch, E., Koeppe, J.K. & Mills, R.R. (1978). Vitellogenins of the American cockroach: electrophoretic and antigenic characterisation of haemolymph and oocyte proteins. *Journal of Insect Physiology,* 24, 45–51.

Cohn, Z.A. & Benson, B. (1965). The *in vitro* differentiation of mononuclear phagocytes. II. The influence of serum on granule formation, hydrolase production and pinocytosis. *Journal of Experimental Medicine,* 121, 835–48.

Cole, G.T. & Wynne, M.J. (1973). Nuclear pore arrangement and structure of the Golgi complex in *Ochromonas danica* (*Chrysophyceae*). *Cytobios,* 8, 161–73.

Cone, M.V. & Eschenberg, K.M. (1966). Histochemical localisation of acid phosphatase in the ovary of *Gerris remigis* (Hemiptera). *Journal of Experimental Zoology,* 161, 337–52.

Contractor, S.F. & Krakauer, K. (1976). Pinocytosis and intracellular digestion of ^{125}I-labelled haemoglobin by trophoblastic cells in tissue culture in the presence and absence of serum. *Journal of Cell Science,* 21, 595–607.

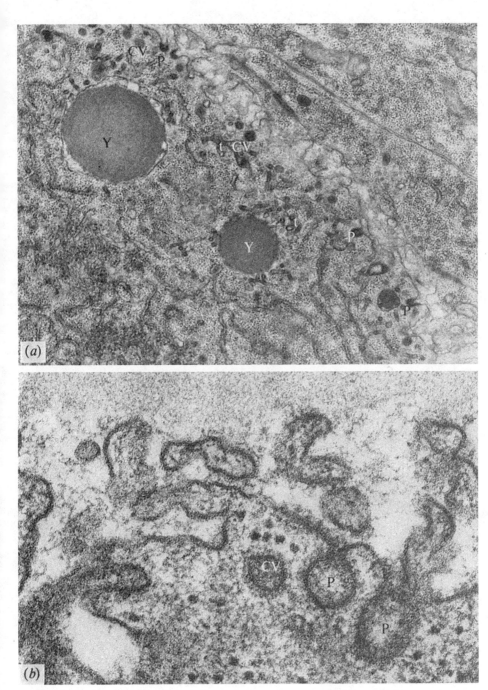

For explanation of plates see p. 176

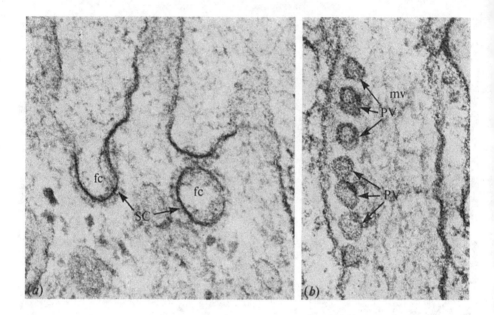

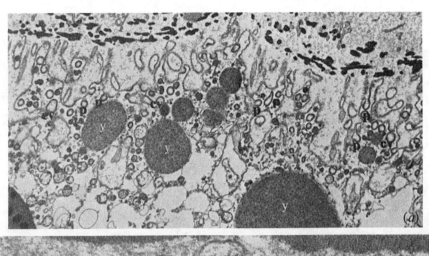

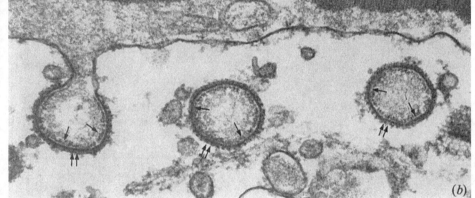

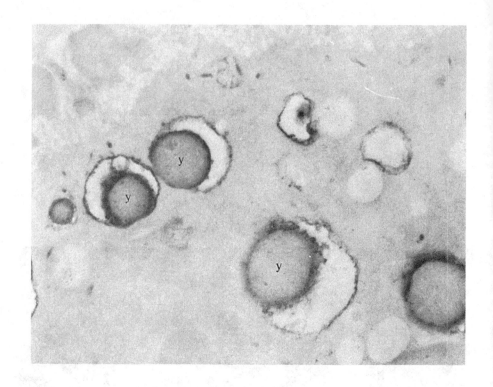

F. Giorgi   Plate 5

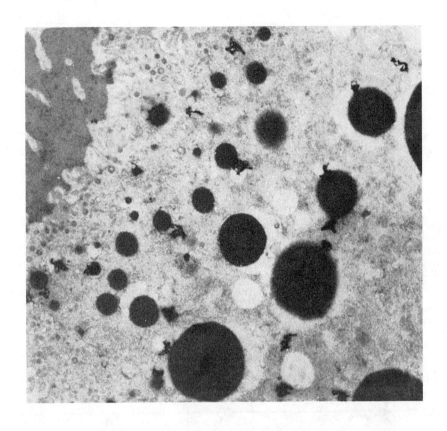

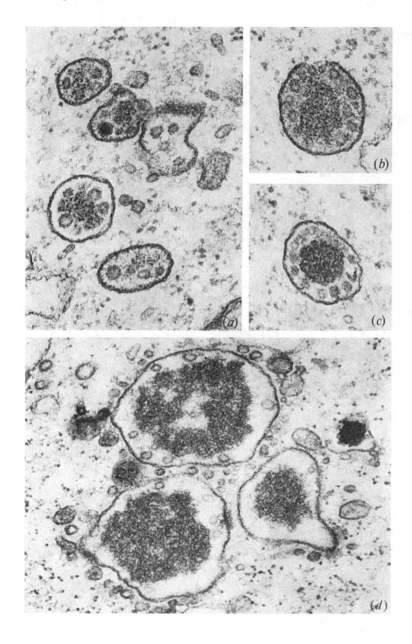

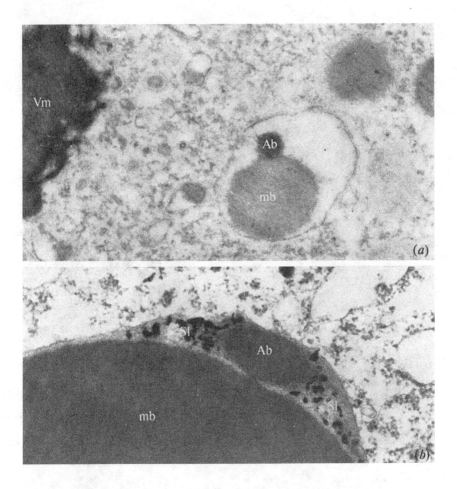

F. Giorgi   Plate 8

Crowther, R.A., Finch, J.T. & Pearse, B.M.F. (1976). On the structure of coated vesicles. *Journal of Molecular Biology*, 103, 785–98.

Cummings, M.R. & King, R.C. (1969). The cytology of the vitellogenic stages of oogenesis in *Drosophila melanogaster*. I. General staging characteristics. *Journal of Morphology*, 128, 427–42.

Cummings, M.R. & King, R.C. (1970). The cytology of the vitellogenic stages of oogenesis in *Drosophila melanogaster*. II. Ultrastructural investigation on the origin of protein yolk spheres. *Journal of Morphology*, 130, 467–78.

Cutting, J.A. & Roth, T.F. (1970). Changes in macromolecular transport specificity during development of the oocyte. *Journal of Cell Biology*, 47, 44A.

Cutting, J.A. & Roth, T.F. (1973). Changes in specific sequestration of protein during transport into the developing oocyte of the chicken. *Biochimica et Biophysica Acta*, 298, 951–5.

Davey, K.G. & Huebner, E. (1974). The response of the follicle cells of *Rhodnius prolixus* to juvenile hormone and antigonadotropin *in vitro*. *Canadian Journal of Zoology*, 52, 1407–12.

Davis, P., Allison, A.C. & Haswell, A.D. (1973). The quantitative estimation of pinocytosis using radioactive colloidal gold. *Biochemical and Biophysical Research Communications*, 52, 627–34.

De Duve, C. & Wattiaux, R. (1966). Function of lysosomes. *Annual Review of Physiology*, 28, 435–93.

Dehn, P.F. & Wallace, R.A. (1973). Sequestered and injected vitellogenin. Alternative routes of protein processing in *Xenopus laevis*. *Journal of Cell Biology*, 58, 721–4.

Dejmal, R.G. & Brookes, V.J. (1972). Chemical and physical characteristics of a yolk protein from the ovaries of *Leucophaea maderae*. *Journal of Biological Chemistry*, 247, 869–74.

De Loof, A. & Lagasse, A. (1970). The ultrastructure of the follicle cells of the ovary of the Colorado beetle in relation to yolk formation. *Journal of Insect Physiology*, 16, 211–20.

Denis, H. (1964). Phosphoprotéine phosphatase et résorption du vitellus chez amphibiens: une étude cytochimique, électrophoretique et immunologique. *Journal of Embryology and Experimental Morphology*, 12, 197–217.

Doira, H. & Kawaguchi, Y. (1972). Changes in haemolymph and egg protein by the castration and implantation of the ovary in *Bombyx mori*. *Journal of the Faculty of Agriculture, Kyushu University*, 17, 119–27.

Droller, M.J. & Roth, T.F. (1966). An electron microscope study of yolk formation during oogenesis in *Lebistes reticulatus* guppy. *Journal of Cell Biology*, 28, 209–25.

Dufour, D., Taskar, S.P. & Perron, J.M. (1970). Ontogenesis of a female specific protein from the locust *Schistocerca gregaria*. *Journal of Physiology*, 16, 1369–77.

Dumont, J.N. (1969). Oogenesis in the annelid *Enchytraeus albidus* with special reference to the origin and cytochemistry of yolk. *Journal of Morphology*, 129, 317—44.

Dumont, J.N. (1972). Oogenesis in *Xenopus laevis* (Daudin). I. Stages of oocyte development in laboratory maintained animals. *Journal of Morphology*, 136, 153—79.

Dumont, J.N. & Anderson, E. (1967). Vitellogenesis in the horse shoe crab *Limulus polyphemus*. *Journal of Microscopie*, 6, 791—806.

Dumont, J.N. & Wallace, R.A. (1972). The effects of vinblastine on isolated *Xenopus* oocytes. *Journal of Cell Biology*, 53, 605—10.

Elliott, R.H. & Gillott, G. (1976). Histological changes in the ovary in relation to yolk deposition, allatectomy and destruction of the median neurosecretory cells in *Melanopus sanguinipes*. *Canadian Journal of Zoology*, 54, 185—92.

Emanuelsson, H. (1969). Electronmicroscopical observations on yolk and yolk formation in *Ophryotrocha labronica* La Greca & Bacci. *Zeitschrift für Zellforschung und Mikroskopische Anatomie*, 95, 19—36.

Emanuelsson, H. (1971). Metabolism and distribution of yolk DNA in embryos of *Ophryotrocha labronica* La Greca & Bacci. *Zeitschrift für Zellforschung und Mikroskopische Anatomie*, 113, 450—60.

Emmerson, B. & Kjaer, K. (1974). Seasonal and hormonally induced changes in the serum level of the precursor protein vitellogenin in relation to ovarian vitellogenic growth in the toad *Bufo bufo bufo* (L.). *General and Comparative Endocrinology*, 22, 261—7.

Engelmann, F. (1969). Female specific protein: biosynthesis controlled by corpus allatum in *Leucophaea maderae*. *Science*, 165, 407—9.

Engelmann, F., Friedel, T. & Ladduwahetty (1976). The native vitellogenin of the cockroach *Leucophaea maderae*. *Insect Biochemistry*, 6, 211—20.

Engelmann, F., Hill, L. & Wilkens, J.L. (1971). Juvenile hormone control of specific female protein synthesis in *Leucophaea maderae, Schistocerca vaga* and *Sarcophaga bullata*. *Journal of Insect Physiology*, 17, 2179—91.

Engelmann, F. & Penney, D. (1966). Studies on the endocrine control of metabolism in *Leucophaea maderae* (Blattaria). I. The haemolymph protein during egg maturation. *General and Comparative Endocrinology*, 7, 314—25.

Engels, W. (1972). Quantitative Untersuchungen zum Dotterprotein Haushalt der Honigbiene (*Apis mellifica*). *Wilhelm Roux' Archiv für Entwicklungsmechanik der Organismen*, 171, 55—86.

Favard, P. & Favard-Sereno, C. (1969). Electron microscope study of polysaccharides in the amphibian oocyte. *Journal of Submicroscopic Cytology*, 1, 91—111.

Favard-Sereno, C. (1964). Phénomène de pinocytose au cours de la vitellogenèse protéique chez le grillon (Orthoptère). *Journal de Microscopie*, 3, 323—38.

Favard-Sereno, C. (1969). Capture de polysaccharides par micropinocytose dans l'ovocyte du grillon en vitellogenèse. *Journal de Microscopie*, 8, 401—14.

Fluri, P., Wille, H., Gerig, L. & Luscher, M. (1977). Juvenile hormone, vitellogenin and haemocyte composition in worker honey bees (*Apis mellifica*). *Experientia,* 33, 1240–1.

Follett, B.K. & Redshaw, M.R. (1974). The physiology of vitellogenesis. In *Physiology of the Amphibia,* ed. B. Loft, vol. 2, pp. 219–308. New York & London: Academic Press.

Franke, W.W. & Herth, W. (1974). Morphological evidence for *de novo* formation of plasma membrane from coated vesicles in exponentially growing cultured plant cells. *Experimental Cell Research,* 89, 447–51.

Friend, D.S. & Farquhar, M.G. (1967). Functions of coated vesicles during protein adsorption in the rat vas deferens. *Journal of Cell Biology,* 35, 357–75.

Ganion, L.R. & Kessel, R.G. (1972). Intracellular synthesis, transport and packaging of proteinaceous yolk in oocytes of *Orconectus immunis.* *Journal of Cell Biology,* 52, 420–37.

Gavin, J.A. & Williamson, J.H. (1976*a*). Synthesis and deposition of yolk protein in adult *Drosophila melanogaster. Journal of Insect Physiology,* 22, 1457–64.

Gavin, J.A. & Williamson, J.H. (1976*b*). Juvenile hormone-induced vitellogenesis in apterous four, a non-vitellogenic mutant in *Drosophila melanogaster. Journal of Insect Physiology,* 22, 1737–42.

Gelissen, G., Wajo, E., Cohen, E., Emmerich, H., Applebaum, S.W. & Flossdorf, J. (1976). Purification and properties of oocyte vitellin from the migratory locust. *Journal of Comparative Physiology,* 108B, 207–301.

Gelti-Douka, H., Gingeras, T.R. & Kambysellis, M.P. (1974). Yolk protein in *Drosophila*: identification and site of synthesis. *Journal of Experimental Zoology,* 187, 167–72.

Gillott, G. & Elliott, R.H. (1976). Reproductive growth in normal allatectomised median-neurosecretory-cell-cauterised and ovariectomised females of *Melanopus sanguinipes. Canadian Journal of Zoology,* 54, 162–71.

Giorgi, F. (1975). Ultrastructural and cytochemical studies of oogenesis in *Drosophila melanogaster* with special reference to the formation of yolk. Ph.D Thesis, University of Edinburgh.

Giorgi, F., Bucci Innocenti, S. & Ragghianti, M. (1976). Osmium/zinc iodide staining of Golgi elements in oocytes of *Triturus cristatus. Cell and Tissue Research,* 172, 121–31.

Giorgi, F. & Jacob, J. (1977*a*). Recent findings on oogenesis of *Drosophila melanogaster.* I. Ultrastructural observations on the developing ooplasm. *Journal of Embryology and Experimental Morphology,* 38, 115–24.

Giorgi, F. & Jacob, J. (1977*b*). Recent findings on oogenesis of *Drosophila melanogaster.* II. Further evidence on the origin of yolk platelets. *Journal of Embryology and Experimental Morphology,* 38, 125–38.

Giorgi, F. & Jacob, J. (1977*c*). Recent findings on oogenesis of *Drosophila melanogaster.* III. Lysosomes and yolk platelets. *Journal of Embryology and Experimental Morphology,* 39, 45–57.

Glass, L.E. (1959). Immunohistological localisation of serum-like molecules in frog oocytes. *Journal of Experimental Zoology,* 141, 257–90.

Glass, L.E. & Cons, J.M. (1968). Stage dependent antigens and radiolabel into mouse ovarian follicles. *Anatomical Record*, 162, 139—56.

Graham, J.M., Sumner, M.C.B., Curtis, D.H. & Pasternak, C.A. (1973). Sequence of events in plasma membrane assembly during the cell cycle. *Nature, London*, 246, 291—5.

Green, C.D. & Tata, J.R. (1976). Direct induction by estradiol of vitellogenin synthesis in organ cultures of male *Xenopus laevis* liver. *Cell*, 7, 131—9.

Gwadz, R.W. & Spielman, A. (1973). Corpus allatum control of ovarian development in *Aedes aegypti. Journal of Insect Physiology*, 19, 1441—8.

Hagedorn, H.H. & Judson, C.L. (1972). Purification and site of synthesis of *Aedes aegypti* yolk proteins. *Journal of Experimental Zoology*, 182, 367—78.

Hagedorn, H.H., O'Connor, J.D., Fuchs, M.S., Sage, B., Schlaeger, D.A. & Bohm, M.K. (1975). The ovary as a source of alpha-ecdysone in adult mosquito. *Proceedings of the National Academy of Sciences, USA*, 72, 3255—9.

Hagedorn, H.H., Turner, S., Hagedorn, E.A., Pontecorvo, D., Greenbaum, P., Pferffer, D., Wheelock, G. & Flanagan, T.R. (1977). Postemergence growth of the ovarian follicles of *Aedes aegypti. Journal of Insect Physiology*, 23, 203—6.

Handler, A.M. & Postlethwait, J.H. (1977). Endocrine control of vitellogenesis in *Drosophila melanogaster*: effects of the corpus allatum. *Journal of Experimental Zoology*, 202, 389—401.

Hanocq-Quertier, J., Baltus, E., Ficq, A. & Brachet, J. (1968). Studies on the DNA of *Xenopus laevis* oocytes. *Journal of Embryology and Experimental Morphology*, 19, 273—82.

Hausman, S.J., Anderson, L.M. & Telfer, W.H. (1971). The dependence of yolk formation *in vitro* on specific blood proteins. *Journal of Cell Biology*, 48, 303—13.

Heesen te, D. & Engels, W. (1973). Elektrophoretische Untersuchungen zur Vitellogenese von *Brachydanio rerio* (*Cyprinidae*, Teleostei). *Wilhelm Roux' Archiv für Entwicklungsmechanik der Organismen*, 173, 46—59.

Herbaut, C. (1972). Nature et origine des réserve vitellines dans l'ovocyte de *Lithobius forficatus* L. (Myriapode, Chilopode). *Zeitschrift für Zellforschung und Mikroskopische Anatomie*, 130, 18—27.

Herman, W.S. & Barker, J.F. (1976). Ecdysterone antagonism mimicry and synergism of juvenile hormone action on the monarch butterfly reproductive tract. *Journal of Insect Physiology*, 22, 643—8.

Herzog, V. & Farquhar, M.G. (1977). Luminal membrane retrieved after exocytosis reaches most Golgi cisternae in secretory cells. *Proceedings of the National Academy of Sciences, USA*, 74, 5073—7.

Holland, C.A. & Dumont, J.N. (1975). Oogenesis in *Xenopus laevis* (Daudin). IV. Effects of gonadotropin, estrogen and starvation on endocytosis in developing oocytes. *Cell and Tissue Research*, 162, 177—84.

Hope, J., Humphreys, A.A. Jr & Bourne, G.H. (1964). Ultrastructural studies on developing oocytes of the salamander *Triturus viridescens*. II. Formation of yolk. *Journal of Ultrastructure Research*, 10, 547–56.

Hopkins, C.R. & King, P.E. (1966). An electron microscope and histochemical study on the oocyte periphery in *Bombus terrestris* during vitellogenesis. *Journal of Cell Science*, 1, 201–16.

Huebner, E. & Anderson, E. (1970). The effects of vinblastine sulfate on the microtubular organisation of the ovary of *Rhodnius prolixus*. *Journal of Cell Biology*, 46, 191–8.

Huebner, E. & Anderson, E. (1972). Cytological study on the ovary of *Rhodnius prolixus*. Oocyte differentiation. *Journal of Morphology*, 137, 385–416.

Huebner, E., Tobe, S.S. & Davey, K.G. (1975). Structural and functional dynamics of oogenesis in *Glossina austeni*: vitellogenesis with special reference to the follicular epithelium. *Tissue and Cell*, 7, 535–58.

Humbert, F., Montesano, R., Grosso, A., de Sousa, R.C. & Orci, L. (1977). Particle aggregates in plasma and intracellular membranes of toad bladder (granular cell). *Experientia*, 33, 1364–7.

Ittycheriah, P.I. & Stephanos, S. (1969). *In vitro* culture of ovary of the plant bug, *Iphita limbata* Stal. *Indian Journal of Experimental Biology*, 7, 17–19.

Jacques, P.J. (1969). Endocytosis. In *Lysosomes in Biology and Pathology*, ed. J.T. Dingle & H.B. Fell, vol. 1, pp. 395–420. New York: Wiley.

Jaffe, L.F. (1977). Electrophoresis along cell membranes. *Nature, London*, 265, 600–2.

Jared, D.W., Dumont, J.N. & Wallace, R.A. (1973). Distribution of incorporated and synthesised protein among cell fractions of *Xenopus laevis* oocytes. *Developmental Biology*, 35, 19–28.

Jollie, W.P. & Triche, T.J. (1971). Ruthenium labelling of micropinocytotic activity in the rat visceral yolk sac placenta. *Journal of Ultrastructure Research*, 35, 541–53.

Kageyama, T., Takahashi, S.Y. & Ohnishi, E. (1973). Acid phosphatase in the eggs of the silkworm, *Bombyx mori*: its purification and its properties. *Insect Biochemistry*, 3, 373–88.

Kambysellis, M.P. & Heed, W.B. (1974). Juvenile hormone induces ovarian development in diapausing cave-dwelling *Drosophila* species. *Journal of Insect Physiology*, 20, 1779–86.

Kanaseki, T. & Kadota, K. (1969). The 'vesicle in a basket'. A morphological study of the coated vesicles isolated from the nerve endings of the guinea pig brain, with special reference to the mechanism of membrane movements. *Journal of Cell Biology*, 42, 202–20.

Karasaki, S. (1963). Studies on amphibian yolk. I. The ultrastructure of the yolk platelet. *Journal of Cell Biology*, 18, 135–51.

Karasaki, S. (1967). An electron microscope study on the crystalline structure of the yolk platelets of the lamprey egg. *Journal of Ultrastructure Research*, 18, 377–90.

Kawaguchi, Y. & Doira, H. (1973). Gene-controlled incorporation of haemolymph protein into the ovaries of *Bombyx mori*. *Journal of Insect Physiology*, 19, 2083–96.

Kelly, T.J. & Davenport, R. (1976). Juvenile hormone induced ovarian uptake of a female specific blood protein in *Oncopeltus fasciatus*. *Journal of Insect Physiology*, 22, 1381–93.

Kerr, M.S. (1969). The hemolymph proteins of the blue crab, *Callinectes sapidus*. II. A lipoprotein serologically identical to oocyte lipovitellin. *Developmental Biology*, 20, 1–17.

Knight, P.F. & Schechtman, A.M. (1954). The passage of heterologous serum proteins from the circulation into the ovum of the fowl. *Journal of Experimental Zoology*, 127, 271–304.

Koeppe, J. & Ofengand, J. (1976). Juvenile hormone-induced biosynthesis of vitellogenin in *Leucophaea maderae*. *Archives of Biochemistry and Biophysics*, 173, 100–13.

Korfsmeier, K.H. (1966). Zur Genese des Dottersystems in der Oocyte von *Brachydanio rerio*. Autoradiographische Untersuchungen. *Zeitschrift für Zellforschung und Mikroskopische Anatomie*, 71, 283–96.

Kress, A. & Spornitz, U.M. (1972). Ultrastructural studies of oogenesis in some European amphibians. I. *Rana esculenta* and *Rana temporaria*. *Zeitschrift für Zellforschung und Mikroskopische Anatomie*, 128, 438–56.

Krolak, J.M., Clore, J.N., Petrovitch, E. & Mills, R. (1977). Vitellogenesis by the American cockroach: haemolymph and follicle protein patterns during vitellogenin synthesis. *Journal of Insect Physiology*, 23, 381–5.

Kunkel, J.G., Johnson, M.E., Haggerty, W.T. & Sargent, T.D. (1976). Conservation of an active site for oocyte recognition in rapidly evolving vitellogenins. *American Zoologist*, 16, 388.

Kunkel, J.G. & Pan, M.L. (1976). Selectivity of yolk protein uptake: comparison of vitellogenins of two insects. *Journal of Insect Physiology*, 22, 809–18.

Kunz, W. & Petzelt, C. (1970). Synthese und Einlagerung der Dotterproteine bei *Gryllus domesticus*. *Journal of Insect Physiology*, 16, 941–7.

Lagueux, M., Hirn, M. & Hoffmann, J.A. (1977). Ecdysone during ovarian development in *Locusta migratoria*. *Journal of Insect Physiology*, 23, 109–19.

Lanzrein, B. (1974). Influence of a juvenile hormone analogue on vitellogenin synthesis and oogenesis in larvae of *Nauphoeta cinerae*. *Journal of Physiology*, 20, 1871–85.

Laverdure, A.M. (1975). Culture *in vitro* des ovaries des *Tenebrio molitor*, hormone juvenile, vitellogenèse et suivii des jeunes ovocytes. *Journal of Insect Physiology*, 21, 33–8.

Lazarovici, P. & Pener, M.P. (1977). Juvenile hormones (JHs) and completion of oocyte development in the African migratory locust: a comparative and quantitative study. *General and Comparative Endocrinology*, 33, 434–52.

Legay, J.M., Calvez, B., Hirn, M. & De Reggi, M.L. (1976). Ecdysone and oocyte morphogenesis in *Bombyx mori*. *Nature, London*, 262, 489–90.

Lender, T. & Laverdure, A.M. (1967). Culture *in vitro* des ovaries de *Tenebrio molitor* (Coléoptère). Croissance et vitellogenèse. *Compte Rendus de L'Académie des Sciences, Paris,* 265, 451–4.

Leonard, R., Deamer, D.W. & Armstrong, P. (1972). Amphibian yolk platelets ultrastructure visualised by freeze-etching. *Journal of Ultrastructure Research,* 40, 1–24.

Liu, T.P. (1973). Ultrastructure of the yolk protein granules in the frozen-etched oocyte of an insect. *Cytobiologie,* 7, 33–41.

Luft, J.H. (1976). The structure and properties of the cell surface coat. *International Review of Cytology,* 45, 291–382.

Mahowald, A.P. (1972*a*). Oogenesis. In *Developmental Systems: Insects,* ed. S.J. Counce & C.H. Waddington, vol. 1, pp. 1–47. New York & London: Academic Press.

Mahowald, A.P. (1972*b*). Ultrastructural observations on oogenesis in *Drosophila. Journal of Morphology,* 137, 29–48.

Masner, P. (1968). The inductors of differentiation of prefollicular tissue and the follicular epithelium in ovarioles of *Pyrrochoris apterus* (Heteroptera). *Journal of Embryology and Experimental Morphology,* 20, 1–13.

Massover, W.H. (1971). Nascent yolk platelets of anuran amphibian oocytes. *Journal of Ultrastructure Research,* 37, 574–91.

Matsuzaki, M. (1971). Electron microscopic studies on the oogenesis of dragonfly and cricket with special reference to the panoistic ovaries. *Development, Growth and Differentiation,* 13, 379–98.

Matsuzaki, M. (1975). Ultrastructural changes in developing oocytes, nurse cells and follicular cells during oogenesis in the telotrophic ovarioles of *Bothrogonia japonica* Ishihara (Homoptera, Tettigellidae). *Kontyu, Tokyo,* 43, 75–90.

Melius, M.E. & Telfer, W.H. (1969). An autoradiographic analysis of yolk deposition in the cortex of the cecropia moth oocyte. *Journal of Morphology,* 129, 1–16.

Mjeni, A.M. & Morrison, P.E. (1975). Delayed allatectomy and feeding in *Phormia regina* (Meig.): effects on follicular development. *Journal of Experimental Zoology,* 194, 547–51.

Mjeni, A.M. & Morrison, P.E. (1976). Juvenile hormone analogue and egg development in the blow fly, *Phormia regina* (Meig.). *General and Comparative Endocrinology,* 28, 17–23.

Mulherkar, L., Kothari, R.M. & Vaidya, V.G. (1972). Study of catheptic and acid phosphatase activities during development and metamorphosis of *Drosophila melanogaster. Wilhelm Roux' Archiv für Entwicklungsmechanik der Organismen,* 171, 195–9.

Nagahama, Y., Chan, K. & Hoar, W.S. (1976). Histochemistry and ultrastructure of pre- and post-ovulatory follicles in the ovary of the goldfish, *Carassius auratus. Canadian Journal of Zoology,* 54, 1128–39.

Neaves, W.B. (1972). The passage of extracellular tracers through the follicular epithelium of lizard ovaries. *Journal of Experimental Zoology,* 179, 339–64.

Nicholls, T.J. & Maple, G. (1972). Ultrastructural observations on possible sites of steroid biosynthesis in the ovarian follicular epithelium of two species of cichlid fish *Cichlasoma nigrofasciatum* and *Haplochromis multicolor*. *Zeitschrift für Zellforschung und Mikroskopische Anatomie,* 128, 317–35.

Nicolson, G.L. & Yanagimachi, R. (1974). Mobility and the restriction of mobility of plasma membrane lectin-binding components. *Science,* 184, 1294–6.

Nielsen, D.J. & Mills, R.R. (1968). Changes in the electrophoretic properties of haemolymph and terminal oocyte proteins during vitellogenesis in the American cockroach. *Journal of Insect Physiology,* 14, 163–70.

Nijhout, M.M. & Riddiford, L.M. (1974). The control of egg maturation by juvenile hormone in the tobacco hornworm moth, *Manduca sexta. Biological Bulletin,* 146, 377–92.

Nishimura, M., Urakawa, N. & Iwata, M. (1976). An electron microscope study on ^{203}Hg transport in the ovarian tissue of laying Japanese quail. *Japanese Journal of Veterinary Science,* 38, 83–92.

Ohlendorf, D.H., Collins, M.L., Puronen, E.O., Banaszak, L.J. & Harrison, S.C. (1975). Crystalline lipoprotein–phosphoprotein complex in oocytes from *Xenopus laevis*: determination of lattice parameters by X-ray crystallography and electron microscopy. *Journal of Molecular Biology,* 99, 153–65.

Ohno, S., Karasaki, S. & Takata, K. (1964). Histo- and cytochemical studies on the superficial layer of yolk platelets in the *Triturus* embryo. *Experimental Cell Research,* 33, 310–18.

Opresko, L., Wiley, H.S. & Wallace, R.A. (1978). The origin of yolk-DNA in *Xenopus laevis. Journal of Cell Biology,* 75, 163a.

Orci, L. & Perrelet, A. (1973). Membrane-associated particles: increase at sites of pinocytosis demonstrated by freeze-etching. *Science,* 181, 868–9.

Pan, M.L., Bell, W.L. & Telfer, W.H. (1969). Vitellogenic blood protein synthesis by insect fat body. *Science,* 165, 393–4.

Pan, M.L. & Wallace, R.A. (1974). Cecropia vitellogenin: isolation and characterisation. *American Zoologist,* 14, 1239–42.

Pan, M.L. & Wyatt, G.R. (1976). Control of vitellogenin synthesis in the monarch butterfly by juvenile hormone. *Developmental Biology,* 54, 127–34.

Pappas, C. & Fraenkel, G. (1978). Hormonal aspects of oogenesis in the flies *Phormia regina* and *Sarcophaga bullata. Journal of Insect Physiology,* 24, 75–80.

Pasteels, J.J. (1969). L'activité phosphatasique acide étudiée au microscope electronique dans des oeufs de *Barnea candida* (Mollusque, Bivalve). *Archives of Biology,* 80, 1–17.

Pasteels, J.J. (1973). Yolk and lysosomes. In *Lysosomes in Biology and Pathology,* ed. J.T. Dingle & H.B. Fell, vol. 3, pp. 238–56. New York: Wiley.

Paulson, J. & Rosenberg, M.D. (1972). The function and transposition of lining bodies in developing avian oocytes. *Journal of Ultrastructure Research*, **40**, 25–43.

Pearse, B.M.F. (1975). Coated vesicles from pig brain: purification and biochemical characterisation. *Journal of Molecular Biology*, **97**, 93–8.

Pearse, B.M.F. (1976). Clathrin, a unique protein associated with intracellular transfer of membrane by coated vesicles. *Proceedings of the National Academy of Sciences, USA*, **73**, 1254–5.

Perry, M.M., Gilbert, A.B. & Evans, A.J. (1978a). Electron microscope observations on the ovarian follicle of the domestic fowl during the rapid growth phase. *Journal of Anatomy*, **125**, 481–97.

Perry, M.M., Gilbert, A.B. & Evans, A.J. (1978b). The structure of the germinal disc region of the hen's ovarian follicle during the rapid growth phase. *Journal of Anatomy*, **127**, 379–92.

Postlethwait, J.H. & Gray, P. (1975). Regulation of acid phosphatase activity in the ovary of *Drosophila melanogaster*. *Developmental Biology*, **47**, 196–205.

Postlethwait, J.H. & Handler, A.M. (1978). Non-vitellogenic female sterile mutants and the hormonal control of vitellogenesis in *Drosophila melanogaster*. *Developmental Biology*, **67**, 202–13.

Postlethwait, J.H., Handler, A.M. & Gray, P.W. (1976). A genetic approach to the study of juvenile hormone control of vitellogenesis in *Drosophila melanogaster*. In *The Juvenile Hormones*, ed. L.I. Gilbert, pp. 449–69. New York: Plenum Press.

Postlethwait, J.H. & Weiser, K. (1973). Vitellogenesis induced by juvenile hormone in the female sterile mutant apterous-four in *Drosophila melanogaster*. *Nature, London*, **244**, 284–5.

Prabhu, V.K.K. & Nayar, K.K. (1972). Haemolymph protein electrophoretic pattern in *Periplaneta americana* after administration of farnesyl methyl ether. *Journal of Insect Physiology*, **18**, 1435–40.

Rahil, K.S. & Narbaitz, R. (1973). Ultrastructural studies on the relationship between follicular cells and growing oocytes in the turtle *Pseudemy scripta elegans*. *Journal of Anatomy*, **115**, 175–86.

Ramamurty, P.S. (1964). On the contribution of the follicle epithelium to the deposition of yolk in the oocyte of *Panorpa communis* (Mecoptera). *Experimental Cell Research*, **33**, 601–5.

Ramamurty, P.S. (1968). The route of haemolymph protein transport into the oocytes of *Panorpa communis* (Mecoptera-Insecta). *Journal of Animal Morphology and Physiology*, **15**, 188–90.

Ramamurty, P.S. & Majumdar, U. (1967). Heterosynthetic origin of protein yolk on *Delias eucharis* (Drury). *Indian Journal of Experimental Biology*, **5**, 250–2.

Rankin, M.A. & Riddiford, L.M. (1978). Significance of haemolymph juvenile hormone titer changes in timing of migration and reproduction in adult *Oncopeltus fasciatus*. *Journal of Insect Physiology*, **24**, 31–8.

Redshaw, M.R. & Follett, B.K. (1971). The crystalline yolk platelet proteins and their soluble plasma precursor in an amphibian ovary, *Xenopus laevis*. *Biochemical Journal*, **124**, 759–66.

Redshaw, M.R. & Nicholls, T.J. (1971). Oestrogen biosynthesis by ovarian tissue of the South African clawed toad, *Xenopus laevis* Daudin. *General and Comparative Endocrinology*, 16, 85–96.

Roseler, P.F. (1977). Juvenile hormone control of oogenesis in bumblebee workers, *Bombus terrestris*. *Journal of Insect Physiology*, 23, 985–92.

Roth, T.F., Cutting, J.A. & Atlas, S.B. (1976). Protein transport: a selective membrane mechanism. *Journal of Supramolecular Structure*, 4, 527–48.

Roth, T.F. & Porter, K.R. (1962). Specialised sites on the cell surface for protein uptake. In *Electron Microscopy*, ed. S.S. Breese, pp. 11–14. New York & London: Academic Press.

Roth, T.F. & Porter, K.R. (1964). Yolk protein uptake in the oocyte of the mosquito *Aedes aegypti* L. *Journal of Cell Biology*, 20, 313–32.

Sakurai, H. (1977). Endocrine control of oogenesis in the housefly *Musca domestica vicina*. *Journal of Insect Physiology*, 23, 1295–302.

Sawicki, J.A. & MacIntyre, R.J. (1977). Synthesis of ovarian acid phosphatase-1 in *Drosophila melanogaster*. *Developmental Biology*, 60, 1–13.

Schechtman, A.M. (1955). Ontogeny of the blood and related antigens and their significance for the theory of differentiation. In *Biological Specificity and Growth*, ed. E.G. Butler, pp. 3–31. Princeton, NJ: Princeton University Press.

Schellens, J.P.M., Brunk, U.T. & Lindgren, A. (1976). Influence of serum on ruffling activity, pinocytosis and proliferation of *in vitro* cultivated human glia cells. *Cytobiologie*, 13, 93–106.

Scheuer, R. (1969). Haemolymph proteins and yolk formation in the cockroach *Leucophaea maderae*. *Journal of Insect Physiology*, 15, 1673–82.

Schjeide, O.A., Briles, W.E., Holshouser, S. & Jones, D.G. (1976). Effect of 'restricted ovulator' gene on uptake of yolk precursor protein. *Cell and Tissue Research*, 166, 109–16.

Schjeide, O.A., Lin, R.I., Grellert, E.A., Galey, F.R. & Mead, J.F. (1969). Isolation and preliminary chemical analysis of coated vesicles from chicken oocytes. *Physiological Chemistry and Physics*, 1, 141–63.

Schneider, Y.J., Tulkens, P. & Trouet, A. (1977). Recycling of fibroblast plasma membrane antigens internalised during endocytosis. *Biochemical Society Transactions*, 5, 1164–7.

Schuel, H., Wilson, W.L., Wilson, J.R. & Bressler, R.S. (1975). Heterogenous distribution of lysosomal hydrolases in yolk platelets isolated from unfertilised sea urchin eggs by zonal centrifugation. *Developmental Biology*, 46, 404–12.

Schuetz, A.W., Wallace, P.A. & Dumont, J.N. (1974). Steroid inhibition of protein incorporation by isolated amphibian oocytes. *Journal of Cell Biology*, 61, 26–34.

Shacley, S.E. & King, P.E. (1977). Oogenesis in a marine teleost, *Blennius pholis* L. *Cell and Tissue Research*, 181, 105–28.

Silverstein, S.C., Steinman, R.M. & Cohn, Z.A. (1977). Endocytosis. *Annual Review of Biochemistry*, 46, 669–722.

Singer, S.J. & Nicolson, G.L. (1972). The fluid mosaic model of the structure of cell membrane. *Science,* 175, 720–31.
Slaughter, D. & Triplett, E. (1975). Amphibian embryo protease inhibitor. II. Biological properties of the inhibitor and its associated protease. *Cell Differentiation,* 4, 23–33.
Slaughter, D. & Triplett, E. (1976). Amphibian embryo protease inhibitor. III. Binding studies on the trypsin inhibitor and properties of its yolk-bound form. *Cell Differentiation,* 4, 429–40.
Spornitz, U.M. (1972). Some properties of crystalline inclusion bodies in oocytes of *Rana temporaria* and *Rana esculenta. Experientia,* 28, 66–7.
Spornitz, U.M. & Kress, A. (1973). Ultrastructural studies of oogenesis in some European amphibians. II. *Triturus vulgaris. Zeitschrift für Zellforschung und Mikroskopische Anatomie,* 143, 387–407.
Stacey, D.W. & Allfrey, V.G. (1977). Evidence for the autophagy of micro-injected proteins in HeLa cells. *Journal of Cell Biology,* 75, 807–17.
Stay, B. (1965). Protein uptake in the oocyte of the cecropia moth. *Journal of Cell Biology,* 26, 49–62.
Steinman, R.M., Silver, J.M. & Cohn, Z.A. (1974). Pinocytosis in fibroblasts. Quantitative studies *in vitro. Journal of Cell Biology,* 63, 949–69.
Stockem, W. (1977). Endocytosis. In *Mammalian Cell Membranes. Responses of Plasma Membranes,* ed. G.A. Jamieson & D.M. Robinson, pp. 151–95. London: Butterworth.
Takashima, Y. (1971). Cytochemical studies on the components of yolk granules and cortical granules in sea urchin eggs. *Medical Journal of Osaka University,* 22, 109–27.
Tanaka, A. & Ishizaki, H. (1974). Immunohistochemical detection of vitello-genin in the ovary of the cockroach, *Blattella germanica. Development, Growth and Differentiation,* 16, 247–55.
Tandler, C.J. & La Torre, J.L. (1967). An acid polysaccharide in the yolk platelets of *Bufo arenarum* oocytes. *Experimental Cell Research,* 45, 491–4.
Tata, J.R. (1976). The expression of the vitellogenin gene. *Cell,* 9, 1–14.
Telfer, W.H. (1954). Immunological studies of insect metamorphosis. II. The role of a sex-limited blood protein in egg formation by the cecropia silkworm. *Journal of General Physiology,* 37, 539–58.
Telfer, W.H. (1960). The selective accumulation of blood protein by the oocytes of saturniid moths. *Biological Bulletin,* 118, 338–51.
Telfer, W.H. (1961). The route of entry and localisation of blood proteins in the oocytes of saturniid moths. *Journal of Biophysical and Biochemical Cytology,* 9, 747–59.
Telfer, W.H. (1965). The mechanism and control of yolk formation. *Annual Review of Entomology,* 10, 161–84.
Telfer, W.H. & Anderson, M.L. (1968). Functional transformation accom-panying growth phase in the cecropia moth oocyte. *Developmental Biology,* 17, 512–35.
Tesoriero, J.V. (1977). Formation of the chorion (zona pellucida) in the

teleost, *Oryzias latipes*. I. Morphology of early oogenesis. *Journal of Ultrastructure Research*, 59, 282–91.

Tsukahara, J. (1970). Formation and behaviour of pinosomes in the sea urchin oocyte during oogenesis. *Development, Growth and Differentiation*, 12, 53–64.

Tsukahara, J. & Sugiyama, M. (1969). Structural changes in the surface of the oocyte during oogenesis of the sea urchin *Hemicentrotus pulcherrimus*, *Embryologia*, 10, 343–55.

Ulrich, E. (1969). Etude des ultrastructures au cours de l'ovogenèse d'un poisson téleosteen le danio, *Brachydanio rerio* (Hamilton-Buchnan). *Journal de Microscopie*, 8, 447–78.

Vernier, J.M. & Sire, M.F. (1977). Plaquettes vitellines et activité hydrolasique acid au cours du développement embryonnaire de la Truite arc-en-ciel. Etude ultrastructurale et biochimique. *Biologie Cellulaire*, 29, 99–112.

Wallace, R.A. (1963). Studies on amphibian yolk. IV. An analysis of the main body components of the yolk platelets. *Biochimica et Biophysica Acta*, 74, 505–18.

Wallace, R.A. (1965). Resolution and isolation of avian and amphibian yolk-granule proteins using TEAE–cellulose. *Analytical Biochemistry*, 11, 297–311.

Wallace, R.A. (1970). Studies on amphibian yolk. IX. *Xenopus* vitellogenin. *Biochimica et Biophysica Acta*, 215, 176–83.

Wallace, R.A. (1978). Oocyte growth: non-mammalian vertebrates. In *Evolution of the Vertebrate Ovary*, ed. R.E. Jones. pp. 469–502. New York: Plenum Press.

Wallace, R.A. & Dumont, J.N. (1968). The induced synthesis and transport of yolk proteins and their accumulation by the oocyte in *Xenopus laevis. Journal of Cell Physiology*, 72 (Suppl.), 73–89.

Wallace, R.A. & Ho, T. (1972). Protein incorporation by isolated amphibian oocytes. II. A survey of inhibitors. *Journal of Experimental Zoology*, 181, 303–18.

Wallace, R.A., Ho, T., Salter, D.W. & Jared, D.W. (1973). Protein incorporation by isolated amphibian oocytes. IV. The role of follicle cells and calcium during protein uptake. *Experimental Cell Research*, 82, 287–95.

Wallace, R.A. & Jared, D.W. (1969). Studies on amphibian yolk. VIII. The estrogen-induced hepatic synthesis of a serum lipophosphoprotein and its selective uptake by the ovary and transformation into yolk platelet proteins in *Xenopus laevis. Developmental Biology*, 19, 498–526.

Wallace, R.A. & Jared, D.W. (1976). Protein incorporation by isolated amphibian oocytes. V. Specificity for vitellogenin incorporation. *Journal of Cell Biology*, 69, 345–51.

Wallace, R.A., Jared, D.W. & Nelson, B.L. (1970). Protein incorporation by isolated amphibian oocytes. I. Preliminary studies. *Journal of Experimental Zoology*, 175, 259–70.

Wallace, R.A., Nickol, J.M. & Jared, D.W. (1972). Studies on amphibian yolk. X. The relative roles of autosynthetic and heterosynthetic processes during yolk protein assembly by isolated amphibian oocytes. *Developmental Biology*, 29, 255–72.

Wangh, L.J. & Knowland, J. (1975). Synthesis of vitellogenin in cultures of male and female frog liver regulated by estradiol treatment *in vitro*. *Proceedings of the National Academy of Sciences, USA*, 72, 3172–5.

Wartenberg, H. (1964). Experimentelle Untersuchungen über die Stoffausnahme durch Pinocytose wahrend der Vitellogenese des Amphibienoocyten. *Zeitschrift für Zellforschung und Mikroskopische Anatomie*, 63, 1004–19.

Went, D.F. (1978). Ecdysone stimulates and juvenile hormone inhibits follicle formation in a gall midge ovary *in vitro*. *Journal of Insect Physiology*, 24, 53–9.

Westin, M. (1975). Phosphatase active antigens in sea urchin eggs and embryos. II. A comparison between the activities in unfertilised eggs and plutei. *Journal of Experimental Zoology*, 192, 315–22.

Whitmore, E. & Gilbert, L.I. (1974). Haemolymph proteins and lipoproteins in Lepidoptera. A comparative electrophoretic study. *Comparative Biochemistry and Physiology*, 47B, 63–78.

Wiley, H.S. & Dumont, J.N. (1978). Stimulation of vitellogenin uptake in stage IV *Xenopus* oocytes by treatment with chorionic gonadotropin *in vitro*. *Biology of Reproduction*, 18, 762-71.

Wilhelm, R. & Luscher, M. (1974). On the relative importance of juvenile hormone and vitellogenin for oocyte growth in the cockroach *Nauphoeta cinerae*. *Journal of Insect Physiology*, 20, 1887–94.

Wittliff, J.L. & Kenney, F.T. (1972). Regulation of yolk protein synthesis in amphibian liver. I. Induction of lipovitellin synthesis by estrogen. *Biochimica et Biophysica Acta*, 269, 485–92.

Woods, J.W., Woodward, M.P. & Roth, T.F. (1978). Common features of coated vesicles from dissimilar tissues: composition and structure. *Journal of Cell Science*, 30, 87–97.

Yamamoto, K. & Oota, I. (1967). Fine structure of yolk globules in the oocyte of the zebrafish *Brachydanio rerio*. *Annotationes Zoologicae Japonenses*, 40, 20–7.

Yonge, C. & Hagedorn, H.H. (1977). Dynamics of vitellogenin uptake in *Aedes aegypti* as demonstrated by trypan blue. *Journal of Insect Physiology*, 23, 1199–203.

Yusoko, S.C. & Roth, T.F. (1976). Binding to specific receptors on oocyte plasma membranes by serum 'phosvitin–lipovitellin'. *Journal of Supramolecular Structure*, 4, 89–97.

Zamboni, L. (1970). Ultrastructure of mammalian oocytes and ova. *Biology of Reproduction* (Suppl.), 2, 44–63.

Plates

Plates 1 to 8 are between pp. 162 and 163.

Plate 1. (a) A low-magnification picture of the cortical ooplasm from a late stage eight oocyte of *Drosophila melanogaster*. Note the presence of several pits (*P*), coated vesicles (*CV*), tubules (t) and forming yolk spheres (*Y*). X 14 400. (b) A high-magnification picture of the cortical ooplasm from a late stage eight oocyte of *D. melanogaster* showing the irregular profile of the oolemma. A coated vesicle (*CV*) and two pits (*P*) are evident. X 90 000. (By courtesy of A.P. Mahowald.)

Plate 2. (a) Section through the cortical region of an oocyte of *Triturus* showing the formation of coated vesicles. The outer coat (*SC*) and the fibrous inner coat (fc) of the pit are clearly visible. X 121 000. (b) Seven pinocytic vesicles (*PV*) with a diameter of 45 nm inside a microvillus (mv) of a *Triturus* oocyte are apparently entering the cytoplasm of the oocyte. X 121 000. (From Spornitz & Kress, 1973; by courtesy of Springer-Verlag.)

Plate 3. (a) Electron micrograph of a 25–30 nm oocyte of the chicken *Gallus domesticus* in the final growth phase of yellow yolk deposition. The oolemma is very irregular and is indented by numerous coated pits (p). In the cortical ooplasm are coated vesicles (cv), 250–350 nm in diameter, and yolk spheres (y). X 5000. (b) A high-magnification picture of a 25–30 mm oocyte of *G. domesticus* showing coated vesicles with a lining of dense granules (single arrows) originating from the external surface of the oolemma and a projecting coat on the ooplasmic side (double arrows). X 58 100. (By courtesy of Perry, Gilbert & Evans.)

Plate 4. The cortical ooplasm from a stage nine ovarian chamber of *Drosophila melanogaster* fixed 30 minutes after injection of peroxidase. The tracer is clearly visible over three forming yolk spheres (y). Section unstained. X 13 300. (From Giorgi, 1975.)

Plate 5. An EM autoradiographic preparation of the cortical ooplasm from a late stage nine ovarian chamber of *Drosophila melanogaster* fixed six hours after injection of tritiated lysine. Note the preferential association of silver grains over the yolk spheres. X 13 300.

Plate 6. (a) Multivesicular bodies formed mainly through coalescence of pinocytic vesicles in the central ooplasm of a previtellogenic oocyte of *Rana*. X 62 300. (b) Pinocytic vesicles dissolve inside a multivesicular body in the central ooplasm of a previtellogenic oocyte of *Rana*. X 62 300. (c) An electron-dense centre with a homogeneous granular appearance is formed inside a multivesicular body in the central ooplasm of a previtellogenic oocyte of *Rana*. X 62 300. (d) Through continuous growth, the multivesicular bodies in the central ooplasm of a previtellogenic oocyte of *Rana* have formed into yolk precursors. X 49 800. (From Kress & Spornitz, 1972; by courtesy of Springer-Verlag.)

Plate 7. The cortical ooplasm from a stage 10 ovarian chamber of *Drosophila melanogaster* embedded in glycolmethacrylate and stained with phosphotungstic acid (PTA). Note that PTA staining is clearly visible over the associated body and the inner lining of the limiting membrane of the forming yolk sphere. Also shown are: Vm, vitelline membrane; mb, mainly dense body; Ab, associated body. X 19 900. (*b*) Part of a mature yolk sphere from the central ooplasm of a stage 10 ovarian chamber of *D. melanogaster*. Note the presence of several osmium zinc iodide (OZI) deposits in the cap-like region of the superficial layer (Sl). The associated body (Ab) itself is devoid of OZI deposits. X 13 300. (Part (*b*) from Giorgi & Jacob, 1977; by courtesy of Cambridge University Press.)

Plate 8. The central ooplasm from an early stage eight ovarian chamber of *Drosophila melanogaster*. The Golgi apparatus (G) contains heavy deposits of osmium zinc iodide (OZI) complex. The organelle on the right of the micrograph exhibits a mainly dense body (mb) embedded with an asymmetrical layer (Sl) which contains scattered OZI deposits. X 12 000. (From Giorgi & Jacob, 1977; by courtesy of Cambridge University Press.)

7

Adsorptive and passive pinocytic uptake

MARGARET K. PRATTEN, RUTH DUNCAN
& JOHN B. LLOYD

Introduction

Many different terms have been employed to describe different types
of endocytosis, and these are listed by Chapman-Andresen (1962) and Jacques
(1969), but in general endocytic phenomena fall into two broad categories,
phagocytosis and pinocytosis. Phagocytosis describes the ingestion of particulate
matter such as bacteria, latex beads and erythrocytes by specialised cells such
as macrophages and certain unicellular organisms, whereas the more universal
process of pinocytosis describes the engulfment of small droplets of extracellular fluid.

Sequence of events in pinocytosis

All types of pinocytosis show a common sequence of events (see
Fig. 1).

1. Internalisation of plasma membrane. This may be triggered by attach-
    ment of some substance to the plasma membrane or by some other
    mechanism.

2. Translocation. Once the pinosome 'pinches off' from the plasma
    membrane, it migrates towards the perinuclear region. During this
    time many fusion events may occur. Initially these may be pinosome—
    pinosome fusions but subsequently pinosome—lysosome fusions take
    place, producing a secondary lysosome compartment which can also
    participate in the fusion sequence. The exposure of the pinosome
    contents to lysosomal enzymes results in the catabolism of any
    degradable material.

3. Lysosomal regression. Small molecules produced as a result of degra-
    dation escape through the lysosomal membrane, whereas any large
    non-biodegradable material remains trapped in the secondary lysosome
    compartment. Certain types of cell, especially unicellular organisms,

have the ability to regurgitate material, but many mammalian cells accumulate material they cannot digest as residual bodies within the cell.

Mechanism of pinocytosis

During pinocytosis substances may be captured in two possible ways. They may be taken up in solution, so called fluid-phase pinocytosis, or alternatively they can be carried into the cell attached to the invaginating plasma membrane (Jacques, 1969). Thus substrates with no affinity for the membrane are taken up solely in solution and conversely substrates with membrane

Fig. 1. Membrane events during pinocytosis: a diagrammatic representation of the dynamic flow of membrane associated with the pinocytic process.

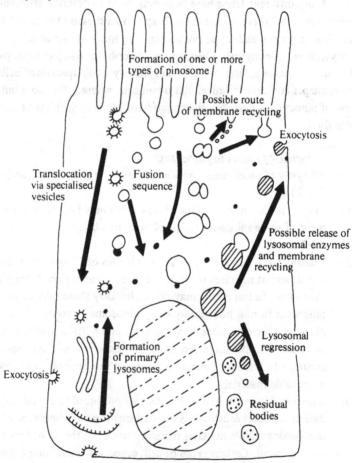

affinity can be taken up either entirely membrane-bound, with total exclusion
of fluid, or by a combination of the fluid and adsorptive modes. Fig. 2 demonstrates these three mechanisms of pinocytic capture. Electron microscopy
shows that pinosome morphology can be variable, although the majority of
intracellular vesicles are spherical. In many cells the vesicles observed near
the cell membrane are of smaller diameter than those in the perinuclear region
and this gradation in size has been interpreted as being the consequence of
many fusion events. The smallest vesicles have a diameter of approximately
70 nm and the larger vesicles, including secondary lysosomes, have a diameter
of between 0.5 and 2.0 μm. Some of the small pinocytic vesicles are electron-
dense, while others are electron-lucent and also have a 'fuzzy coating'. The

Fig. 2. Mechanisms of pinocytic capture: the three theoretical modes
of pinocytosis which determine the substrate selectivity of the
process.

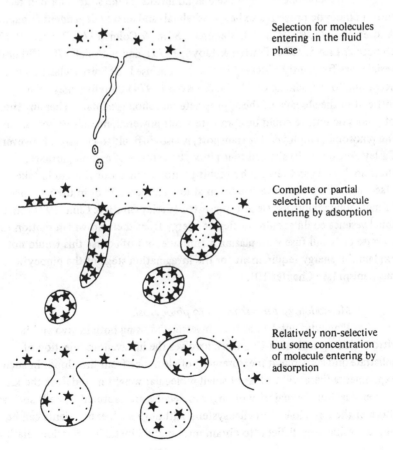

Selection for molecules
entering in the fluid
phase

Complete or partial
selection for molecule
entering by adsorption

Relatively non-selective
but some concentration
of molecule entering by
adsorption

latter have been named 'coated vesicles' and it has been claimed (see Chapter
12) that their filamentous encapsulation is polysaccharide in nature. It has
been suggested (see Chapter 4) that this type of vesicle is responsible for the
selective uptake of immunologically important molecules and that the
vesicles' structure prevents fusion with lysosomes, affording protection to
their contents and possibly facilitating transmission across the cell. Other
chapters of this volume are devoted to discussion of the functional significance
of the specialised morphology of this particular category of vesicle.

Further classification of pinocytic mechanisms, according to their metabolic
and cytoskeletal requirements as well as vesicle size, has been proposed by
Allison & Davies (1974). Macropinocytosis is defined as requiring metabolic
energy and microfilament–microtubule function and involves vesicles of
diameter 0.3–2.0 μm. Micropinocytosis, in contrast, is defined as involving
smaller vesicles (diameter 70–100 nm) and being able to continue in the
presence of metabolic and cytoskeletal inhibitors. Some studies have in fact
shown pinocytic processes to be cytoskeletal- and energy-dependent (Chapman-
Andresen, 1977; Ryser, 1970; Steinman, Silver & Cohn, 1974; Bowers, 1977;
Duncan & Lloyd, 1978; Pratten & Lloyd, 1979), and yet small (70–100 nm)
vesicles are frequently observed in the systems used. Definitive classification
according to the scheme of Allison & Davies (1974) therefore may seem
difficult in the absence of the appropriate morphological data. That any mode
of pinocytic uptake could be completely self-powered, no matter how small
the pinosomes employed for transport, is also difficult to envisage. However
Casley-Smith (1969) showed that pinocytic capture of carbon particles,
thorium dioxide and ferritin by rabbit peritoneal macrophages could take
place in small pinocytic vesicles even at low temperature or in the presence
of metabolic inhibitors. He also suggested (Casley-Smith & Chin, 1971) that
small vesicles could acquire sufficient energy from their Brownian motion to
traverse cells and fuse with plasma membrane, but of course this would not
explain the energy requirement for the invagination stage of the pinocytic
mechanism (see Chapter 10).

Methodology for research into pinocytosis

Pinocytic uptake has been investigated using both in-vivo and in-
vitro techniques. Most in-vivo work involves the intravenous injection of
substrate and subsequent measurement of clearance from the blood. In such
experiments there will be loss of small molecular weight material via the kidney
glomerulus, but the majority of larger molecules are captured by cells such as
those of the reticulo-endothelial system. Although such experiments can be
of great value, it is difficult to obtain meaningful kinetic information relating

to the uptake of specific substrates by a defined cell type, owing to continuous decline in the concentration of substrate in the blood, the large variety of cell types to which the substrate is exposed and the different locations of cells of a single type in relation to the circulatory system. The environment of the substrate is also such that it is potentially subject to uncontrolled modifications such as metabolism and interactions with blood proteins.

An in-vitro culture system can be more precisely defined, thus facilitating the accurate quantitation of both fluid-phase and adsorptive pinocytosis of a specific substrate by a given cell type. Of course, one has to be very careful to minimize changes in the physiological and physical environment of the cell, or extrapolation of data to the in-vivo situation may be totally invalid. Two main approaches have been used to quantitate pinocytosis *in vitro*.

(1) *Morphological methods*. The number of pinocytic vacuoles within a cell, the number of particles of substrate within a cell or the number of cells within a population that have taken up substrate, visualised using light or electron microscopy, have all been used as a measure of pinocytic activity. Although there are advantages in being able to observe the morphological integrity of pinocytically active cells, the interpretation of morphometric data is not always simple. It is difficult to assess whether membrane-bounded profiles are intracellular vesicles or transverse sections of canaliculi that are really extracellular. Only by studying serial sections or focussing through a cell can one of these alternatives be confirmed. The extrapolation from vesicle number to pinocytic rate also has its pitfalls. The number of vesicles present within a cell at any one time depends not only on the rate of formation of those vesicles but also on their rate of disappearance and their rate of fusion. Another determinant of the number of vesicles visible within a cell is the resolution of the microscope used and this could be particularly important if different mechanisms of pinocytosis involved vesicles in discrete size ranges.

(2) *Direct measurement of substrate capture*. Many techniques employ a radiolabelled substrate or exploit the chemical detectability of a macromolecule. The use of this type of marker is usually a more efficient method of quantitation but these techniques are not without problems. One must be sure that the probe employed is in fact penetrating the cell by pinocytosis and not just progressively binding to the exterior of the cell or lodging within intracellular spaces. In some cases, such as when using an electron-dense marker, this question can be easily resolved by parallel microscopical examination of the system. Certain criteria have been suggested to confirm that accumulation is by pinocytic uptake. These include continuous progressive accumulation of substrate over long time periods (binding would be a rapidly saturable process), inhibition of uptake at low temperature or in the presence of metabolic

inhibitors (binding is not an energy-consuming process), and a low rate of release of substrate from the system when the cells are preincubated in the presence of substrate and then reappearance of substrate in the culture medium is subsequently monitored (externally disposed substrate would be rapidly exchangeable). Subcellular fractionation can also confirm the intracellular location of marker macromolecules.

Studies on the uptake of degradable substances have the added complication that interiorisation is followed by lysosomal digestion. Consequently, to calculate the pinocytic uptake of such markers, it is necessary to sum both the tissue level of the substance and the release of digestion products from the cells. Most studies have failed to take account of possible catabolism, but nevertheless the initial rate of uptake of a degradable substrate probably approximates in most cases to the true rate of capture. A relatively rapid plateauing of substrate accumulation is one indication that the cell may be metabolising the substrate, although this is not the only possible explanation for such data.

Theoretically the ideal method for quantitation of pinocytosis would attempt to correlate biochemical and morphological observations. This correlation is particularly important where potential modifiers are introduced into the pinocytic system, since it is necessary to monitor both physiological function and cellular integrity simultaneously.

The concept of Endocytic Index

It is of great importance when studying pinocytosis that the comparison of results should be facilitated by the use of units that take account of the effect of inter-experimental variables. Thus, expressing the uptake of a radiolabelled substrate in counts per minute is only of use if all experiments are performed with precisely the same concentration of substrate at the same stage of radioactive decay. Likewise, most attempts to relate the pinocytic activities of different tissues by using data reported in the literature are defeated by the widespread use of non-comparable units and by the failure to provide sufficient ancillary information to allow further calculations.

The concept of Endocytic Index as proposed by Williams, Kidston, Beck & Lloyd (1975a) would seem to answer these criticisms. Endocytic Index is defined as the volume of culture medium (μl) whose contained substrate is captured per milligram of cell protein (or per 10^6 cells: Pratten, Williams & Lloyd, 1977) per hour. Expressing uptake in this way has a number of advantages. It eliminates variability arising within an experiment from variations in the amount of radioactivity or chemical marker added to each culture vessel, and normalises for the effects of variation in specific radioactivity or specific enzymic activity of substrates and for the amount of tissue or number of cells present. Direct numerical comparison of rates of uptake, both in experiments with the same radioactive substrate or chemical marker and in experiments with different substrates, thus becomes possible. Uptake by different cell types can also be compared, assuming the culture conditions employed make such comparisons meaningful. Although the units of Endocytic Index are microlitres per unit tissue per hour, this does not imply that an equivalent volume of fluid is actually ingested. Obviously, in the case of a substrate that enters a cell entirely in the fluid phase, the volume of fluid taken up by the cell is the same as that indicated by the Endocytic Index. However, in the case of a substrate that enters mainly adsorbed to the membrane, the actual volume of fluid ingested will be far smaller than the calculated Endocytic Index. In such cases it is often quite helpful to calculate not only the Endocytic Index but also uptake in terms of nanograms of substrate internalised per unit of tissue per unit time. In this way it is possible to estimate the concentration of a pinocytosed substrate within the cell.

The units of Endocytic Index are sufficiently versatile to describe the uptake of degradable substrates. In this case both the tissue level of substrate and the release of digestion products can be estimated in microlitres per milligram protein or 10^6 cells, the values summed and a net rate of uptake derived (Williams, Kidston, Beck & Lloyd, 1975b; Moore, Williams & Lloyd, 1977). Very recently a new technique has been devised (Pittman & Steinberg, 1978) whereby the rate of uptake of a degradable substrate can be assessed by covalently coupling a small non-degradable molecule to the substrate and following the rate of accumulation of the secondary marker. The feasibility of the method was tested using serum albumin doubly labelled with ^{125}I and [14C]sucrose, uptake by normal fibroblasts in terms of ^{14}C accumulated in the cells being correlated with uptake measured in a more conventional way.

Differentiation between adsorptive and fluid-phase pinocytosis

The following criteria may be used to establish the mode of uptake of a substrate as adsorptive.

(1) In a carefully defined system where the uptake of a number of substrates has been measured, the magnitude of the Endocytic Index can give a good indication of the mechanism of capture. If it is found that more than one substrate enters the system at the same (low) rate, thus giving a minimal Endocytic Index which is less than that obtained for any other substrate, it may be postulated that the uptake of these substrates is entirely fluid-phase. However, deviation from the established fluid-phase rate could have several explanations. A higher Endocytic Index is most likely to be a function of the

degree of membrane adsorption, but would also result if the substrate were stimulating pinosome formation at the plasma membrane. Likewise, a lower Endocytic Index would suggest that the assumed fluid-phase rate was invalid, unless the substrate could inhibit the rate of membrane invagination. These alternatives could be distinguished by assessing the effect of such substrates on the uptake of an accepted fluid-phase marker, assuming that there is no vesicular discrimination against certain substrates.

(2) In the case of a substrate that enters entirely in the fluid phase, an increase in concentration in the extracellular medium is without effect on uptake expressed in terms of microlitres per unit of tissue per hour, although uptake expressed in nanograms increases linearly with concentration. A substrate entering by adsorption, however, does so in a concentration-dependent manner: as the concentration increases and the sites to which it binds become saturated, the Endocytic Index will decrease. As concentration rises, the binding component of uptake becomes less significant and the Endocytic Index will tend to the fluid-phase value.

(3) When cells ingesting a substrate that enters to a large extent in a membrane-bound fashion are subsequently incubated in substrate-free medium, any externally bound material will tend to be released into the medium. Any internalised material can only be released by an exocytic mechanism, and so a fluid-phase marker is only released by this route. Thus if the tissue-association of a substrate has a substantial readily reversible component, it may be indicative of an adsorptive mechanism. Similarly, trypsin may be used to remove externally disposed material. The advantage of such treatment is that more tightly bound material can be released, but the harshness of the treatment requires careful control to eliminate errors caused by cell damage.

(4) Because the binding of substrate to the cell surface does not require a metabolic energy supply or the cytoskeletal system, inhibitors of these should be without effect on the binding component of the association of substrate with cells. However, as mentioned previously, some pinocytic processes are also thought to be independent of energy and microtubules and microfilaments, so a persistence of 'uptake' can be interpreted either as a continuation of pinocytosis or as an indication of an adsorptive mechanism.

One or more of these criteria has often been used to discriminate between fluid-phase and adsorptive pinocytosis. Ideally all four criteria should be used.

Experimental systems used to quantitate pinocytosis

Three main types of cell have been used to study endocytosis. The large unicellular protozoans such as *Amoeba* and *Tetrahymena* have been investigated extensively and these cells are particularly interesting, as their normal feeding cycle involves the capture of food particles by a phagocytic mechanism, although they are also able to pinocytose. A second category comprises mammalian cells in culture, chiefly mononuclear phagocytes and fibroblasts. More recently work has begun which investigates pinocytosis of mammalian epithelial cells. Owing to their fixed position and microvillous apical plasma membrane, these cells can only engulf small particles of a more limited size range and they are probably restricted to pinocytosis. It was not until the advent of electron microscopy that pinocytic vesicles were seen within these cells and this explains why quantitation in these systems is a more recent innovation.

Unicellular animals

The periodicity of pinocytosis in many unicellular organisms restricts the quantitation of substrate capture to relatively short time intervals. Pinocytosis in *Amoeba proteus* is induced by solutions such as 0.125 M sodium chloride, with maximum channel formation after 10–15 minutes and almost complete cessation of channel formation after 30 minutes (Chapman-Andresen, 1965). Similarly the high initial rate of uptake of peptone by *Tetrahymena pyriformis* drops markedly after 30 minutes to an hour (Ricketts, 1971). By means of a dual-label technique, the uptake of ^{131}I-labelled albumin and [14C]glucose has been studied in *Amoeba proteus* (Chapman-Andresen & Holter, 1964). The rate of uptake of protein was some 10 times greater than that of glucose (which is known to be a non-inducer and is only taken up extremely slowly by amoebae that do not pinocytose), and this so-called 'enrichment factor' for the protein component was attributed to the uptake of protein adsorbed to hair-like extensions of the plasmalemma. Negligible amounts of glucose were adsorbed to the cell and so it was concluded that glucose entered only in the fluid phase. Pinocytosis in freshwater amoebae has recently been reviewed by Chapman-Andresen (1977).

The smaller *Acanthamoeba castellanii* showed different pinocytic properties. Here ^{131}I-labelled albumin, [3H]inulin, [14C]-L-leucine and [14C]-D-glucose were used as pinocytic markers (Bowers & Olszewski, 1972) and were found to be captured at a constant rate for the duration of the experiment, the rates of capture being similar when calculated as volume of fluid ingested per unit time (approximately 2 μl/10^6 cells/hour). This result and the unsaturability of the uptake mechanism for albumin and leucine led to the conclusion that all the tracers are taken up in the liquid phase. Unlike the two large amoebae, *Chaos chaos* and *Amoeba proteus, Acanthamoeba castellanii* does not have a demonstrable surface coat and furthermore pinocytosis appears to be a continuous process. These differences might explain why no evidence was found for surface binding of molecules by this organism.

Ricketts & Rappit (1975) examined the endocytic uptake of three radio-labelled substrates into the digestive food vacuoles of *Tetrahymena pyriformis*, and concluded that ^{125}I-labelled albumin, ^{125}I-labelled polyvinylpyrrolidone ([^{125}I] PVP) and [^{3}H] - or [^{14}C] sucrose were all concentrated within this compartment to as much as 20–50 times the concentration in the extracellular medium. Substrate capture was measured in terms of intravacuolar concentrations by estimating the total volume of the vacuolar compartment, using phase-contrast microscopy to determine the number and size of vacuoles within the cell. The authors admit that their estimated concentrations may be too high as the calculation makes no allowance for the uptake, which may well be considerable, of substrate into smaller unresolved pinocytic vesicles. In any case intravacuolar concentration of substrate gives little indication of the specificity of the capture mechanism as it depends on the vacuole volume, which can decrease greatly during passage through the cell.

Mammalian cells in culture

Mononuclear phagocytes. Although these cells are best known for their phagocytic activity, interest in the mechanism and pharmacology of pinocytosis has led to many attempts to quantify pinocytosis by macrophages in culture. Morphometric methods have been used (Cohn, 1966; Cohn & Parks, 1967*a*, *b*, *c*; Westwood & Longstaff, 1976), but these are of limited value since the number of vacuoles visible within the cell depends not only on their rate of formation, but also on their longevity and size. Recently it was reported that only 16 % of the pinosomes within mouse peritoneal macrophages are detectable with the phase-contrast microscope (Steinman, Brodie & Cohn, 1976).

Radiolabelled or enzymic markers afford a method for quantifying uptake more directly. Cohn and co-workers have examined pinocytosis by mouse peritoneal macrophages using a wide range of probes, including ^{131}I-labelled human serum albumin ([^{131}I] HSA) (Ehrenreich & Cohn, 1967), [^{3}H] haemo-globin (Ehrenreich & Cohn, 1968), [^{14}C] sucrose (Cohn & Ehrenreich, 1969), horseradish peroxidase (HRP) and [^{125}I] HRP (Steinman & Cohn, 1972*a*; Edelson, Zweibel & Cohn, 1975), HRP:anti-HRP aggregates (Steinman & Cohn, 1972*b*), [^{3}H] double-stranded RNA and [^{3}H] poly-L-lysine (Seljelid, Silverstein & Cohn, 1973). They demonstrated the ability of macrophages to digest endocytosed protein molecules, but only when examining the effect of concanavalin A on uptake of albumin and HRP (Edelson & Cohn, 1974*a*, *b*) was the intracellular accumulation of protein and degradative loss summed to give a net uptake. An Endocytic Index of 0.3 μl/10^6 cells/hour may be deduced from the initial rate for [^{3}H] haemoglobin uptake, which when compared with an Endocytic Index of 0.045 μl/10^6/hour for HRP, believed to be a fluid-phase marker, suggests that haemoglobin is pinocytosed to a large extent in close association with the plasma membrane. HRP:anti-HRP aggregates also bind avidly to the macrophage membrane utilising a trypsin-insensitive Fc receptor, and this results in a rate of capture some 4000-fold greater than that of soluble HRP (Steinman & Cohn, 1972*b*). The size of such aggregates may also be an important factor in determining their rate of capture and could initiate a phagocytic rather than pinocytic response.

Mehl & Lagunoff (1975) studied the uptake of aggregated albumin by rat peritoneal macrophages cultured *in vitro,* measuring the number of binding sites and affinity for this substrate, using Scatchard plots. They found that there were fewer binding sites for large aggregates than for the monomer, but that the binding capacity for the aggregates was greater by a factor of two.

Colloidal [^{198}Au] gold has also been used as a pinocytic marker, and Davies, Allison & Haswell (1973) quantified its uptake by mouse peritoneal macrophages. The rate of uptake varied with concentration of newborn calf serum in the culture medium but was in the range 3.9–8.9 $\mu l/10^6$ cells/hour. They assumed that uptake of colloidal gold was entirely in the fluid phase, although they were aware that their calculated rate of uptake of liquid exceeded values obtained for other markers for endocytosis. Pratten *et al.* (1977) also used colloidal [^{198}Au] gold as a tracer, concurrently investigating the rate of capture of [^{125}I] PVP, [^{14}C] sucrose and ^{125}I-labelled denatured bovine serum albumin ([^{125}I] dBSA). The measured Endocytic Indices for [^{125}I] PVP and [^{14}C] sucrose using rat peritoneal macrophages were two orders of magnitude lower than the value obtained for colloidal [^{198}Au] gold, suggesting that colloidal [^{198}Au] gold binds strongly to the plasma membrane. The other plausible explanation of these data, that colloidal [^{198}Au] gold is a powerful stimulator of pinosome formation, was eliminated because colloidal gold was unable to stimulate the uptake of [^{125}I] PVP. Although [^{125}I] dBSA was captured at a lower rate than colloidal [^{198}Au] gold, its Endocytic Index was markedly higher than that of [^{125}I] PVP and [^{14}C] sucrose, indicating that adsorption plays a major role in its pinocytic uptake. The values reported for [^{125}I] dBSA are markedly similar to those that can be calculated from the data of Ehrenreich & Cohn (1968) for [^{3}H] haemoglobin.

Liver cells. Early studies on endocytosis by liver were confined to extrapolations from data obtained from blood clearance experiments (e.g. Straus, 1964, 1971; Ashwell & Morell, 1971; Normann, 1973). It was assumed that to a large extent the disappearance of a substance from the blood stream could be accounted for by its uptake into liver cells, especially those cells of the mononuclear phagocyte series. The type of cell involved in the capture of

HRP was extensively investigated by Creemers & Jacques (1971). They found that HRP was taken up by endothelial cells, hepatocytes and non-parenchymal cells and established that HRP entered the lysosomal compartment after internalisation. Likewise, Regoeczi (1976) has used blood clearance of [^{131}I]-PVP as a measure of reticulo-endothelial activity, and has demonstrated that quite a large proportion of the administered dose is found in the liver. An attempt was made to quantitate pinocytosis by liver cells *in vitro* using Chang-strain adult human liver cells incubated in the presence of [^{3}H]sucrose (Wagner, Rosenberg & Estensen, 1971). This substrate was deemed to be a suitable marker for fluid-phase endocytosis, since it was taken up in a linear fashion over four hours and since increasing the concentration of sucrose did not stimulate or inhibit its uptake except at concentrations that caused excessive vacuolation. It is possible to calculate an Endocytic Index of 13.9 μl/10^6 cells/hour.

More recently, sophisticated techniques have emerged which not only permit the culture of liver cells under controlled conditions *in vitro*, but also allow their separation into parenchymal and non-parenchymal cells (Berg & Boman, 1973). Nilsson & Berg (1977) used this technique to measure the uptake and degradation of formaldehyde-treated [^{125}I]HSA. They found that this substrate was taken up rapidly by non-parenchymal cells (the data yield an Endocytic Index of 100.0 μl/10^6 cells/hour), but relatively slowly by parenchymal cells, the Endocytic Index here being 8.3 μl/10^6 cells/hour. The degradation of the substrate to produce TCA-soluble radioactivity occurred between 20 and 30 minutes after addition of substrate. In contrast it was found that parenchymal cells took up asialofetuin preferentially (Tolleshaug, Berg, Nilsson & Norum, 1977), but the Endocytic Index was so high (we calculate, approximately 320 μl/10^6 cells/hour) that the medium must have become depleted of substrate very rapidly. Non-parenchymal cells, however, have an estimated Endocytic Index of only 4.4 μl/10^6 cells/hour for this substrate. Other studies on non-parenchymal cells by Munthe-Kaas (1977, and personal communication) show that the uptake of colloidal [^{198}Au] gold (Endocytic Index approximately 2 μl/10^6 cells/hour) and [^{3}H]sucrose (Endocytic Index approximately 0.01 μl/10^6 cells/hour) is very similar to the uptake of the same substrates by rat peritoneal macrophages (Pratten *et al.*, 1977), indicating the similar membrane–substrate affinities of these two closely related cell types. It seems probable that [^{3}H]sucrose is a fluid-phase marker in non-parenchymal cells, but that colloidal gold, albumin and asialo-fetuin enter by adsorptive pinocytosis. The latter two substrates also enter parenchymal cells by an adsorptive mechanism, although there is clearly a different range of specific cell–substrate affinities.

Furbish and co-workers have combined the techniques of blood clearance and in-vitro cell culture to study the uptake of lysosomal hydrolases by liver cells (Furbish, Steer & Barranger, 1978). They injected glucocerebrosidase in both its native and desialated form into rats and then separated the liver into parenchymal and non-parenchymal cells. They found that the former removed a greater quantity of both native and asialo-enzyme than the non-parenchymal cells and also that they took up asialoglucocerebrosidase preferentially. This indicated yet again that parenchymal cells have a receptor for galactose residues which is lacking in non-parenchymal cells, and thus that the characteristics of adsorptive pinocytosis in the two cell types are different.

Fibroblasts. Fibroblasts used for pinocytic measurements are normally cultured in monolayer, and both normal and transformed cells have been used. Steinman *et al.* (1974) showed that L-cells maintained at high density after reaching confluence pinocytosed more rapidly (up to four times faster) than growing preconfluent cells. Using 1 mg/ml HRP as a solute marker they calculated an uptake of $32-35$ ng/10^6 cells/hour, a concentration that can be converted into an Endocytic Index of 0.035 μl/10^6 cells/hour. They point out that uptake in terms of ng/mg protein/hour for fibroblasts is approximately tenfold lower than the pinocytic activity of unstimulated mouse peritoneal macrophages, but in fact when calculated on a per cell basis the values are very similar owing to the fibroblasts studied containing 10 times as much protein per cell. Others have used either [^{3}H] - or [^{14}C] sucrose as fluid-phase markers. Miller, Weinstein & Steinberg (1977) cultured normal skin fibroblasts and estimated an Endocytic Index of 0.19 μl/mg protein/hour, which is equivalent to 0.086 μl/10^6 cells/hour and remarkably similar to the value quoted above for HRP. Becker & Ashwood-Smith (1973) used [^{3}H] sucrose as a probe and showed that serum, PVP, ATP, insulin and cyclic AMP (cAMP) all had no effect on fluid-phase pinocytosis.

Proteins, particularly low-density (LDL) and high-density (HDL) lipoproteins, lysosomal enzymes, mucopolysaccharides and ferritin have all been used to investigate adsorptive pinocytosis. Normal human fibroblasts regulate their cholesterol content by controlling the rate of uptake of LDL and extensive studies have been carried out to investigate the nature of the LDL receptor (Goldstein & Brown, 1976, 1977). This work is described in detail in Chapter 12. A recent study on the kinetics of binding, internalisation and degradation of ^{125}I-labelled HDL showed only slightly lower binding, on a molar basis, than for ^{125}I-labelled LDL, but the rates of internalisation and degradation of ^{125}I-labelled HDL were considerably lower. Although HDL does bind to the membrane, it is suggested that HDL uptake could almost completely be accounted for by uptake of medium and invagination of surface membrane, whereas the uptake of LDL is 10-fold greater than this, implying concentration on the plasma membrane by some specific receptor site, which is internalised in a LDL-specific class of vesicle that forms more rapidly than other pinosomes. Since LDL had no stimulatory effect on the uptake of [^{14}C] sucrose it was postulated that LDL is internalised by some mechanism that does not involve the uptake of extracellular fluid, unless the LDL-binding sites represent such a small proportion of the cell surface area that any increment in fluid uptake is undetectable.

The incorporation of ^{35}S-labelled chondroitin sulphate into Chinese hamster cells displayed kinetics that indicate that surface adsorption is involved (Saito & Uzman, 1971). Uptake was depressed by 10 % calf serum and showed characteristic saturation kinetics, with Endocytic Indices of, we estimate, between 0.15 and 3.75 μl/mg protein/hour. Ryser & Hancock (1965) measured the uptake, over periods from 30 seconds to two hours, of ^{131}I-labelled albumin, fluorescein-labelled histones and basic polyamino acids by sarcoma-180 cells. They found that the basic proteins were taken up some 3000 times faster than serum albumin, but unfortunately no estimate of intracellular degradation was made in this study. The differences in uptake rate may be partially explained by different catabolic rates, but the discrepancies are so large that it seems likely that certain proteins are captured selectively, probably due to high membrane affinity. Several basic compounds, such as poly-L-ornithine, poly-L-lysine and DEAE-dextran, appeared to stimulate the uptake of serum albumin, and later work (Ryser, 1967) showed a correlation between stimulation and molecular weight of the polymer and it was suggested that multiple attachment of a large basic polymer to the membrane components results in conformational changes that may induce pinocytosis. Polynucleotides were shown to have similar effects when used as large aggregates (Ryser, Termini & Barnes, 1976). Recently it has been shown (Shen & Ryser, 1978) that conjugation of low molecular weight fragments of poly-L-lysine to radiolabelled HSA and HRP enhances their pinocytosis by mouse fibroblasts approximately 11-fold and 200-fold respectively, despite the fact that addition of comparable amounts of free poly-L-lysine had no effect.

Petitpierre-Gabathuler & Ryser (1975) used soluble and aggregated ferritin to investigate endocytosis in sarcoma-180 cells and found that they were taken up differently, the larger aggregates being captured more rapidly and by a mechanism that is susceptible to metabolic inhibition. It was inferred that small aggregates of ferritin (0.1 μm and less), together with serum albumin, are taken up by pinocytosis, whereas the larger aggregates are phagocytosed, the criterion proposed for the distinction being the energy requirement.

The hypothesis that hydrolytic enzymes may be transferred to lysosomes via a pathway that involves secretion and recapture by a receptor-mediated endocytosis (Hickman & Neufeld, 1972) has led to exhaustive studies investigating pinocytosis of various lysosomal enzymes by fibroblasts (Hickman & Neufeld, 1972; Hieber *et al.*, 1977; Von Figura, Kresse, Meinhard & Holtfrerich, 1978). There is now clear evidence that a specific receptor for certain enzymes is present (Kaplan, Achord & Sly, 1977; Kaplan, Fischer, Achord & Sly, 1977) but the physiological significance of the apparent specificity is still unclear, as an obligatory pinocytic step in the translocation of lysosomal enzymes has not been confirmed. From the data of Von Figura *et al.* (1978) on normal skin fibroblasts, it is possible to calculate Endocytic Indices for the uptake of β-N-acetylglucosaminidase and β-glucuronidase of 5.8 μl/mg protein/ hour and 27.0 μl/mg protein/hour respectively. These relatively high values, compared with those of the fluid-phase markers discussed earlier, are concordant with uptake by an adsorptive mechanism.

Blood vessels. Modification of the pinocytic rate of smooth muscle cells may occur in the diseased state of arterial walls known as atherosclerosis, where lesions are formed by the extensive deposition of lipid. Recently, attempts have been made to investigate the pinocytic activity in these cells using HRP, [^{14}C]sucrose and [^{125}I]PVP as probes (Leake & Bowyer, 1977; Davies & Ross, 1978). [^{125}I]PVP was captured with an Endocytic Index of approximately 0.1 μl/mg protein/hour by pig aortic smooth muscle cells, and corresponding values for [^{14}C]sucrose and HRP in monkey cells were 0.06 and 0.055 μl/10^6 cells/hour. It was suggested that these substrates are all fluid-phase markers, so possibly the rate of vesicle formation is very similar in cells derived from these two different species. Factors derived from whole blood serum and from platelets apparently stimulate the rate of vesicle formation by monkey aortic smooth muscle cells by approximately twofold (Davies & Ross, 1978).

Trophoblast. Interest in the nutritive and digestive roles of the human placenta led to studies on trophoblast cells cultured in monolayer (Contractor & Krakauer, 1976). Uptake and digestion of ^{125}I-labelled haemoglobin were studied in the presence and absence of serum. By summing cell-associated and TCA-soluble radioactivity it is possible to calculate an Endocytic Index of 4.30 μl/10^6 cells/hour for cells cultured in the absence of serum, but in the presence of serum there was no intracellular digestion and the Endocytic Index was 3.68 μl/10^6 cells/hour.

Vertebrate organ cultures

Kidney. The tubular endothelial cells of the kidney are highly endocytic. It is, however, a complicated organ to study because of the difficulty of obtaining a homogeneous sample of cells with which to quantitate pinocytosis in a meaningful way.

Miller and his co-workers studied renal pinocytosis using kidney slices (Miller, Hale & Alexander, 1965). They used HRP as an endocytic marker and measured its uptake using histochemical techniques. The process was found to be energy-dependent and they suggested that an ouabain-sensitive ATPase could be part of the control mechanism for the uptake of proteins. The technique they employed to quantitate uptake was rather subjective, and neither could it distinguish adequately between different modes of substrate capture.

Others have used radiolabelled substrates. Maunsbach (1966) injected ^{125}I-labelled rat albumin into the proximal tubules of rat kidney and after various time intervals examined the tubules using electron microscopic auto-radiography. After 30 minutes most of the label was found in cytoplasmic bodies and there was no evidence that label crossed the wall of the proximal tubule by passing between cells or that vacuoles containing ingested albumin could exocytose their contents into the peritubular space (i.e. no evidence for transport of proteins across the cell). Intravenously injected ^{125}I-labelled bovine pancreatic ribonuclease A is specifically removed from the circulation by the kidneys, 55 % being present in this location seven minutes after injection, whilst the liver and gut each absorbed less than 50 % (Davidson, Hughes & Barnwell, 1971). Subsequent fractionation of the tissue showed the presence of label in pinocytic vesicles and lysosomes, and degradation was demonstrated because incubation of the lysosomal fraction at pH 5.5 caused the release of phosphotungstic-acid-soluble radioactivity.

Christensen (1976) has studied the uptake and digestion of proteins by proximal tubules, but in a slightly more quantitative way. He injected ^{125}I-labelled cytochrome *c in vivo* as a tracer and subsequently studied its distribution in the kidney using autoradiography. At first the label was seen adsorbed to the brush border membrane, but after 30 minutes approximately 40 % had transferred to the lysosomal system of the proximal tubule cells. He also measured the catabolism of captured protein by incubating renal cortex slices. Preloaded renal tissue was found to release TCA-soluble material which was identified as monoiodo- and diiodotyrosine, indicating that the digestion of cytochrome *c* is intralysosomal (any plasma membrane hydrolysis of the sub-strate would be expected to yield a range of high molecular weight fragments). The morphological data indicate that uptake is probably by adsorptive pinocytosis.

A different approach to the study of pinocytosis in kidney was taken by Bode, Baumann & Kinne (1976). In order to investigate the movement of pinocytic vesicles across rat kidney cortex cells they isolated different membrane types and analysed them for enzyme activities and protein, glycoprotein and lipid composition. In comparison with brush border microvilli, the pinocytic membranes were found to have a different composition with a high concentration of acid phospholipids and glycoproteins, both of which may be of importance in non-specific binding sites during adsorptive pinocytosis. This suggested difference in composition of pinocytic vesicle membrane is not, however, without difficulties when viewed in the light of the origin of the vesicles and routes of membrane recycling which have been suggested for other types of endocytic process (Steinman *et al.*, 1976; Duncan & Pratten, 1977).

Intestine. Interest in uptake of macromolecules by the gut probably originated from the early observations (e.g. Halliday & Kekwick, 1960) that one of the primary routes for passive transfer of immunity in many species is by way of the neonatal gut. It was thought for a long time that the ability to take up IgG and other immunologically important molecules was lost completely after 'closure', but it is now becoming apparent that the adult intestinal wall is also pinocytically active. Much of the work that has been published in this area does not attempt to quantitate pinocytosis or indeed study the mode of capture of the substrate employed. It has however been suggested that the uptake of IgG into neonatal gut is by a route that avoids the lysosomal digestion of these molecules and may involve a specialised form of adsorptive pinocytosis by way of coated vesicles. This aspect of gut pinocytosis is dealt with at greater length in Chapter 4 and therefore the discussion here will be confined to experiments that study the more usual pinocytic process.

Early studies by Isselbacher's group were mainly based on the histochemical visualisation of HRP (Cornell, Walker & Isselbacher, 1971). Using everted gut sacs from both adult and neonatal rats it was found that HRP was taken up into intracellular vesicles by pinocytosis and that the rate of uptake increased proportionally with increase in concentration and was energy-dependent. Since there was no evidence of saturability of the process it is likely that HRP uptake is fluid-phase in the gut, as in other tissues so far studied. Subsequently Walker, Cornell, Davenport & Isselbacher (1972) measured uptake of HRP into adult and neonatal rat gut sacs by assaying enzymic activity, and expressed uptake into the serosal fluid as μmol enzyme /mg protein/hour, but did not state the original concentration of enzyme in the medium, thus making it impossible to calculate an Endocytic Index and compare the value with that of other tissues. They found absorption to be higher in adult than neonatal rat and that the jejunum was the most pinocytically active part of the intestine. This was borne out by data obtained using ^{125}I-labelled HRP, although the stated presence of intestinal iodinases make the interpretation of these results somewhat difficult.

In 1971 Clarke & Hardy studied uptake of [^{125}I] PVP by neonatal rat gut. They administered the substrate orally and measured radioactivity taken up by the cells of the intestinal wall. Uptake was found to be dose-dependent at low concentrations but some saturation effect was observed at oral doses above 1 mg. They also observed that the uptake of [^{125}I] PVP was dramatically decreased once the rats reached 18 days of age, and suggest that this correlates with the time at which maternofoetal transmission of immunoglobulins ceases to be of importance. This implies that the transfer of immunity by this route is closely linked to the pinocytic process by which a fluid-phase marker such as PVP enters the cell. It is possible that a mechanism that permits entry of PVP into the cells is present in both adults and neonatal rats and that the difference between the two age-groups is the cessation at around 18 days of a completely different process which in the neonate permits entry and passage across the cell of both immunoglobulins and PVP.

The recent work by Bridges & Woodley (1979) using adult rat small intestinal sacs cultured *in vitro* certainly indicates that there is uptake of [^{125}I] PVP by the gut and also that it is transferred into the serosal fluid. This is presumably a pinocytic mechanism which persists in adulthood. Uptake was found to be linear with time and the Endocytic Index obtained for [^{125}I] PVP cultured in medium 199 containing 10 % calf serum was 0.79 μl/mg protein/ hour into the tissue and 0.17 μl/mg protein/hour into the serosal fluid. ATP was found to enhance this uptake, especially the transfer of the macromolecules across the tissue. Further studies (Bridges, Millard & Woodley, 1978) indicate that the uptake of [^{125}I] PVP can be greatly enhanced by liposomal entrapment. This study of the mechanism of transfer of liposomally entrapped material across the gut stemmed from many earlier observations that liposomally entrapped insulin administered orally appeared to bring about changes in blood glucose levels (Dapergolas & Gregoriadis, 1976; Patel & Ryman, 1976, 1977; Patel *et al.*, 1978). These observations are of great interest because of their implications in the field of drug administration. Many therapeutic agents when administered orally are broken down by gut enzymes and so never reach the blood stream in active form. If it were possible to protect such agents by administering them inside a liposome, and subsequently be certain they would be taken up and transferred across the enterocyte cell into the circulatory system, many drugs that at present can only be given parenterally could be administered by the oral route.

Williams (1978) found that adult gut would also take up and transfer ferritin molecules. When ferritin was injected into ligated segments of duodenum and ileum *in vivo*, it was possible after three hours to find ferritin located in the lymphatic and other body organs. Ferritin was observed both in cytoplasmic membrane-bounded vesicles and free in the cytoplasm of enterocytes, so a pinocytic process is quite likely to be occurring. These data gave no indication of the rate of pinocytosis but the intracellular location of the substrate confirms the ability of adult intestine to take up macromolecules from the gut lumen.

Yolk sac. The yolk sac epithelium is an intensely pinocytic tissue which is probably involved in embryotrophic nutrition or embryonic protection. Like the intestine, the yolk sac is also thought to play a vital role in the transfer of immunoglobulins from mother to foetus in certain animal types. There is a considerable volume of data concerning this process in both rabbits and rats which suggests that transfer involves a mechanism by which proteins can be ingested pinocytically but escape lysosomal hydrolysis, possibly by means of coated vesicles that do not fuse with lysosomes. The question of maternofoetal transmission is discussed in Chapter 4, so this section will be confined to the uptake of substrates that is followed by conventional entry into the vacuolar system.

Over the last 10 years a quantitative technique has been developed in our laboratory for the measurement of pinocytosis in the rat visceral yolk sac. An early interest was the degradative capacity of the yolk sac with regard to its role in embryotrophic nutrition and this was investigated by intravenous injections of HRP into pregnant rats, followed by subsequent histochemical and biochemical investigations of the embyro and its membranes (Beck, Lloyd & Griffiths, 1967). In later experiments explants of the mesometrial pole of yolk sacs were taken and cultured using a raft technique (Beck, Parry & Lloyd, 1970). This series of experiments showed that HRP was taken up by the yolk sac epithelium and to some extent degraded, but no peroxidase reached the embryo. To study the kinetics of the pinocytic process in more detail, an organ culture technique was devised (Williams *et al.*, 1975*a, b*) whereby the pinocytic capture of both degradable and non-degradable macromolecules could be measured as an Endocytic Index.

Initially a non-degradable macromolecule, [^{125}I] PVP, was used as a marker (Williams *et al.*, 1975*a*) and the mean Endocytic Index for this substrate was 1.71 μl/mg protein/hour. More recently [^{14}C] sucrose and invertase have also been found to be captured at a similar rate, having mean Endocytic Indices of 2.04 μl/mg protein/hour (Roberts, Williams & Lloyd, 1977) and 1.14 μl/mg protein/hour (Brown & Segal, 1977) respectively. In contrast [^{125}I] dBSA and colloidal [^{198}Au] gold have higher Endocytic Indices and also show a marked batch variation in the rate of uptake which was not observed with the other substrates (Williams *et al.*, 1975*b*; Roberts *et al.*, 1977). That the two latter substrates do not increase the rate of vesicle formation was shown by their inability to stimulate the rate of uptake of [^{125}I] PVP. The conclusion has been drawn that [^{125}I] PVP, [^{14}C] sucrose and invertase all enter the cell by fluid-phase pinocytosis whilst [^{125}I] dBSA and colloidal [^{198}Au] gold enter to varying extents by adsorption to the membrane. It can be postulated that the different Endocytic Indices observed are directly related to the proportion of substrate entering in association with the plasma membrane. For example, assuming that 2.0 μl/mg protein/hour is the Endocytic Index ascribed to fluid-phase uptake alone, an Endocytic Index of 80 μl/mg protein/hour for [^{125}I] dBSA would indicate that 97.5 % is entering by binding and 2.5 % is taken up in the fluid phase. Similarly 50 % of colloidal gold, with Endocytic Index 4 μl/mg protein/hour is taken up attached to the plasma membrane.

The inter-batch variation seen with [^{125}I] dBSA was attributed to differences in the extent of denaturation during preparation. Likewise the variation observed with colloidal gold may be caused by alterations in the gelatin stabiliser, which is an integral part of the substrate and is probably the cause of the binding. These explanations are compatible with the results of Moore *et al.* (1977) who studied the effect of different physical and chemical treatments on the uptake of [^{125}I] dBSA and ^{125}I-labelled orosomucoid and found that chemical modification generally caused an increase in the rate of capture (see Table 1). Several explanations were possible for the higher rates of uptake. An increase in the rate of pinosome formation was adequately discounted because the modified proteins failed to affect uptake of [^{125}I] PVP. Several modes of endocytosis might exist, each one involved in the capture of a specific molecule, but this seems implausible, if not completely impossible, in view of the immense variation in the type of molecules presented to the cell. However, ultrastructural studies have indicated that more than one class of vesicle exists in the yolk sac, one type being the 'coated' vesicle which has been implicated in the specific uptake of homologous γ-globulin (Moxon, Wild & Slade, 1976). The third possibility is the original postulate that denaturation changes the substrate—membrane affinity by aggregation or conformational change in the protein.

Embryonic skeletal tissue. Dingle and co-workers extensively studied the endocytosis of sucrose by chick limb-bone rudiments using histological (Fell & Dingle, 1969), electron-microscopical (Glauert, Fell & Dingle, 1969) autoradiographical (Appleton, Pelc, Dingle & Fell, 1969) and biochemical approaches (Dingle, Fell & Glauert, 1969). Autoradiography showed that sucrose is taken up by the perichondrium and cartilage, whereas chondrocytes were unlabelled, probably due to their weak endocytic activity. Biochemically it was shown that 50 % of the administered $[^{14}C]$ sucrose was recoverable in the cells after two days and microscopy showed the cells to be heavily vacuolated. To test whether sucrose stimulated pinosome formation, the effect of unlabelled sucrose on the uptake of $[^{14}C]$ dextran (a substrate that had been shown to accumulate linearly with time over six days), $[^{14}C]$ sucrose and iron-dextran was measured. There was no effect on uptake of any of these substrates and it was further concluded that sucrose was taken up by fluid-phase pinocytosis. The vacuolation was attributed to persistence of pinosomes owing to the absence of enzymes able to hydrolyse the disaccharide. The large size of these vacuoles could be due not only to vesicle fusions but also to subsequent hydration of hyperosmotic vesicles.

Other metazoan cells

Although much of the recent work on pinocytosis has been confined to higher vertebrates, cells from many phyla have been shown to display this phenomenon.

The foot epithelium of the slug *Agriolimax reticulatus* has been shown by morphological techniques to endocytose ferritin and peroxidase, the former being visible in larger, probably phagocytic, vacuoles and the latter in smaller pinocytic vesicles (Ryder & Bowen, 1977). This is believed to be the first report of endocytosis in a terrestrial mollusc and may be of importance in the design of molluscicides.

Recently phagocytes have been isolated from the coelomic fluid of the sea urchin *Stronglyocentrotus droebachiensis* (Bertheussen & Seljelid, 1978). These cells have been likened to mammalian macrophages because they ingest carbon particles, adhere to glass and form syncytia similar to macrophage giant cells. Because they can be cultured in very simple media without the addition of complex protein sources, these cells may prove to be useful for the study of endocytosis.

It is believed that, during oogenesis in the mosquito *Aedes aegypti,* yolk deposition is accompanied by concomitant removal of proteins from the blood. Roth & Porter (1962) studied oocyte development morphometrically and found a correlation between the number of coated pits observed at the surface of the oocyte and the stage of development, there being a 15-fold increase in the number observed seven hours after feeding. It was suggested that these coated pits were engaged in the specific binding of protein prior to pinocytic uptake. More recently the kinetics of uptake of yolk protein in the oocyte have been studied in both the mosquito *Culex fatigans* and the chicken (Roth, Cutting & Atlas, 1976). In the mosquito uptake displayed saturation kinetics, was abrogated by glycolytic inhibitors and not much reduced by the presence of increasing concentrations of BSA. Since uptake of ferritin could be stimulated by a female-specific phosphoglycolipoprotein, it was inferred that yolk deposition is regulated by this substance. Electron-microscopical observations showed that ferritin—IgG complexes were located in coated vesicles forming from chicken oocyte plasma membrane. Although coated vesicles may mediate the specific capture of yolk protein, morphological observations indicated that this substrate was not sufficiently large to fill completely the central volume of the vesicle and so exclude fluid-phase uptake.

Functional significance of pinocytosis

The physiological function of the pinocytic process may be twofold. Firstly, it affords a mechanism for the internalisation of macromolecules that normally cannot penetrate the cell membrane by either diffusion or active transport. Secondly, in some situations the formation of pinosomes at the plasma membrane may simply be a membrane retrieval system: membrane recycling may be essential to maintain an appropriate distribution of membranes between different compartments of the cell, and the consequent internalisation of extracellular material could be merely incidental. Membrane recycling also has a potential for membrane surveillance and elimination of defective components. Teleological thinking would suggest that fluid-phase uptake, where there is no substrate selectivity, is less likely to have functional significance than adsorptive pinocytosis involving relatively specific membrane binding sites.

The concept of a universal pinocytic rate

The previous section has shown that many different experimental systems, in combination with many types of substrate, have been used in attempts to quantitate pinocytosis. We have summarised these data in Table 1, using Endocytic Index as a universal unit. This has involved the recalculation, wherever possible, of results reported in the original texts so that values for Endocytic Index are obtained in terms of either μl/mg protein/hour or μl/10^6 cells/hour. Some systems lend themselves to one set of units rather than the other but for fuller comparability we have calculated the Endocytic Index in both terms, using values from Steinman *et al.* (1976) of 80 ± 12 μg protein equalling 10^6 macrophage cells and 455 ± 20 μg protein equalling 10^6 L-cells. Of course this can only be an approximation since the relationship between cell number and total protein varies according to cell type, culture conditions and time in culture. In the calculations for Table 1 we have used the conversion factor more appropriate to the cell type involved to give the most meaningful comparison available at the present time.

As can be seen from the table, a large range of Endocytic Indices can be obtained according to cell type and pinocytic marker employed. It is notable that [^{125}I] PVP, sucrose and invertase have a similar and apparently minimal rate of capture by rat yolk sac, probably indicating that in this system they are fluid-phase markers. This postulate apparently applies to several other cell types where the same markers, along with HRP, have a comparable low Endocytic Index − e.g. the uptake of HRP by mouse peritoneal macrophages, [^{125}I] PVP by intestine and [^{14}C] sucrose by endothelial cells and smooth muscle cells. Other markers have higher Endocytic Indices, presumably attributable to uptake in association with the plasma membrane, and some of these also show startling similarities in different cell types (e.g. the uptake of serum albumin by fibroblasts, yolk sac and rat peritoneal macrophages). In contrast colloidal gold is taken up approximately 100 times faster than PVP by the rat macrophage, but only twice as quickly as PVP by the rat yolk sac. In these systems it is known that this difference is not attributable to a substrate-induced change in the rate of pinosome formation, so there are different membrane−substrate affinities in operation.

One might suppose that cells of different types would have widely different basal pinocytic activities. However it is surprising that the estimated rates of pinocytic uptake of liquid for many cell types on which data are available are of the same order of magnitude, with the exception of fibroblasts, which appear to have lower activity, and Chang-strain liver cells for which a very high Endocytic Index may be calculated. From Table 1 it can be seen that values for fluid phase pinocytosis for six different cell types fall in the range 0.4−2.1 µl/mg protein/hour. In fact over 60 % of the values estimated for various fluid-phase markers fall in this range, a remarkable similarity when the variation in culture techniques employed is taken into account. It is theoretically possible that many cells have a similar rate of pinocytic membrane internalisation and that observed specificities for substrates entering by an adsorptive mechanism are entirely due to differences in the densities of the various receptors and their relative affinities for the substrate. This would enable cells to take up a range of molecules appropriate to their requirements.

Selectivity in pinocytosis

The degree of binding of a substrate to the exterior of a cell depends on the composition of the plasma membrane and the chemical nature of the

Table 1. *Quantitative data of rates of pinocytosis in various cell types*
Values have been calculated by the authors from published data. Figures in parentheses have been derived by taking 10^6 cells as equivalent to 80 μg protein.

Cell type	Tracer (concentration if stated)	Calculated Endocytic Index μl/mg protein/hour	μl/10^6 cells/hour	Reference
Unicellular animals				
Amoeba proteus	[125I]dBSA (1 mg/ml)	30–100	(13.69–45.66)	Chapman-Andresen (1977)
Acanthamoeba castellanii	[131I]HSA (1 mg/ml)	(3.23)[a]	1.47	Bowers & Olszewski (1972)
	[3H]inulin (0.9 mM)	(7.88)[a]	3.6	
	[14C]leucine (3.8 mM)	(5.47)[a]	2.5	
	1[14C]glucose	(4.12)[a]	1.88	
Mononuclear phagocytes				
Rat peritoneal macrophages	[125I]PVP (10 μg/ml)	(0.43)	0.034	Pratten *et al.* (1977)
	[14C]sucrose (1 μg/ml)	(0.25)	0.020	
	Colloidal [198Au]gold (8 μg/ml)[b]	(15.13–104.12)	1.21–8.33	
	[125I]dBSA (10 μg/ml)[b]	(5.5–13.0)	0.44–1.04	
Mouse peritoneal macrophages	[3H]haemoglobin	(3.75)	0.3	Ehrenreich & Cohn (1968)
	HRP (2 μg/ml)	(0.56)	0.045	
	HRP:anti-HRP aggregates (2 μg/ml)			Steinman & Cohn (1972b)
	added directly	(132.75)	10.62	
	stored 30 min at 37 °C	(164.06)	13.12	
	2 days at 4 °C	(484.37)	38.75	
	4 days at 4 °C	(853.12)	68.25	
	7 days at 4 °C	(1337.50)	107.00	

HRP (1 mg/ml)	0.56	(0.045)	Edelson *et al.* (1975)
Endotoxin-stimulated cells, HRP (1 mg/ml)	0.85	(0.068)	
Thioglycollate-stimulated cells, HRP (1 mg/ml)	1.90	(0.152)	
Colloidal [^{198}Au] gold (0.1–1.0 µCi)			
10 % foetal calf serum	(48.75)	3.9	Davies *et al.* (1973)
50 % foetal calf serum	(111.25)	8.9	
Liver cells			
Rat parenchymal cells			
[^{125}I]asialofetuin (10^{-7} M)	(700)[a]	320	Tolleshaug *et al.* (1977)
[^{125}I]dHSA (20 µg/ml)	(18.24)[a]	8.3	Nilsson & Berg (1977)
Rat non-parenchymal cells			
[^{125}I]asialofetuin (10^{-7} M)	(55)	4.4	Tolleshaug *et al.* (1977)
[^{3}H] sucrose	(0.13)	~0.01	
[^{131}I]albumin	(3.75)	~0.30	Munthe-Kaas (1977)
Colloidal [^{198}Au]gold (1.75 µg/ml)	(25)	~2.0	
[^{125}I]dHSA (25 µg/ml)	(1250)	100.0	Nilsson & Berg (1977)
Chang-strain human liver cells			
[^{3}H] sucrose (2×10^{-5} mM)	(173.75)	13.9	Wagner *et al.* (1971)
Fibroblasts			
Fibroblasts (L-cells)			
HRP (1 mg/ml)	0.08	(0.035)[a]	Steinman *et al.* (1974)
Fibroblasts (L929-cells)			
[^{125}I]HSA (50 µg/ml)	10.2	(4.64)[a]	Shen & Ryser (1978)
[^{125}I]HSA –poly-L-lysine	112.4	(51.14)[a]	
Normal human skin fibroblasts			
[^{125}I]LDL (5–250 µg/ml)	33.06–1.42	(15.09–0.65)[a]	Miller *et al.* (1977)
[^{125}I]HDL (1–250 µg/ml)	1.38–0.32	(0.63–0.15)[a]	
[^{14}C] sucrose (1.88 µg/ml)	0.19	(0.086)[a]	
β-*N*-acetylglucosaminidase	5.80	(2.65)[a]	Von Figura *et al.* (1978)
β-glucuronidase	27.00	(12.33)[a]	

Table 1 (*cont.*).

Cell type	Tracer (concentration if stated)	Calculated Endocytic Index		Reference
		μl/mg protein/hour	μl/10⁶ cells/hour	

Cell type	Tracer (concentration if stated)	μl/mg protein/hour	μl/10^6 cells/hour	Reference
Fibroblasts				
Sarcoma-180 cells	[131I] HSA (0.7 mg/ml)	0.046	(0.021)[a]	Petitpierre-Gabathuler & Ryser (1975)
	Soluble ferritin (5.4 mg/ml)	0.398	(0.182)[a]	
	Aggregated ferritin (5.4 mg/ml)	2.29	(1.044)[a]	
Chinese hamster cells	35S-labelled chondroitin sulphate			Saito & Uzman (1971)
	no serum	0.25	(0.11)[a]	
	10 % foetal calf serum	0.104	(0.05)[a]	
3T3 cells	[14C] sucrose	(0.107)[a]	0.049	Davies & Ross (1978)
	HRP	(0.059)[a]	0.027	
Normal human fibroblasts	125I-labelled albumin + [14C] sucrose	17.6	(8.04)[a]	Pittman & Steinberg (1978)
Aorta blood vessels				
Monkey smooth muscle cells	[14C] sucrose	(0.75)	0.06	Davies & Ross (1978)
	HRP	(0.69)	0.055	
Monkey endothelial cells	[14C] sucrose (serum-factor-dependent)	(1.86–0.625)	0.15–0.05	
Pig smooth muscle cells	[125I] PVP (30 μg/ml)	0.082–0.123	(0.007–0.01)	Leake & Bowyer (1977)
Pig endothelial cells	[125I] PVP (30 μg/ml)	0.034	(0.003)	
Trophoblast				
Human trophoblast	125I-labelled haemoglobin (5–12 μg/ml)			Contractor & Krakauer (1976)
	no serum	(53.75)	4.30	
	serum	(46.00)	3.68	

Intestine

Rat jejunum

[125I]-PVP (2 μg/ml)			
10% calf serum	0.96	(0.077)	} Bridges & Woodley (1979)
10% calf serum and ATP	1.48	(0.12)	

Yolk sac

Rat yolk sac

[125I]-PVP (3 μg/ml)	1.71	(0.14)	Williams et al. (1975*a*)
[14C]sucrose (0.1 μg/ml)	2.04	(0.16)	
Colloidal [198Au]gold (1 μg/ml)[b]	1.45–4.83	(0.12–0.39)	} Roberts et al. (1977)
[125I]-BSA (1 μg/ml) (untreated)	4.8	(0.38)	
[125I]-dBSA (1 μg/ml)			
frozen, −20 °C	8.9	(0.71)	
acetic acid, pH 3.5	16.8	(1.34)	
acetic acid, pH 3.0	23.4	(1.87)	
acetic acid, pH 2.5	14.5	(1.16)	
formaldehyde, pH 10	65.0	(5.20)	
buffer, pH 10	8.10	(0.65)	
urea, pH 5.5	73.3	(5.86)	
125I-labelled orosomucoid (untreated)	6.1	(0.49)	Moore et al. (1977)
125I-labelled denatured orosomucoid			
urea, pH 5.5	5.5	(0.44)	
buffer, pH 10	4.8	(0.38)	
formaldehyde, pH 10	2.2	(0.18)	
125I-labelled asialo-orosomucoid	12.5	(1.0)	
Invertase (44 units/ml)	1.14	(0.09)	Brown & Segal (1977)

[a] Fibroblast conversion factor used: 455 μg protein = 10^6 cells.
[b] Batch-dependent.

substrate. The area of plasma membrane to which the substrate molecule binds has been termed a receptor or binding site, and these receptors may be lipid, protein or glycoprotein in composition. They may be highly specific, with affinity for one molecule or group of molecules, or they may accept a very wide range of types of molecule (the so-called non-specific receptor).

Recognition systems. Certain specific mechanisms have been studied in detail and one example is the LDL receptor of human fibroblasts (see Chapter 12). Another system that has been studied in detail is the asialoglycoprotein re-cognition system of liver parenchymal cells. The in-vivo injection of modified glycoproteins and subsequent measurement of clearance has shown that the significant area on the plasma membrane has a sialic acid residue which recog-nises a terminal galactose unit on the substrate molecule (Morell *et al.*, 1968; Gregoriadis, Morell, Sternlieb & Scheinberg, 1970; Furbish *et al.*, 1978). Asialo-transferrin does not have the typical rapid clearance of an asialoglycoprotein and it has recently been shown (Regoeczi *et al.*, 1978) that binding of this molecule by the rat hepatocyte does not necessarily lead to endocytosis. Although the asialotransferrin was readily bound by the hepatic lectin, the individual molecules were thought to lack either the quantity or spacing of terminal galactose residues necessary for triggering endocytosis, but several molecules could act synergistically to initiate pinosome formation. Other relatively specific systems that have been studied are the phosphohexose recognition system of fibroblasts (Von Figura & Kresse, 1974; Kaplan *et al.*, 1977; Kaplan, Fischer & Sly, 1978) and the *N*-acetyl-glucosamine/mannose system of Kupffer cells and macrophages (Stahl *et al.*, 1975, 1976; Achord *et al.*, 1977; Stahl, Rodman, Miller & Schlesinger, 1978). All these are discussed at greater length by Lloyd & Griffiths (1979).

Non-specific chemical modifications of substrates cause quite dramatic changes in their rates of capture. Moore and co-workers (1977) used a variety of physical and chemical treatments to modify proteins such as bovine serum albumin and orosomucoid. The effect of such modifications on the pinocytic uptake by rat yolk sacs is shown in Table 1. Simple modifications by treatments such as urea, acetic acid, formaldehyde or freezing and thawing are thought to cause changes in the tertiary structure of the molecule which alter its affinity for the non-specific receptor. Currently little is known at the mole-cular level of the nature of such interactions but it is possible that specific charged or polar amino acids at the surface of the protein may interact with protein or glycoprotein components of the plasma membrane, or alternatively hydrophobic regions on the protein may bind to lipid components.

Role of adsorptive pinocytosis. The existence of specific and non-specific binding sites on the cell membrane facilitates the selection of certain molecules for internalisation. This type of selectivity permits the control of relative concentrations of metabolites between the extracellular and intracellular compartments. The vectorial flow of material into the cell could either serve to decrease the extracellular concentration (this might be important in the removal of unwanted debris such as denatured macromolecules whose useful lifespan is at an end) or alternatively permit an increase in the intracellular concentration of material that is needed by the cell (this could be of nutritional or immunological importance). An increase in the intracellular concentration may be only transient if internalisation is followed either by digestion or subsequent translocation of the material.

The overall density and spatial location of binding sites determines the rate of uptake of a substrate that can interact with membrane. The quantity of substrate internalised per unit area of membrane could vary according to the local distribution of receptors. It is possible that receptors for a particular substrate are anchored in the membrane in groups, although for maximum efficiency of binding there are steric considerations that would determine the optimum separation required between these sites. Or it is possible that receptors are distributed at random throughout the membrane but, even so, post-binding migration may occur in order to economise on membrane and fluid internalisation (see Chapter 8). Such clumping of receptors is observed in the capping phenomenon reported in relation to the initial stages of internalisation of concanavalin A (Yahara & Edelman, 1973; Zagyansky & Edidin, 1976; Williams, Boxer, Oliver & Bachner, 1977; Albertini, Berlin & Oliver, 1977).

All types of pinocytosis involve the internalisation of large quantities of membrane (and fluid). A compensation for this influx must be carefully controlled by the cell, and an efficient recycling mechanism must exist, since in many systems pinocytosis can be maintained for long periods. The possible routes for such membrane recycling have been discussed by Duncan & Pratten (1977). Obviously the internalisation of substrate bound to special sites on the membrane, especially when the sites are grouped, can be a major economy in the amount of membrane involved, but for the rate of capture to be maintained in this case there must be a constant replenishment of the receptor pool at the surface, either by de-novo synthesis or regeneration. If, as Duncan & Pratten (1977) suggest, a major part of membrane regeneration is by a process of recycling of some sort of intact membrane unit following a fusion event prior to fusion with lysosomes, perhaps the receptors could be regenerated too. However, this is only practicable if the substrate−receptor complex dissociates within the pinosome; otherwise the recycling mechanism would be futile. It has been postulated by Lloyd (1977) that lysosomal enzymes may reach the plasma membrane by such a membrane recycling mechanism after lysosomal fusion and that their loss to the exterior is prevented by membrane binding. He proposes that a faulty enzyme–receptor linkage may explain the autosomal recessive human condition known as I-cell disease (or mucolipidosis type II), in which several lysosomal enzymes are absent or deficient from connective tissue cells. If lysosomal enzymes do circulate in this fashion it is noteworthy that they could be present in newly formed pinosomes although their activity would remain low until the internal pH was lowered by fusion with a lysosome. These ideas are further discussed by Lloyd & Williams (1978). If the extracellular fluid were excluded during pinosome formation, either by the spatial relationship of substrate molecules or the geometry of the pre-pinosome channel, grouping of receptors prior to internalisation would also allow formation of pinosomes containing one substrate exclusively. It is quite possible that a pinosome that selectively excludes extracellular fluid during its formation step would become turgid due to osmotic effect once it is within the cytoplasm. Coated vesicles have been proposed as such a class of specialised vesicle and it is suggested that they may carry exclusively immunoglobulin molecules and that their structure protects them from fusion with the lysosomal system.

To summarise, adsorptive pinocytosis has the following advantages:
1. enhancement of the rate of substrate capture;
2. selectivity between substrates taken up;
3. the ability to direct substrates into a certain vesicle type;
4. cell specificity in uptake of different substrates;
5. maximisation of the uptake of substrate per unit of membrane internalised and thus assistance in membrane dynamics within the cell.

The control of pinocytosis

The rate-limiting step in all pinocytic phenomena is the number of pinosomes formed per unit time and as yet little is known about how this is controlled. The metabolic status and extent of polymerisation of the cyto-skeletal components exert some control and it has been suggested that various secondary messengers exist which act as a fine control on pinocytic rate. By analogy with the skeletal muscle and secretory systems (Rasmussen, 1970), candidates for these are calcium and cAMP but, as they are soluble compounds, they might produce a diffuse response throughout the whole cell. In order to control the localised internalisation of grouped substrate–receptor complexes, a more specific mechanism must operate, but the involvement of molecules such as calcium cannot be totally excluded since it has recently been postulated

(Matthews, 1979) that such substances might move through cells relatively slowly owing to physical and physiological restraints on their diffusibility. The idea that a group of secondary messengers exists which link external and internal pinocytic triggers to actual changes in the rate of pinosome formation at the plasma membrane has been termed 'the transduction mechanism of pinocytosis', and this aspect of endocytosis has recently been reviewed by Stossel (1977). There is no reason to suppose that the same control processes govern pinocytosis in all cell types and inevitably some cell types will be more suitable for the study of these mechanisms. A cell type that pinocytoses continuously throughout its lifespan, such as the yolk sac, may not be the ideal choice, whereas cells involved in the immune system may be more subject to control.

Many attempts have been made to discover substances that change, and in particular enhance, the pinocytic rate. It was Cohn (1966) who first observed that a variation in the serum concentration employed in his culture system for mouse peritoneal macrophages caused changes in the number of vesicles visible with the phase-contrast microscope. He extensively investigated this phenomenon and tried to identify the active component within the serum, and found that the negatively charged bovine plasma albumin and fetuin were particularly active. Further studies with the synthetic polymers poly-L-glutamic acid and dextran sulphate confirmed this apparent anion effect. However, subsequent studies have shown polycations to be potent stimulators of pinocytosis (Ryser & Hancock, 1965; Seljelid *et al.*, 1973; Pratten, Duncan & Lloyd, 1978; Shen & Ryser, 1978). The apparent contradiction of these data with the early reports of Cohn could be explicable on the basis of the quantitation methods employed.

Charged polymers could stimulate uptake of pinocytic markers by increasing the rate of pinosome formation, by initiating a new pinocytic mechanism, or by increasing the proportion of substrate entering in association with the plasma membrane. So far no studies have described an effect of known polymeric stimulators on the uptake of a fluid-phase marker. Ryser & Hancock (1965) attributed the poly-L-lysine stimulation of uptake of ^{131}I-labelled human serum albumin by sarcoma-180 cells to the induction of a different endocytic mechanism, whereas the majority of other studies ascribe the stimulation to an electrostatic interaction between substrate and modifier leading to an increased affinity for the cell membrane. (Seljelid *et al.*, 1973; Shen & Ryser, 1978; Duncan, Pratten & Lloyd, 1979).

Substances that stimulate fluid-phase pinocytosis are relatively uncommon. Concanavalin A (Con A) apparently stimulates the uptake of HRP and [125I]-BSA by mouse peritoneal macrophages (Edelson & Cohn, 1974a) but this

effect is combined with the Con A inhibition of pinosome—lysosome fusion
(Edelson & Cohn, 1974*b*), the stimulation of lysosomal enzyme synthesis
(Goldman & Raz, 1975) and the inhibition of phagocytosis of latex particles
by mouse peritoneal macrophages (Friend, Ekstedt & Duncan, 1975). Obviously the membrane perturbations caused by Con A are complex, but they are
dependent on the interaction of the lectin with saccharide residues on the
membrane, as shown by the action of the competitive inhibitor, α-methyl
mannoside. This subject is discussed at greater length by Silverstein, Steinman
& Cohn (1977).

Very recently Davies & Ross (1978) reported that factors derived from
whole blood serum and from platelets stimulate (approximately twofold) the
rate of vesicle formation by monkey aortic smooth muscle cells.

To the future

Over the last 10 years understanding of pinocytosis has increased
enormously but many areas are as yet virgin territory. Now that the methodology for adequate quantitation of pinocytosis is available, investigations can
begin in a manner that will give a comprehensive understanding of all the
physiological functions of this intriguing process. Attempts can be made to
discover whether fluid-phase pinocytosis is truly common to all cell types and
indeed proceeds at a universal rate.

Many areas require urgent consideration. Although several recognition
systems have already been described, the determinants of membrane affinities
for many diverse substances are still a mystery. The existence of discrete
categories of pinocytic uptake has not been adequately confirmed and the
uptake of substrates by specific types of vesicles remains hypothetical. There
is a need for definitive experimental design to clarify these issues. Finally,
very little is understood about the control mechanisms of pinocytosis. The
feasibility of an ongoing pinocytic process that does not require a trigger must
surely be in doubt, but, if stimuli are necessary, they need to be identified,
as does the way in which the stimulation process is linked to the membrane
invagination.

M.K.P. and R.D. are supported by research grants from the Science Research
Council and the Cancer Research Campaign.

References

Achord, D., Brot, F., Gonzalez-Noriega, A., Sly, W. & Stahl, P. (1977). Human
    β-glucuronidase. II. Fate of infused human placental β-glucuronidase
    in the rat. *Pediatric Research*, 11, 816—22.
Albertini, D.F., Berlin, R.D. & Oliver, J.M. (1977). The mechanism of con-
    canavalin A cap formation in leukocytes. *Journal of Cell Science*, 26,
    57—75.

Allison, A.C. & Davies, P. (1974). Mechanisms of endocytosis and exocytosis. In *Transport at the Cellular Level*, ed. M.A. Sleigh & D.H. Jennings, *Society for Experimental Biology Symposium* 28, pp. 419–46. Cambridge University Press.

Appleton, T.C., Pelc, S.R., Dingle, J.T. & Fell, H.B. (1969). Endocytosis of sugars in embryonic skeletal tissues in organ culture. III. Radioautographic distribution of [^{14}C] sucrose. *Journal of Cell Science*, 4, 133–8.

Ashwell, G. & Morell, A.G. (1971). Galactose: a cryptic determinant of glycoprotein catabolism. In *Glycoproteins of Blood Cells and Plasma*, ed. G.A. Jameson & T.J. Greenwalt, pp. 173–89. Philadelphia: J.B. Lippincott.

Beck, F., Lloyd, J.B. & Griffiths, A. (1967). A histochemical and biochemical study of some aspects of placental function in the rat using maternal injection of horseradish peroxidase. *Journal of Anatomy*, 101, 461–78.

Beck, F., Parry, L.M. & Lloyd, J.B. (1970). In vitro studies on the degradation of horseradish peroxidase by rat yolk-sac. *Cytobiologie*, 1, 331–42.

Becker, G. & Ashwood-Smith, M.J. (1973). Endocytosis in Chinese hamster fibroblasts. Inhibition by glucose. *Experimental Cell Research*, 82, 310–14.

Berg, T. & Boman, D. (1973). Distribution of lysosomal enzymes between parenchymal and Kupffer cells of rat liver. *Biochimica et Biophysica Acta*, 321, 585–96.

Bertheussen, K. & Seljelid, R. (1978). Echinoid phagocytes *in vitro. Experimental Cell Research*, 111, 401–12.

Bode, F., Baumann, K. & Kinne, R. (1976). Analysis of the pinocytic process in rat kidney. II. Biochemical composition of pinocytic vesicles compared to brush border microvilli, lysosomes and basolateral plasma membranes. *Biochimica et Biophysica Acta*, 433, 294–310.

Bowers, B. (1977). Comparison of pinocytosis and phagocytosis in *Acanthamoeba castellanii. Experimental Cell Research*, 110, 409–17.

Bowers, B. & Olszewski, T.E. (1972). Pinocytosis in *Acanthamoeba castellanii*. Kinetics and morphology. *Journal of Cell Biology*, 53, 681–94.

Bridges, J.F., Millard, P.C. & Woodley, J.F. (1978). The uptake of liposome-entrapped ^{125}I-labelled poly(vinylpyrrolidone) by rat jejunum *in vitro. Biochimica et Biophysica Acta*, 544, 448–51.

Bridges, J.F. & Woodley, J.F. (1979). The uptake of [^{125}I] polyvinylpyrrolidone by adult rat gut *in vitro*. In *Maternofoetal Transmission*, ed. W.A. Hemmings, vol. 2, pp. 249–57. Amsterdam: Elsevier.

Brown, J.A. & Segal, H.L. (1977). Effect of glucagon on pinocytosis by the yolk sac of the rat. *Journal of Biological Chemistry*, 252, 7151–5.

Casley-Smith, J.R. (1969). Endocytosis: the different energy requirements for the uptake of particles by small and large vesicles into peritoneal macrophages. *Journal of Microscopy*, 90, 15–30.

Casley-Smith, J.R. & Chin, J.C. (1971). The passage of cytoplasmic vesicles across endothelial and mesothelial cells. *Journal of Microscopy*, 93, 167–89.

Chapman-Andresen, C. (1962). Studies on pinocytosis in *Amoebae*. *Comptes Rendus des Travaux du Laboratoire Carlsberg*, 33, 73–264.

Chapman-Andresen, C. (1965). The induction of pinocytosis in *Amoebae*. *Archives de Biologie*, Liège, 76, 189–207.

Chapman-Andresen, C. (1976). Studies on endocytosis in *Amoebae*. The distribution of pinocytically ingested dyes in relation to food vacuoles in *Chaos chaos*. I. Light microscopic observations. *Comptes Rendus des Travaux du Laboratoire Carlsberg*, 36, 161–207.

Chapman-Andresen, C. (1977). Endocytosis in freshwater amoebas. *Physiological Reviews*, 57, 371–85.

Chapman-Andresen, C. & Holter, H. (1964). Differential uptake of protein and glucose by pinocytosis in *Amoeba proteus*. *Comptes Rendus des Travaux du Laboratoire Carlsberg*, 34, 211–26.

Christensen, E.I. (1976). Rapid protein uptake and digestion in proximal tubule lysosomes. *Kidney International*, 10, 301–10.

Clarke, R.M. & Hardy, R.N. (1971). Factors influencing the uptake of [^{125}I]-polyvinylpyrrolidone by the intestine of the young rat. *Journal of Physiology*, 212, 801–17.

Cohn, Z.A. (1966). The regulation of pinocytosis in mouse macrophages. I. Metabolic requirements as defined by the use of inhibitors. *Journal of Experimental Medicine*, 124, 557–71.

Cohn, Z.A. & Ehrenreich, B.A. (1969). The uptake, storage, and intracellular hydrolysis of carbohydrates by macrophages. *Journal of Experimental Medicine*, 129, 201–19.

Cohn, Z.A. & Parks, E. (1967a). The regulation of pinocytosis in mouse macrophages. II. Factors inducing vesicle formation. *Journal of Experimental Medicine*, 125, 213–30.

Cohn, Z.A. & Parks, E. (1967b). The regulation of pinocytosis in mouse macrophages. III. The induction of vesicle formation by nucleosides and nucleotides. *Journal of Experimental Medicine*, 125, 457–66.

Cohn, Z.A. & Parks, E. (1967c). The regulation of pinocytosis in mouse macrophages. IV. The immunological induction of pinocytic vesicles, secondary lysosomes, and hydrolytic enzymes. *Journal of Experimental Medicine*, 125, 1091–104.

Contractor, S.F. & Krakauer, K. (1976). Pinocytosis and intracellular digestion of ^{125}I-labelled haemoglobin by trophoblastic cells in tissue culture in the presence and absence of serum. *Journal of Cell Science*, 21, 595–607.

Cornell, R., Walker, W.A. & Isselbacher, K.J. (1971). Small intestine absorption of horseradish peroxidase. A cytochemical study. *Laboratory Investigation*, 25, 42–8.

Creemers, J. & Jacques, P.J. (1971). Endocytic uptake and vesicular transport of injected horseradish peroxidase in the vacuolar apparatus of rat liver cells. *Experimental Cell Research*, 67, 188–203.

Dapergolas, G. & Gregoriadis, G. (1976). Hypoglycaemic effect of liposome-entrapped insulin administered intragastrically into rats. *Lancet*, ii, 824–7.

Davidson, S.J., Hughes, W.L. & Barnwell, A. (1971). Renal protein absorption into subcellular particles. I. Studies with intact kidneys and fractionated homogenates. *Experimental Cell Research*, 67, 171–87.

Davies, P., Allison, A.C. & Haswell, A.D. (1973). The quantitative estimation of pinocytosis using radioactive colloidal gold. *Biochemical and Biophysical Research Communications*, 52, 627–34.

Davies, P.F. & Ross, R. (1978). Mediation of pinocytosis in cultured arterial smooth muscle and endothelial cells by platelet-derived growth factor. *Journal of Cell Biology*, 79, 663–71.

Dingle, J.T., Fell, H.B. & Glauert, A.M. (1969). Endocytosis of sugars in embryonic skeletal tissues in organ culture. IV. Lysosomal and other biochemical effects. General discussion. *Journal of Cell Science*, 4, 139–154.

Duncan, R. & Lloyd, J.B. (1978). Pinocytosis in the rat visceral yolk sac: effect of temperature, metabolic inhibitors and some other modifiers. *Biochimica et Biophysica Acta*, 544, 647–55.

Duncan, R. & Pratten, M.K. (1977). Membrane economics in endocytic systems. *Journal of Theoretical Biology*, 66, 727–35.

Duncan, R., Pratten, M.K. & Lloyd, J.B. (1979). Mechanism of polycation stimulation of pinocytosis. *Biochimica et Biophysica Acta*, in press.

Edelson, P.J. & Cohn, Z.A. (1974a). Effects of concanavalin A on mouse peritoneal macrophages. I. Stimulation of endocytic activity and inhibition of phago-lysosome formation. *Journal of Experimental Medicine*, 140, 1364–86.

Edelson, P.J. & Cohn, Z.A. (1974b). Effects of concanavalin A on mouse peritoneal macrophages. II. Metabolism of endocytozed proteins and reversibility of the effects by mannose. *Journal of Experimental Medicine*, 140, 1387–403.

Edelson, P.J., Zweibel, R. & Cohn, Z.A. (1975). The pinocytic rate of activated macrophages. *Journal of Experimental Medicine*, 142, 1150–64.

Ehrenreich, B.A. & Cohn, Z.A. (1967). The uptake and digestion of iodinated human serum albumin by macrophages *in vitro*. *Journal of Experimental Medicine*, 126, 941–58.

Ehrenreich, B.A. & Cohn, Z.A. (1968). Fate of hemoglobin pinocytosed by macrophages *in vitro*. *Journal of Cell Biology*, 38, 244–8.

Fell, H.B. & Dingle, J.T. (1969). Endocytosis of sugars in embryonic skeletal tissues in organ culture. I. General introduction and histological effects. *Journal of Cell Science*, 4, 89–103.

Friend, K., Ekstedt, R.D. & Duncan, J.L. (1975). Effect on concanavalin A on phagocytosis by mouse peritoneal macrophages. *Journal of the Reticuloendothelial Society*, 17, 10–19.

Furbish, F.S., Steer, C.J. & Barranger, J.A. (1978). The uptake of native and desialylated glucocerebrosidase by rat hepatocytes and Kupffer cells. *Biochemical and Biophysical Research Communications*, 81, 1047–53.

Glauert, A.M., Fell, H.B. & Dingle, J.T. (1969). Endocytosis of sugars in embryonic skeletal tissues in organ culture. II. Effect of sucrose on cellular fine structure. *Journal of Cell Science*, 4, 105–31.

Goldman, R. & Raz, A. (1975). Concanavalin A and the *in vitro* induction in macrophages of vacuolation and lysosomal enzyme synthesis. *Experimental Cell Research,* 96, 393–405.

Goldstein, J.L. & Brown, M.S. (1976). The LDL pathway in human fibroblasts: a receptor-mediated mechanism for the regulation of cholesterol metabolism. In *Current Topics in Cellular Regulation,* ed. B.L. Horecker & E.R. Stadtman, vol. 11, pp. 147–81. New York & London: Academic Press.

Goldstein, J.L. & Brown, M.S. (1977). The low density lipoprotein pathway and its relation to atherosclerosis. *Annual Review of Biochemistry,* 46, 897–930.

Gregoriadis, G., Morell, A.G., Sternlieb, I. & Scheinberg, I.H. (1970). Catabolism of desialylated ceruloplasmin in the liver. *Journal of Biological Chemistry,* 245, 5833–7.

Halliday, R. & Kekwick, R.A. (1960). The selection of antibodies by the gut of the young rat. *Proceedings of the Royal Society of London, Series B,* 153, 279–86.

Hickman, S. & Neufeld, E.F. (1972). A hypothesis for I-cell disease: defective hydrolases that do not enter lysosomes. *Biochemical and Biophysical Research Communications,* 49, 992–9.

Hieber, V., Distler, J., Myerowitz, R., Schmickel, R.D. & Jourdian, G.W. (1977). The role of glycosidically bound mannose in the assimilation of β-galactosidase by generalised gangliosidosis fibroblasts. *Biochemical and Biophysical Research Communications,* 73, 710–17.

Jacques, P.J. (1969). Endocytosis. In *Lysosomes in Biology and Pathology,* ed. J.T. Dingle & H.B. Fell, vol. 2, pp. 395–420. Amsterdam: North-Holland.

Kaplan, A., Achord, D.T. & Sly, W.S. (1977). Phosphohexosyl components of a lysosomal enzyme are recognised by pinocytosis receptors on human fibroblasts. *Proceedings of the National Academy of Sciences, USA,* 74, 2026–30.

Kaplan, A., Fischer, D., Achord, D. & Sly, W. (1977). Phosphohexosyl recognition is a general characteristic of pinocytosis of lysosomal glycosidases by human fibroblasts. *Journal of Clinical Investigation,* 60, 1088–93.

Kaplan, A., Fischer, D. & Sly, W.S. (1978). Correlation of structural features of phosphomannans with their ability to inhibit pinocytosis of human β-glucuronidase by human fibroblasts. *Journal of Biological Chemistry,* 253, 647–50.

Leake, D.S. & Bowyer, D.E. (1977). A method to measure the pinocytic uptake of ^{125}I-labelled polyvinylpyrrolidone by pig aortic smooth muscle and endothelial cells in culture. *Progress in Biochemistry and Pharmacology,* 14, 78–83.

Lloyd, J.B. (1977). Cellular transport of lysosomal enzymes. An alternative hypothesis. *Biochemical Journal,* 164, 281–2.

Lloyd, J.B. & Griffiths, P.A. (1979). Enzyme replacement therapy of lysosome storage disease. In *Lysosomes in Biology and Pathology,* ed. J.T. Dingle, P.J. Jacques & I.H. Shaw, vol. 6. Amsterdam: Elsevier (in press).

Lloyd, J.B. & Williams, K.E. (1978). Lysosomal digestion of endocytosed proteins: opportunities and problems for the cell. In *Protein Turnover and Lysosome Function*, ed. H.L. Segal & D.J. Doyle, pp. 395–416. New York & London: Academic Press.

Matthews, E.K. (1979). Calcium translocation and control mechanisms for endocrine secretion. In *Secretory Mechanisms*, ed. C.R. Hopkins & C.J. Duncan. *Society for Experimental Biology Symposium 33*, pp. 225–49. Cambridge University Press.

Maunsbach, A.B. (1966). Absorption of I^{125}-labelled homologous albumin by rat kidney proximal tubule cells. A study of microperfused single proximal tubules by electron microscopic autoradiography and histochemistry. *Journal of Ultrastructure Research*, 15, 197–241.

Mehl, T.D. & Lagunoff, D. (1975). Uptake of aggregated albumin by rat macrophages *in vitro*: affinities of cells for monomeric and aggregated bovine serum albumin. *Journal of the Reticuloendothelial Society*, 18, 125–35.

Miller, A.T., Hale, D.M. & Alexander, K.D. (1965). Histochemical studies on the uptake of horseradish peroxidase by rat kidney slices. *Journal of Cell Biology*, 27, 305–12.

Miller, N.E., Weinstein, D.B. & Steinberg, D. (1977). Binding, internalisation, and degradation of high density lipoprotein by cultured normal human fibroblasts. *Journal of Lipid Research*, 18, 438–50.

Moore, A.T., Williams, K.E. & Lloyd, J.B. (1977). The effect of chemical treatments of albumin and orosomucoid on rate of clearance from the rat bloodstream and rate of pinocytic capture by rat yolk sac cultured *in vitro*. *Biochemical Journal*, 164, 607–16.

Morell, A.G., Irvine, R.A., Sternlieb, I., Scheinberg, I.H. & Ashwell, G. (1968). Physical and chemical studies on ceruloplasmin. *Journal of Biological Chemistry*, 243, 155–9.

Moxon, L.A., Wild, A.E. & Slade, B.S. (1976). Localisation of proteins in coated micropinocytotic vesicles during transport across rabbit yolk sac endoderm. *Cell and Tissue Research*, 171, 175–93.

Munthe-Kaas, A.C. (1977). Uptake of macromolecules by rat Kupffer cells *in vitro*. *Experimental Cell Research*, 107, 55–62.

Nilsson, M. & Berg, T. (1977). Uptake and degradation of formaldehyde-treated ^{125}I-labelled human serum albumin in rat liver cells *in vivo* and *in vitro*. *Biochimica et Biophysica Acta*, 497, 171–82.

Normann, S.J. (1973). The kinetics of phagocytosis. I. A study on the clearance of denatured bovine albumin and its competitive inhibition by denatured human albumin. *Journal of the Reticuloendothelial Society*, 14, 587–98.

Patel, H.M., Harding, N.G.L., Logue, F., Kesson, C., Maccuish, A.G., McKenzie, J.C., Ryman, B.E. & Scobie, I. (1978). Intrajejunal absorption of liposomally entrapped insulin in normal man. *Biochemical Society Transactions*, 6, 784–5.

Patel, H.M. & Ryman, B.E. (1976). Oral administration of insulin by encapsulation within liposomes. *Federation of European Biochemical Societies Letters*, 62, 60–3.

Patel, H.M. & Ryman, B.E. (1977). The gastrointestinal absorption of liposomally entrapped insulin in normal rats. *Biochemical Society Transactions*, 5, 1054–5.

Petitpierre-Gabathuler, M.-P. & Ryser, H. J.-P. (1975). Cellular uptake of soluble and aggregated ferritin: distinction between pinocytosis and phagocytosis. *Journal of Cell Science*, 19, 141–56.

Pittman, R.C. & Steinberg, D. (1978). A new approach for assessing cumulative degradation of proteins or other macromolecules. *Biochemical and Biophysical Research Communications*, 81, 1254–9.

Pratten, M.K., Duncan, R. & Lloyd, J.B. (1978). A comparative study of the effects of polyamino acids and dextran derivatives on pinocytosis in the rat yolk sac and the rat peritoneal macrophage. *Biochimica et Biophysica Acta*, 540, 455–62.

Pratten, M.K. & Lloyd, J.B. (1979). A comparative study of effect of temperature, metabolic inhibitors and some other factors on fluid phase and adsorptive pinocytosis by rat peritoneal macrophages. *Biochemical Journal*, 180, 567–71.

Pratten, M.K., Williams, K.E. & Lloyd, J.B. (1977). A quantitative study of pinocytosis and intracellular proteolysis in rat peritoneal macrophages. *Biochemical Journal*, 168, 365–72.

Rasmussen, H. (1970). Cell communication, calcium ion, and cyclic adenosine monophosphate. *Science*, 170, 404–12.

Regoeczi, E. (1976). Labelled polyvinylpyrrolidone as an *in vivo* indicator of reticuloendothelial activity. *British Journal of Experimental Pathology*, 57, 431–42.

Regoeczi, E., Taylor, P., Hatton, M.W.C., Wong, K.-L. & Koj, A. (1978). Distinction between binding and endocytosis of human asialo-transferrin by the rat liver. *Biochemical Journal*, 174, 171–8.

Ricketts, T.R. (1971). Periodicity of endocytosis in *Tetrahymena pyriformis*. *Protoplasma*, 73, 387–96.

Ricketts, T.R. & Rappit, A.F. (1975). A radioisotopic and morphological study of the uptake of materials into food vacuoles by *Tetrahymena pyriformis* GL-9. *Protoplasma*, 86, 321–37.

Roberts, A.V.S., Williams, K.E. & Lloyd, J.B. (1977). The pinocytosis of ^{125}I-labelled poly(vinylpyrrolidone), [14C]sucrose and colloidal [198Au]gold by rat yolk sac cultured *in vitro*. *Biochemical Journal*, 168, 239–44.

Roth, T.F., Cutting, J.A. & Atlas, S.B. (1976). Protein transport: a selective membrane mechanism. *Journal of Supramolecular Structure*, 4, 527–48.

Roth, T.F. & Porter, K.R. (1962). Yolk protein uptake in the oocyte of the mosquito. *Journal of Cell Biology*, 20, 313–32.

Ryder, T.A. & Bowen, I.D. (1977). Endocytosis and aspects of autophagy in the foot epithelium of the slug *Agriolimax reticulatus* (Müller). *Cell and Tissue Research*, 181, 129–42.

Ryser, H. J.-P. (1967). A membrane effect of basic polymers dependent on molecular size. *Nature, London*, 215, 934–6.

Ryser, H. J.-P. (1970). Transport of macromolecules, especially proteins, into mammalian cells. In *Proceedings of the 4th International Congress of Pharmacology,* vol. 3, pp. 96–132. Basel: Schwabe.

Ryser, H.J.-P. & Hancock, R. (1965). Histones and basic polyamino acids stimulate the uptake of albumin by tumour cells in culture. *Science,* 150, 501–3.

Ryser, H.J.-P., Termini, T.E. & Barnes, P.R. (1976). Polynucleotide aggregates enhance the transport of protein at the surface of cultured mammalian cells. *Journal of Cellular Physiology,* 87, 221–8.

Saito, H. & Uzman, B.G. (1971). Uptake of chondroitin sulfate by mammalian cells in culture. II. Kinetics of uptake and autoradiography. *Experimental Cell Research,* 66, 90–6.

Seljelid, R., Silverstein, S.C. & Cohn, Z.A. (1973). The effect of poly-L-lysine on the uptake of reovirus double-stranded RNA in macrophages *in vitro. Journal of Cell Biology,* 57, 484–98.

Shen, W.-C. & Ryser, H.J.-P. (1978). Conjugation of poly-L-lysine to albumin and horseradish peroxidase: a novel method of enhancing the cellular uptake of proteins. *Proceedings of the National Academy of Sciences, USA,* 75, 1872–6.

Silverstein, S.C., Steinman, R.M. & Cohn, Z.A. (1977). Endocytosis. *Annual Review of Biochemistry,* 46, 669–722.

Stahl, P., Mandell, B., Rodman, J.S., Schlesinger, P. & Lang, S. (1975). Different forms of rat β-glucuronidase with rapid and slow clearance following intravenous injection: selective serum enhancement of slow clearance forms by organophosphate compounds. *Archives of Biochemistry and Biophysics,* 170, 536–46.

Stahl, P., Rodman, J.S., Miller, M.J. & Schlesinger, P.H. (1978). Evidence for receptor mediated binding of glycoprotein, glycoconjugates and lysosomal glycosidases by alveolar macrophages. *Proceedings of the National Academy of Sciences, USA,* 75, 1399–403.

Stahl, P., Six, H., Rodman, J.S., Schlesinger, P., Tulsiani, D.P.R. & Touster, O. (1976). Evidence for specific recognition sites mediating clearance of lysosomal enzymes *in vivo. Proceedings of the National Academy of Sciences, USA,* 73, 4045–9.

Steinman, R.M., Brodie, S.E. & Cohn, Z.A. (1976). Membrane flow during pinocytosis. A stereologic analysis. *Journal of Cell Biology,* 68, 665–87.

Steinman, R.M. & Cohn, Z.A. (1972a). The interaction of soluble horseradish peroxidase with mouse peritoneal macrophages *in vitro. Journal of Cell Biology,* 55, 186–204.

Steinman, R.M. & Cohn, Z.A. (1972b). The interaction of particulate horseradish peroxidase (HRP)-anti-HRP immune complexes with mouse peritoneal macrophages *in vitro. Journal of Cell Biology,* 55, 616–34.

Steinman, R.M., Silver, J.M. & Cohn, Z.A. (1974). Pinocytosis in fibroblasts. Quantitative studies *in vitro. Journal of Cell Biology,* 63, 949–69.

Stossel, T.P. (1977). Endocytosis. In *Receptors and Recognition, Series A,* ed. P. Cuatrecasas & M.F. Greaves, vol. 4, pp. 105–41. London: Chapman & Hall.

Straus, W. (1964). Cytochemical observations on the relationship between lysosomes and phagosomes in kidney and liver by combined staining for acid phosphatase and intravenously injected horseradish peroxidase. *Journal of Cell Biology*, 20, 497–507.

Straus, W. (1971). Comparative analysis of the concentration of injected horseradish peroxidase in cytoplasmic granules of the kidney cortex, in the blood, urine and liver. *Journal of Cell Biology*, 48, 620–32.

Tolleshaug, H., Berg, T., Nilsson, M. & Norum, K.R. (1977). Uptake and degradation of ^{125}I-labelled asialo-fetuin by isolated rat hepatocytes. *Biochimica et Biophysica Acta*, 499, 73–84.

Von Figura, K. & Kresse, H. (1974). Quantitative aspects of pinocytosis and the intracellular fate of N-acetyl-α-D-glucosaminidase in Sanfilippo B fibroblasts. *Journal of Clinical Investigation*, 53, 85–90.

Von Figura, K., Kresse, H., Meinhard, U. & Holtfrerich, D. (1978). Studies on secretion and endocytosis of macromolecules by cultivated skin fibroblasts. *Biochemical Journal*, 170, 313–20.

Wagner, R.C., Rosenberg, M. & Estensen, R. (1971). Endocytosis in Chang liver cells. Quantitation by sucrose-^{3}H uptake and inhibition by cytochalasin B. *Journal of Cell Biology*, 50, 804–17.

Walker, W.A., Cornell, R., Davenport, L.M. & Isselbacher, K.J. (1972). Macromolecular absorption. Mechanism of horseradish peroxidase uptake and transport in adult and neonatal rat intestine. *Journal of Cell Biology*, 54, 195–205.

Westwood, F.R. & Longstaff, E. (1976). Stimulation of cellular ingestion by basic proteins *in vitro*. *British Journal of Cancer*, 33, 392–9.

Williams, D.A., Boxer, L.A., Oliver, J.M. & Bachner, R.L. (1977). Cytoskeletal regulation of concanavalin A capping in pulmonary alveolar macrophages. *Nature, London*, 267, 255–6.

Williams, E.W. (1978). Ferritin uptake by the gut of the adult rat: an immunological and electron microscopical study. In *Antigen Absorption by the Gut*, ed. W.A. Hemmings, pp. 49–63. Lancaster: MTP Press Ltd.

Williams, K.E., Kidston, E.M., Beck, F. & Lloyd, J.B. (1975a). Quantitative studies of pinocytosis. I. Kinetics of uptake of [^{125}I] polyvinylpyrrolidone by rat yolk sac cultured *in vitro*. *Journal of Cell Biology*, 64, 113–22.

Williams, K.E., Kidston, E.M., Beck, F. & Lloyd, J.B. (1975b). Quantitative studies of pinocytosis. II. Kinetics of protein uptake and digestion by rat yolk sac cultured *in vitro*. *Journal of Cell Biology*, 64, 123–34.

Yahara, I. & Edelman, G.M. (1973). Modulation of lymphocyte receptor motility by concanavalin A and colchicine. *Annals of the New York Academy of Sciences*, 253, 455–69.

Zagyanksy, Y. & Edidin, M. (1976). Lateral diffusion of concanavalin A receptors in the plasma membrane of mouse fibroblasts. *Biochimica et Biophysica Acta*, 433, 209–14.

8

Coated vesicles and receptor biology

A. REES & K. WALLACE

It is a usual assumption that the selective transport of proteins across cell plasma membranes is mediated by receptor molecules located in the membranes. The function of such receptors is to bind those molecules for which they have the correct stereospecificity. The binding process is usually saturable and occurs with a high affinity between receptor and protein. When the receptor–protein complex is subsequently internalised by endocytosis it may be either utilised in some metabolic function or translocated to a site suitable for exocytosis at the basolateral membrane. From there it may be absorbed by the vascular system.

We shall discuss in this chapter the various models which elaborate these basic concepts, with particular emphasis on the role and fate of the receptor molecule. Evidence for and against the involvement of receptors will be presented and an evaluation of the transport models will be attempted in the light of such evidence. Finally we shall propose some experimental approaches which might resolve some of the more ill-understood areas of the subject.

General properties of receptors

The concept of a 'receptor' was introduced about 100 years ago by Langley (1878). Langley's work on the response of certain muscle cells to nicotine led to the idea of a 'receptive substance' at the site of application of the drug. Later work by Ehrlich* allowed a generalisation of the concept and the receptor became identified with the role of a recognition site for a specific region of a drug or ligand. In his later studies of immunity Ehrlich conceived the notion of the antibody as a receptor for its ligand, the antigen. In the transport of antibodies mediated by Fc receptors, to which we will make frequent reference in this chapter, the antibody reverts to the role of ligand

*See *The Collected Papers of Paul Ehrlich* (Pergamon Press, Oxford, 1957).

and its membrane receptor becomes as much an unknown entity as the antibody was in Ehrlich's time.

Our present-day definition of a receptor has changed only to the extent that it can now be described in terms of structural, kinetic and physiological parameters. The interaction between ligand and cell receptor can therefore be defined in more quantifiable terms. The techniques that have made possible the preparation of highly radioactive ligands (hormones, immunoglobulins, etc.) have enabled a direct analysis of the interaction of such ligands with their receptors. This in turn has led to an evaluation of the affinities and specificities of such interactions. Further, progress in the methods of isolating specific receptor substances has enabled a more detailed molecular description of the receptor to be given (e.g. the acetylcholine receptor: Olsen, Mennier & Changeux, 1972; the insulin receptor: Jacobs, Chang & Cuatrecasas, 1975). Although for many hormone and transport systems a considerable quantity of data has been amassed *in vivo* and *in vitro*, the *precise* sequence of events leading from the initial binding to the observed response has not been described for any system.

The relationship between biological activity and binding of ligand has been described by various models, the general equation, shown in (1) below, being the same for all of them:

$$L_{ext} + R \underset{k_{-1}}{\overset{k_1}{\rightleftharpoons}} LR \underset{k_{-2}}{\overset{k_2}{\rightleftharpoons}} L_{int} + R \qquad (1)$$

Response A Response B
(metabolic or
transport)

where R = receptor
L_{ext} = ligand outside the membrane
L_{int} = ligand inside the cell
LR = ligand—receptor complex

The 'occupancy theory', originally proposed by Clark (1926a, b), relates the amount of ligand—receptor complex at any given time to the external response, with a maximum corresponding to total receptor occupancy (response B in (1)). The biological response is mediated either directly by LR and a transducer or by dissociation of LR, possibly after translocation of L to the inner surface of the membrane, to give undissociated ligand which then acts directly on its target system. An alternative description, the 'rate theory' of Paton (1961), focusses attention on the rate of formation of the ligand—receptor complex, such that the process of forming the complex gives rise to the biological response (response A in (1)). Once formed the ligand–receptor complex is inactive and must dissociate before the response sequence can be activated again.

Where the transport of ligand by a specific receptor-mediated mechanism is suspected a third, perfectly acceptable description can be formulated in terms of the Michaelis–Menten equation. Brambell (1970) suggested such a description for the maternofoetal transport of immunoglobulins, according to the scheme in (1), where L_{ext} and L_{int} refer to the immunoglobulin in the maternal blood and that exiting to the foetal circulation respectively (note: LR only exists in this scheme when L is in contact with the membrane).

The crucial factor in a Michaelis–Menten treatment is that the concentration of LR remains constant. Thus it is essential that the process of removal of LR, either by endocytosis or some other method, is slow relative to the rate of dissociation of the complex (i.e. k_2 is small compared to k_{-1}). If this is so then the rate of transport, v, is given by:

$$v = \frac{V_{max}}{1 + K_m/[L]_{ext}} \qquad (2)$$

where V_{max} is equivalent to $k_2[R]_{total}$

and $K_m = \dfrac{k_2 + k_{-1}}{k_1}$, which reduces to $K_m = \dfrac{k_{-1}}{k_1} = K_D$ when k_2 is small.

It should be appreciated that K_m is not an equilibrium constant since the concentration terms must be determined at early times in the reacting system and hence equation (2) is valid only if the velocity is proportional to the concentration of LR. Further assumptions are that there should be no interaction between binding sites (a requirement that is known not to be adequately satisfied by some hormone receptors) and that the ligand is present in a homogeneous form. Provided these criteria are satisfied then the K_m value obtained will approximate to the affinity of the particular macromolecule for its receptor.

It can easily be seen therefore, if equation (2) is a correct description of macromolecular transport kinetics, that when [L] is large relative to K_D the rate should reflect zero order kinetics and that when [L] is small relative to K_D, first order kinetics. Results have been obtained which suggest that such a description is validated for the transport of IgG by what is thought to be a receptor-mediated process. Morris (1964) found that the relationship of IgG concentration in the serum of young mice to the dose was in the form of a rectangular hyperbola. Sonoda & Schlamowitz (1972) demonstrated zero order kinetics *in vivo* for the transfer of rabbit IgG via the rabbit yolk sac membrane when high levels of IgG were used, and Hemmings (1975) has shown that when low levels of IgG are used with young mice first order

kinetics are seen. Wallace & Jared (1976) in studies on the uptake of vitellogenin by *Xenopus laevis* showed that the receptor-mediated transport occurred
with a K_m of 1.5×10^{-6} M. In addition experiments on the receptor-mediated
transport of both IgG and phosvitin–lipovitellin from serum into chicken
oocytes produced an apparent K_m of 10^{-7} M (Roth, Cutting & Atlas, 1976).
These values are close to the K_D that has been observed for Fc receptor interactions.

Although such a kinetic treatment is useful for systems where direct
measurement of macromolecular transport is possible, the more general
equilibrium approach yields directly the affinity constant for the receptor–
ligand interaction. For systems at equilibrium the Langmuir adsorption
isotherm is obeyed. Thus for (1):

$$[LR] = \frac{[R]_T}{1 + K_D/[L]} \tag{3}$$

where $[R]_T$ is the total receptor concentration and is notionally identical to
V_{max} of the Michaelis–Menten equation. The constant K_D is a measure of the
affinity of the ligand for its receptor although more commonly its reciprocal,
K_A, is used since it increases in the same direction as the affinity. It can be
seen that equation (3) is in fact the Michaelis–Menten equation at early times
when the rate of accumulation of ligand is proportional to the concentration
of LR.

The method that is frequently used to obtain affinity constants is the
'displacement' or direct binding method. In order that the affinity constant
obtained is actually a true reflection of the receptor–ligand interaction it is
required that the affinities of both labelled and unlabelled ligand should be
identical. In turn the fraction of labelled ligand displaced is dependent on
both $[L]_T$ and $[R]_T$. In circumstances where, for example, the concentrations
of L and R are close to the value of K_D then the affinity of L for the receptor
would be underestimated (Jacobs *et al.*, 1975).

If these provisos are observed then the data can be analysed by a variety
of methods. The method due to Scatchard (1949) is frequently used since it is
not as susceptible as other analyses to the above-mentioned effects of L and
R on K_D. The Scatchard equation derives easily from (3) to give:

$$\frac{[LR]}{[L]} = -K_A[LR] + K_A[R]_T \tag{4}$$

If the left-hand side, which is the ratio of bound to free ligand, is plotted
against [LR] a straight line with a slope of $-K_A$ is obtained. From the inter-
cept on the LR axis the value of $[R]_T$ can be calculated.

Unkeless & Eisen (1976) have obtained a value of $K_A = 2.2 \times 10^7$ M^{-1} for the binding of monomeric IgG to a mouse macrophage cell line, using the direct binding method and Scatchard analysis. Data obtained by other workers for transporting tissues are compared in Table 1. Also shown are data which suggest the presence in a variety of cell types of Fc receptors for aggregated IgG in addition to those for the monomer. In studies on placental membranes we have been able to distinguish between the affinities of the monomeric and aggregated forms of IgG, as have Johnson, Page-Faulk & Wang (1976). However work by Frøland, Natrig & Michaelson (1974) and by Cooper, Sambray & Friou (1977) on murine leukaemia cells has demonstrated that Fc receptors for antigen—antibody complex are distinct from those for monomeric or heat-aggregated IgG. Clearly there appear to be some anomalies in all these observations, which may be resolved when careful experiments with antigen—antibody binding to the various tissues are performed. The results of Jenkinson, Billingham & Elson (1976) on the binding of antibody-sensitised erythrocytes to placental membranes, and the observations of Contractor (personal communication) on trophoblast cells in culture which exhibit the same effect, suggest that receptors for antigen—antibody complexes are present in this tissue, although no quantitative data are available. If the function of these receptors is to trap and remove by internalisation those antibodies which may be inimical to the sustaining of the foetus as an allograft then this will obviously be advantageous. Whether or not these specialised Fc receptors reside on the syncytial membrane or on the macrophage-like trophoblastic villus cells (Hofbauer cells) remains to be established. In addition it is not clear whether, if they are present on the syncytium, the complexes bound to them are internalised by a coated vesicle system or by some process unrelated to the normal transport route.

Ockleford (1977) proposed that if binding of IgG to syncytiotrophoblast plasma membrane by Fc receptor is more powerful than the binding of (fab)$_2$ to cell surface antigens then antibody will be bound by its Fc end as soon as it comes into contact with the membrane. Thus it would be bound in a configuration in which it would be unable to elicit the normal gamut of rejection responses. There are, however, a number of reasons why this may be an oversimplification of the in-vivo situation. The K_A for antigen—antibody binding is about the same as for the Fc receptor IgG binding, viz. about $10^6 - 10^7 M^{-1}$. Thus the latter interaction would not necessarily be favoured on affinity grounds. Also, the concentration of IgG normally present in maternal blood may, at times, be sufficiently high to saturate the receptor system. The published data indicate that about 50 % of receptor sites would be occupied at an IgG concentration of around $10^{-7} M$. The normal serum level of IgG is close to $10^{-4} M$. Thus there would be considerable levels of unbound IgG some of whose

Table 1. *A comparison of the molecular and kinetic properties of IgG-Fc receptors from different species*

Species and tissue type	Type of IgG receptor found	Experimental observations	Reference
Mouse macrophage	Monomeric	Measured by direct binding assay $K_A = 2.2 \times 10^7 \text{M}^{-1}$; trypsin-sensitive	Unkeless & Eisen (1976)
	Heat-aggregated or antigen–antibody complex	Observed using peroxidase cytochemistry; trypsin-resistant	Steinman & Cohn (1972)
Human placenta	Monomeric	Measured by direct binding $K_A = 4 \times 10^6 \text{M}^{-1}$ $K_A = 5 \times 10^6 \text{M}^{-1}$ $K_A \simeq 4 \times 10^7 \text{M}^{-1}$	Johnson *et al.* (1976); McNabb *et al.* (1976) Rees & Wallace (1979)
	Monomeric Heat-aggregated IgG Antigen–antibody complex	Measured by erythrocyte rosetting	Jenkinson *et al.* (1976)
Rabbit yolk sac	Monomeric	Direct binding $K_A = 5.4 \times 10^4 \text{M}^{-1}$	Tsay & Schlamowitz (1975)
Rat intestine	Monomeric and heat-aggregated	Measured by binding to intestinal loops; no kinetic or equilibrium data; monomeric receptor was trypsin-sensitive, heat-aggregated receptor trypsin-resistant	Borthistle *et al.* (1977)
Mouse leukaemia	Monomeric and antigen–antibody complex		Cooper *et al.* (1977); Frøland *et al.* (1974)

anti-foetal components would be free to pursue their normal antibody activities. In addition, since antigens are shed into the blood during the normal course of membrane turnover, soluble immune complexes would be formed. The binding of these complexes by the Fc receptors for monomeric IgG and subsequent transport in coated vesicles would then, in effect, defeat the system. This latter argument would only apply of course if the receptors do bind both monomer and complex, a supposition for which there is no clear, quantitative evidence as yet. Finally it is equally plausible that the placenta operates in a fashion similar to that of tumour tissues and that blocking factors (shed antigens for example) reduce the accessibility of the syncytial membrane to potentially harmful anti-foetal antibodies. From the foregoing discussion it is clear that there is a need to establish whether or not the placenta transports IgG and antigen complexes by the same route. If this does occur then obviously a second line of defence involving Fc receptors on trophoblast cells would be very important. There is an additional complication to the story which derives from the fact that under some circumstances antibody may be transported by a diffusion process. The observation that IgG is transported in humans during early gestation (8–12 weeks) by a diffusion-controlled mechanism (Gitlin & Biasucci, 1969) frustrates any attempt to describe antibody transport in terms of a single, simple model. This is even more confusing when one considers that early placentae possess both receptors which bind IgG (Gitlin & Gitlin, 1976) and large numbers of coated vesicles (Ockleford & Whyte, 1977). The interpretation of currently favoured transport models will be considered later in the light of such anomalies as these.

Evidence for receptors linked to transport

Much of the evidence for the premise that specific binding of certain macromolecules to receptors must precede their uptake into a specialised transport system has come from studies on the neonatal rat intestine (Jones & Waldmann, 1972; Waldmann & Jones, 1976; Rodewald, 1973) and foetal rabbit yolk sac (Wild, 1970; Schlamowitz, 1976). For example, Jones & Waldmann (1972) showed that unlabelled IgG competed with labelled IgG in the selective transfer of this molecule through the neonatal rat intestine, a system which has been shown to be saturable. In addition, Sonoda & Schlamowitz (1972) have demonstrated the selection of both rabbit serum albumin (RSA) and rabbit IgG (IgG_R) by the rabbit foetal yolk sac and have proposed that binding to separate receptors for these two proteins is the first event in the sequestration process. Roth *et al.* (1976) have reviewed the evidence for the selective uptake of various proteins by the oocyte. Anderson & Spielman (1971) firmly established that uptake of yolk protein was receptor-mediated and that it entered via coated vesicles. The receptors were seen to be located in coated 'pits', specialised regions of the plasma membrane which have been observed by many workers (Roth & Porter, 1964; Fawcett, 1965; Friend & Farquhar, 1967; Anderson, Goldstein & Brown, 1977).

Roth *et al.* (1976), in studies of the chicken oocyte system, have also located receptor-rich coated pits and have observed ferritin-labelled IgG bound to the inner membrane of coated vesicles after incubation of membrane fragments with the conjugate. The fact that specific protein could still bind to the coated pits when transport had been inhibited (by glycolytic inhibitors) and that K_m values of around 10^{-7}M or less were observed for the sequestration process argues strongly for a receptor-mediated transport mechanism. Confirmation of this type of specific uptake has been obtained by Stay (1965) for transport of the 'female protein' into moth oocytes. This was seen to occur against concentration gradients as high as 21 : 1 and coated vesicles were implicated in the transport process. It is relevant to mention that as noted by both Roth *et al.* (1976) and Nagasawa, Douglas & Schulz (1971), naked vesicles, partially coated and fully coated vesicles have been observed adjacent to the receptor pits in oocytes and pituitary glands. It may be that some of what have been observed as membrane-associated vesicles (MAVs) in transporting tissues are in fact coated vesicles that have prematurely lost their coats. If this is the case then it may explain how during transport of a particular protein both undegraded and degraded forms of the protein are produced (Wild, 1970) — those vesicles which have lost their coats would obviously be suitable candidates for fusion with lysosomes.

Further evidence for the existence of receptors localised in coated pits has come from the work of Anderson *et al.* (1977) on the uptake of low-density lipoprotein (LDL) by fibroblasts. These workers have provided probably the first evidence that unless receptors are present internalisation cannot occur. They compared the uptake, into coated vesicles, of LDL by normal human fibroblasts and by mutants from a patient with hypercholesterolaemia. In the latter cells both specific binding and internalisation of other proteins occurred normally but with LDL only binding to the coated pits was observed. The absence of internalisation of LDL by the mutant cells was correlated with a receptor defect. An alternative explanation could be a defect in the ability of the coat protein to form the interlocking hexagonal/pentagonal units (Pearse, 1975) necessary for vesicle formation, although this is less likely if the other proteins that are internalised by the mutant (e.g. epidermal growth factor) themselves enter via coated vesicles.

The presence of coated pits as regions of the plasma membrane whose function may be to localise receptors has also been noted by Rodewald (1973) in the transport of IgG by neonatal rat jejunum. Rodewald's observations using ferritin- or peroxidase-labelled IgG suggested a model for transport (to be elaborated in the next section) which intimately ties together receptor binding and internalisation, although it is not entirely clear whether or not the presence of a coated pit is a prerequisite for coated vesicle formation.

In summary it would be fair to say that for a large number of proteins the criteria for a receptor-mediated transport process have been satisfied. Table 2 reviews the current situation with regard to some of the transport systems that have been well studied.

Molecular aspects of receptors

A large part of our knowledge of receptors for macromolecules involved in transport has derived from studies on IgG-Fc receptors. The characterisation of these receptors, however, is still very much in its early stages, a fact that is attested to by the failure of numerous attempts to solubilise a specific receptor molecule (see for example Waldmann & Jones, 1976, and Hillman, Schlamowitz & Shaw, 1977).

Those results which have been reported suggest that the receptor is a protein (with the exception of Hemmings's (1975) mechanism in which a specific receptor is not implicated at all in the selection of proteins at the plasma membrane but the general 'stickiness' of the glycocalyx is relied on), since it is susceptible to trypsin and papain, but that any carbohydrate present does not affect the capacity of the receptor to bind IgG (Hillman *et al.*, 1977). Also it appears that the interaction of IgG with the membrane-bound receptor involves a degree of ionic bonding since the maximum observable binding is strongly dependent on pH (Balfour & Jones, 1976) and ionic strength (Hillman *et al.*, 1977; Rees & Wallace, 1979).

Fig. 1 illustrates the latter effect on the binding of human IgG to placental plasma membranes at pH 7. Since *in vivo* the tonicity of the maternal blood is likely to be mimicked more closely by Tris-buffered saline (TBS) than by 10 mM Tris buffer, the significance of the very high binding observable at this latter ionic strength has to be interpreted with caution. The fact is however that the binding is specifically competed for by unlabelled IgG and at the lower ionic strengths the affinity is increased. Also it is not possible to measure accurately the ionic strength of the blood in direct contact with the plasma membrane and even less possible to estimate the dielectric constant of this environment; see for example Mehrishi (1976) for a discussion of physico-chemical events at membrane surfaces.

It may be that ionic strength effects such as these are responsible for the differences that have been observed by different workers when measuring the

Table 2. *A summary of some receptor-linked transport systems*

Species and tissue	Labelled protein	Is it transported selectively?	Has a receptor been identified?	Has it been observed in coated vesicles?	References
Rabbit yolk sac	IgG rabbit	Yes	Yes	Yes	Wild (1970); Sonoda & Schlamowitz (1972); Hemmings (1976)
	IgG bovine	No	–	Yes/No	
	Rabbit serum albumin	Yes	Yes	–	
Rat proximal jejunum	IgG[a] rat	Yes	Yes	Yes	Jones & Waldmann (1972); Rees & Wallace (in press); Rodewald (1973); Morris & Morris (1976); Waldmann & Jones (1973)
	IgG bovine	No	No	No	
	Ferritin	No	No	–	
	BSA	No	No	–	
Mouse intestine	IgG1[b]	Yes			Guyer et al. (1976)
	IgG2a[b]	Yes	Yes[c]	Not known	
	IgG2b[b]	Yes			
	IgG3[b]	Yes			
	IgA	No			
	IgM	No			
	Fc	Yes			
	Fab	No			
Human placenta	IgG1[b]	Yes			Mellbye et al. (1970); McNabb et al. (1976); Rees & Wallace (unpublished data)
	IgG2[b]	Yes		Not known	
	IgG3[b]	Yes			
	IgG4[b]	Yes			
	Fc	Yes			
	IgE	No			

Human placenta	IgM	No			Not known	Johnson et al. (1976)
	F(ab)$_2$	Yes/No				Balfour & Jones (1977)
	IgG-coated erythrocytes	–	Yes			Jenkinson et al. (1976)
	Heat-aggregated IgG	–	Yes			Rees & Wallace (1979)
						Gitlin & Gitlin (1976)
Chicken & mosquito (oocytes)	Vitellogenin	Yes	Yes	Yes	$K_m = 10^{-7}$ M	Roth et al. (1976)
	IgG	Yes	Yes	Yes		
Culture of human fibroblasts	Low-density lipoprotein	Yes	Yes	Yes	Yes	Anderson et al. (1977)

[a] Including various separated ionic species.
[b] Myeloma proteins.
[c] The subclasses have different affinities for the same receptor.

binding of IgG subclasses to placental membranes. McNabb, Koh, Dorrington
& Painter (1976) reported the binding of human subclasses in the approximate
ratios (our approximations) $1:0.18:0.96:0.57$ for IgG1, 2, 3 and 4 respectively. Balfour & Jones (1977), however, report ratios for the same subclasses
of $0.19:1.0:0.74:0.2$. The former authors carried out the binding in TBS of
pH 7.4 whereas the latter experiments were performed in 5 mM EDTA buffer
at pH 6.5. The combined effect of pH and ionic strength differences could
have been sufficient to account for the disparity in the observed affinities.

The difficulties of obtaining soluble receptor for monomeric IgG has not
been experienced with similar studies on the Fc receptor from other sources,
e.g. the lymphocyte or macrophage. The Fc receptors on these cells are,
however, in the main, for IgG—antigen complexes and seem to be generally
less susceptible to degradation during preparation. Anderson & Grey (1977)
have reported the solubilisation and partial characterisation of Fc receptors
from macrophage-like cell lines and various other tumour lines. These authors
conclude that the receptor isolated from a mastocytoma (P815) is both trypsin
and phospholipase sensitive, suggesting the presence of a lipid moiety in

Fig. 1. The effect of salt concentration on the binding of IgG to
placental membranes. Percentage binding is that measured for 250 μg
membrane protein incubated with ^{125}I-labelled IgG at approximately
10^{-10} M in Tris buffer, pH 7.0, containing 0.2 % bovine serum albumin.
The effect of salt is produced whether Tris buffers or phosphate
buffers are used and whether the buffer ion itself or sodium chloride
is used to obtain the higher ionic strengths. (Compare the data by
Hillman et al., 1977.)

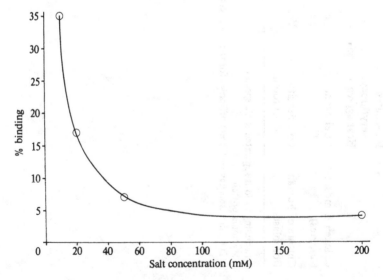

addition to protein. On the macrophage lines there appear to be receptors for both monomeric and antigen-complexed IgG but there is some question regarding the lipid content.

More recently Bourgois, Abney & Parkhouse (1977) have reported the detergent solubilisation of Fc receptors from mouse lymphocytes, macrophages and fibroblasts. A single chain molecule of 120 000 mol. wt was obtained which was highly susceptible to trypsin. Fragments obtained after trypsin treatment varied in molecular weight between 10 000 and 75 000, a range which would encompass some of the other reported values for this molecule (Rask, Klareskog, Ostberg & Peterson, 1975; Wernet & Kunkel, 1975; Cooper & Sambray, 1976). More recently still, Cooper *et al.* (1977) have isolated separate Fc receptors for IgG—antigen complex and monomeric IgG from a murine leukaemia. They extracted four proteins in the molecular weight range 28 000 to 125 000. When the higher molecular weight material was exposed to proteases two further smaller fragments (60 000 and 74 000) were obtained. This result, although apparently consistent with the suggestion of Bourgois *et al.* (1977) that the Fc receptor is trypsin-sensitive and that the degradation products are only seen after reduction and alkylation of cysteine residues, is interpreted by Cooper as indicative of heterogeneous Fc receptor populations. Of the latter worker's four fragments, the largest is proposed as a receptor for antigen—antibody complex whereas the smaller fragments (which might be part of a larger subunit system in the membrane) are specific for both monomeric and heat-aggregated IgG. This latter deduction is in contrast to the results of Borthistle, Kubo, Brown & Grey (1977) on epithelial cells of the neonatal rat intestine where receptors for monomeric IgG were distinguishable from those for aggregated IgG (the former were found to be trypsin-sensitive whereas the latter were not) and Rees & Wallace (1979) who obtained the same result for placental membranes. The ability of heat-aggregated IgG to fix complement is only about 1 % that of antigen—antibody complex, and if this fact is used to index its efficiency in mimicking the complex then it may not be surprising that different results are obtained depending on which ligand is used, especially as the conformational state of the Fc portion of IgG seems to be so critical for the correct receptor interaction.

Clearly then the Fc receptor story is not an uncomplicated one although many of the apparent contradictions look as though they are slowly being resolved. It may be that the monomeric IgG receptor observable on macrophages, leukaemia lines and the like is not related to the receptor in transporting tissues, although this would complicate the situation still further since it would be necessary to postulate three distinct types of Fc receptor. Perhaps the observation that coated vesicles are present in macrophages (Nagura &

Asai, 1976) lends weight to the argument that receptors for monomeric IgG are identical in all tissues within the same species. Whatever the case a method for isolating the transporting Fc receptor is long overdue. Only after its isolation and characterisation will a clearer understanding emerge of the role of the receptor in the induction of coated vesicle formation.

Models for receptor-mediated transport

In the following paragraphs we will attempt to review the current ideas about how the receptor molecule can be of use as a part of a larger mechanism for carrying proteins across interfacial cell layers. Consider again what the system is required to perform: a protein (e.g. IgG) which is highly adapted to maintaining its structure and function in an aqueous environment of particular ionic conditions, i.e. blood, has to traverse two lipid bilayers the composition of which bear little resemblance to the normal vascular milieu. If it were possible to maintain the protein topologically on the outside of the cell within a medium favourable to the maintenance of its structural integrity then this would obviously be advantageous. The process of endocytosis serves just this function and enables proteins bound at the membrane to be deposited by vesicles in the cytoplasm or at whatever intracellular destination is required. Selectivity can then be achieved by the mediation of a receptor which will ensure the entry into the vesicle of specific proteins only.

Three important aspects of this apparently facile process need to be considered:

(i)   How so much membrane, complete with its receptors, can be withdrawn from the plasma membrane pool without depletion and how receptors are returned to it, if at all.

(ii) How the vesicle 'finds its way' across the cell and is able to fuse with a membrane when it is covered by a protein coat.

(iii) Since the transport system is actually pumping selected protein, often against a large gradient, and yet does not seem to be heavily reliant on metabolic energy, where does the energy for the vesiculisation originate.

To gain some insight into the possible solutions to these questions it is necessary to discuss the models of protein transport that are currently in favour. These facts have been presented in other chapters but we will attempt to reinterpret them with an eye to the involvement of receptors. The evidence for and against these models will be summarised later.

The existence of differentiated regions of plasma membrane known as coated pits was proposed by Kanaseki & Kadota (1969) to explain the selective transport of proteins. Their proposal involved the binding of protein by receptors followed by formation of a coated vesicle. The interaction of protein with its receptor results in a conformational change in the receptor which is transmitted to the coat protein underlying the plasma membrane. This in turn induces curvature in the membrane and eventual pinching-off of the vesicle.

Ockleford & Whyte (1977) have proposed that formation of receptor-coated vesicle complexes is essentially a process of patching, analogous to lectin-induced aggregation of antigens in B-lymphocytes. Receptors are normally able to diffuse freely in the plane of the membrane and become associated in some way with the coat protein subunits. Random association between these molecules is short lived until ligand is bound, when large curved patches are formed which ultimately pinch-off from the membrane. The evidence for and against these models for coated vesicle formation will be discussed shortly.

There are many proposed fates for receptor—vesicle complexes, each involving a distinct role for the receptor. These models are rationalised in Fig. 2. The earliest and most complete model is that of Brambell (1970), who proposed that endocytic vesicles when formed at the plasma membrane contain not only the receptor-selected proteins but also non-receptor-bound luminal protein that is fortuitously included during endocytosis. The vesicle once

Fig. 2. The fate of coated vesicles and the role of receptor according to various models of transcellular protein transport.

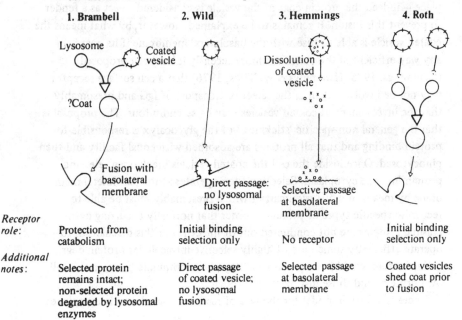

	1. Brambell	2. Wild	3. Hemmings	4. Roth
Receptor role:	Protection from catabolism	Initial binding selection only	No receptor	Initial binding selection only
Additional notes:	Selected protein remains intact; non-selected protein degraded by lysosomal enzymes	Direct passage of coated vesicle; no lysosomal fusion	Selected passage at basolateral membrane	Coated vesicles shed coat prior to fusion

formed traverses the entire cell and expels its contents at the basolateral membrane by exocytosis. Selectivity is achieved *en route* by fusion of the vesicle with a lysosome. The receptor serves to protect the selected protein from degradation by lysosomal enzymes while unbound protein is degraded. The model therefore predicts that, at the basolateral membrane, the vesicles release intact receptor-selected protein, degraded unselected protein and lysosomal enzymes. It also presupposes that these vesicles have a large enough internal diameter to include a significant quantity of unbound protein, a supposition that may only be justified for vesicles larger than coated vesicles. As observed by Woods, Woodward & Roth (1978) coated vesicles have a wide range of sizes (50 nm to greater than 100 nm). If it is the small vesicles that are primarily responsible for selective transport then the internal volume would be too small to permit the inclusion of luminal protein. Again if the protein coat is an important part of the transport process and serves to protect the vesicle during its hazardous passage through the cell then why should a selection be necessary at the lysosomal level when selectivity has already been achieved at the receptor level? A mechanism by which an uncoated vesicle (e.g. a lysosome) fuses with the membrane of a coated vesicle and thus penetrates the protein coat has not been proposed.

Wild (1975) proposes that selectivity is achieved at the level of receptor binding and that no significant quantity of non-receptor-bound protein is included, and thus that fusion with lysosomes need not and does not take place – indeed the protein coat of the vesicle is considered to act as a fender to prevent this fusion. It remains to be explained, however, by what means the coated vesicle is able to fuse with the basolateral membrane if its coat is in any way efficient in this function. More recently it has been proposed (Hemmings, 1975; Hemmings & Williams, 1976) that a cell surface receptor may not be involved at all in the selective transport of IgG and presumably that the involvement of coated vesicles may be serendipitous. The proposal is that the general non-specific 'stickiness' of the glycocalyx is responsible for protein binding and that all proteins are adsorbed with equal facility and then pinocytosed. Once inside the cell the coated vesicles break open releasing protein into the cytoplasm. Selection then operates at the basolateral membrane by means of a diffusion carrier which presumably must be able to recognise specific types of protein in order that normally occurring cytoplasmic proteins are not conducted out of the cell. For this mechanism to operate efficiently some kind of highly selective intracellular protein degradation system would be required which would discriminate between the cells own proteins and those just 'passing through'.

There is a further model for the fate of receptor-coated vesicle complexes which derives from the work of Roth *et al.* (1976) on vitellogenesis of mosquito and chicken oocytes and which states that the coated vesicles containing receptor-bound lipovitellin and phosvitin (LvPv) shed their protein coats prior to fusion with each other to form nascent yolk granules. This model is relevant to the previous discussion concerning the fusion of coated vesicles with other membranes and will be returned to in the next section.

It will be apparent by now that none of the models cited has proposed a fate for the receptor in the transport process. Wild & Dawson (1977) have observed the binding of antibody-coated erythrocytes to sites other than the plasma membrane and have suggested that this may be due to free receptors that are released when fusion of coated vesicles with the basolateral membrane and subsequent loss of bound protein occurs. If this suggestion is followed to its logical conclusion it must be assumed that during continuous priming of the system with a specific protein (e.g. IgG) and transport of it to the baso-lateral membrane, this latter membrane would soon itself become saturated with receptors unless a further receptor-specific translocation mechanism operates. A more complete explanation has been offered by Anderson *et al.* (1977), which suggests that receptors in coated vesicles are recycled after vesicle breakdown in the cytoplasm, although a possible mechanism for the recycling process has not been proposed.

In the next section we will review the evidence that has accumulated for and against the receptor-mediated transport models with particular emphasis on morphological studies.

Evidence for the various models

The pioneering work was that of Rodewald (1973) whose very high quality electron micrographs provided the first real evidence for the vesicular uptake of IgG by the rat intestine. Using ferritin-labelled and antiperoxidase-labelled IgG Rodewald demonstrated the passage of IgG across the proximal intestinal epithelium of the neonatal rat. The transported protein was seen to adhere to the inner surface of the vesicle, an observation that suggested the presence of receptors. In contrast, the uptake of the same proteins into large vacuoles in the distal end of the intestine implied the operation of a non-selective mechanism. These early observations correlate well with the more recent quantitative results of Morris & Morris (1976) on the respective roles of proximal and distal regions of the rat intestine.

However, Rodewald also observed that the initial uptake of label in the proximal intestine took place via complex tubular vesicles with passage into distinct coated vesicles occurring some distance across the cell. Further, no fusion with lysosomes was observed and therefore a 'Wild-type' mechanism may be appropriate. Wild (1976) has provided more recent evidence for coated vesicle uptake in the rabbit yolk sac endothelium using peroxidase-labelled IgG, when he observed the same sequence of events as Rodewald.

In direct contrast to these observations Hemmings (1976), using autoradiography and ferritin-labelled IgG from homologous and heterologous carriers, has seen the labelled molecules distributed throughout the cytoplasmic mass and not in discrete vesicles. The apparent contradictions in the results of these various groups of workers have yet to be resolved.

Unfortunately, due to the impossibility of in-vivo experiments, the route of labelled proteins in the human placenta has not yet been mapped. The observation, that the 8–12 week human placenta possesses both receptors for IgG (Gitlin & Gitlin, 1976) and coated vesicles (Ockleford & Whyte, 1977) but does not 'actively' transport IgG, may cast some doubt on the linked role of these components and provide the proponents of the Hemmings model with some encouragement. Ockleford, Whyte & Bowyer (1977) have estimated the number of coated vesicles in a term placenta as about 10^{12} (the first-trimester placenta though smaller appears to contain similar numbers of coated vesicles per unit area of syncytial surface to the term placenta) and Gitlin & Gitlin (1976) have demonstrated binding to the 60 000g plasma membrane fraction. It would seem therefore that the ability to bind IgG and to form coated vesicles is not of itself conclusive of an ability to transport the molecule, at least in the aforementioned system.

The work of Roth *et al.* (1976) on vitellogenesis has provided evidence that coated vesicles can shed their coats prior to fusion with other membrane organelles, but a much more interesting aspect of the work concerns the release of internal protein from the inner face of the vesicle coincident with loss of the protein coat. This may reflect the intimate interaction between receptor and coat protein subunits. Although Friend & Farquhar (1967) have also observed the shedding of vesicle coats prior to fusion in the rat vas deferens, since this concerned the non-specific uptake of horseradish peroxidase a receptor mediation could not be implicated. It would seem, therefore, that it is possible for coated vesicles to be formed without the receptor–coat protein interaction that is postulated in certain of the models and by Roth's work on vitellogenesis. What is even more perplexing is that the electron micrographs of Friend & Farquhar show labels closely associated with the inner face of the vesicle. The implication is that even in the absence of a receptor, coated vesicles may form and may contain luminal protein apparently bound to the inner plasma membrane. Hence the conclusion that intimate association of label with the inner vesicle face is proof of receptor-bound protein, may require a reappraisal if the same effect can be induced by non-selective uptake (compare Hemmings's model). The work of Anderson *et al.* (1977) on the uptake of cholesterol bound to low-density lipoprotein (LDL) by human fibroblasts is interesting for a number of reasons. Firstly it seems to have provided evidence for a receptor-coated vesicle transport system that is reliant on the integrity of the receptor moiety for vesicle formation. Thus fibroblasts from an individual with a receptor-mislocation mutation were unable to sequester LDL although the ability to bind the carrier protein in coated pits was not impaired. The model of Kanaseki & Kadota (1969) would seem to fit this system although if, as is proposed, a conformational change in the receptor on binding its ligand is required to trigger the coat protein into a conformational state that physically induces curvature in the membrane, then it might be difficult to reconcile the conclusions of Friend & Farquhar with this model.

Finally, with regard to the fate of receptor molecules, what evidence exists is suggestive of a recycling mechanism. In the work of Anderson *et al.* (1977) it was demonstrated that LDL incorporation into vesicles continued for up to six hours after inhibition of protein synthesis had been effected. The membrane turnover time was about 10 minutes and these authors concluded that the only sensible interpretation of their results was the operation of a receptor recycling system. In order for this to occur efficiently the coated vesicles are required to lose their coats within a very short distance from the plasma membrane. The coat proteins (clathrin and others: see Pearse, 1975, and Woods *et al.*, 1978) are then returned to the membrane along with any receptors which are present. Whether or not this hypothesis could be realistically applied to other transport systems remains to be seen, but it is possibly not coincidental that the majority of coated vesicles in the human placenta are observed to be within a region less than 540 nm from the syncytiotrophoblast membrane (Ockleford & Whyte, 1977).

Discussion

From the foregoing text it is clear that the mechanism of specific protein transport is more complex than any one of the existing models proposes. Perhaps the apparent complexity arises because of the variety of different systems being studied, although the almost universal distribution and properties of the coat protein(s) and the similarity of coated vesicle appearance may suggest that a common mechanism is being observed from different aspects.

Some of the functions ascribed to coated vesicles that are relevant to receptor-mediated transport of proteins are depicted in Fig. 3. Whether all of these processes occur in all systems and it is the variance in their relative quantitative importance for a particular metabolic state of the cell that confuses us, or whether some of the observations leading to these functions have been misinterpreted, time will tell. What is of fundamental importance, however, is to establish the existence of a specific receptor for a protein that has been observed in coated vesicles. This receptor could then be used to test some aspects of the various models – for example whether receptor-bound protein complex is less susceptible to lysosomal enzyme degradation than free protein, and perhaps whether an interaction between receptor and isolated coat proteins can be demonstrated. It would also seem fundamental to the whole idea of coated vesicle transport to isolate vesicles which actually contain both receptor and transported protein. Again these could be assessed for their aptitude in undergoing fusion with lysosomes etc. A substantial bonus to such a demonstration would derive from the fact that isolation of a large number of such receptor-containing vesicles would result in a greatly enriched receptor preparation. Thus, for example, perfusion of a rat intestinal segment or a human placenta with IgG might induce the formation of coated vesicles which could then be isolated. A number of workers are engaged in this type of approach and it is to be hoped that their efforts will reveal something of the nature of the link between the receptor and the coated vesicle.

Fig. 3. Some functions ascribed to coated vesicles in receptor biology. (1) Transfer of lysosomal enzymes to cell surface (Lloyd, 1977). (2) Transfer of lysosomal enzymes to secondary lysosomes (Novikoff, 1964). (3) Protein uptake and lysosomal fusion (± coat) (Brambell, 1970). (4) Protein uptake and transcellular passage only (Wild, 1975).

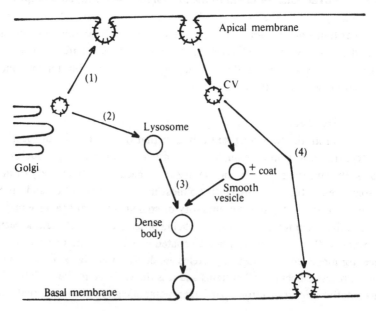

References

Anderson, C.L. & Grey, H.M. (1977). Solubilisation and partial characterisation of cell membrane Fc receptors. *Journal of Immunology*, 118, 819–25.

Anderson, R.G.W., Goldstein, J.L. & Brown, M.S. (1977). A mutation that impairs the ability of lipoprotein receptors to localise in coated pits on the cell surface of human fibroblasts. *Nature, London*, 270, 695–99.

Anderson, W.A. & Spielman, A. (1971). Permeability of the ovarian follicle of *Aedes aegypti* mosquitoes. *Journal of Cell Biology*, 50, 201–21.

Balfour, A. & Jones, E.A. (1976). The binding of plasma proteins to human placental cell membranes. *Clinical Science and Molecular Medicine*, 52, 383–94.

Borthistle, B.K., Kubo, R.T., Brown, W.R. & Grey, H.M. (1977). Studies on receptors for IgG on epithelial cells of the rat intestine. *Journal of Immunology*, 119, 471–76.

Bourgois, A., Abney, E.R. & Parkhouse, R.M. (1977). Structure of mouse Fc receptor. *European Journal of Immunology*, 7, 691–5.

Brambell, F.W.R. (1970). *The Transmission of Passive Immunity from Mother to Young. Frontiers of Biology 18*. Amsterdam: North-Holland.

Clark, A.J. (1926a). The reaction between acetyl choline and muscle cells. *Journal of Physiology*, 61, 530–46.

Clark, A.J. (1926b). The antagonism of acetyl choline by atropine. *Journal of Physiology*, 61, 547–56.

Cooper, S.M. & Sambray, Y. (1976). Isolation of a murine leukaemia Fc receptor by selective release induced by surface redistribution. *Journal of Immunology*, 117, 511–17.

Cooper, S.M., Sambray, Y. & Friou, G.J. (1977). Isolation of separate Fc receptors for IgG complexed to antigen and native IgG from a murine leukaemia. *Nature, London*, 270, 253–5.

David, G.S. (1972). Solid state lactoperoxidase: a highly stable enzyme for simple, gentle iodination of proteins. *Biochemical and Biophysical Research Communications*, 48, 464–71.

Fawcett, D.W. (1965). Surface specialisations of absorbing cells. *Histochemistry and Cytochemistry*, 13, 75–91.

Friend, D.S. & Farquhar, M.A. (1967). Functions of coated vesicles during protein absorption in the rat vas deferens. *Journal of Cell Biology*, 35, 357–76.

Frøland, S.S., Natrig, J.B. & Michaelson, T.E. (1974). Binding of aggregated IgG by human B lymphocytes independant of Fc receptors. *Scandinavian Journal of Immunology*, 3, 375–80.

Gitlin, D. & Biasucci, A. (1969). Development of γG, γA, γM, βIC/βIA, CI esterase inhibitor, ceruloplasmin, transferrin, hemopexin, haptoglobulin, fibrinogen, plasminogen, α_1-antitrypsin, orosomucoid β-lipoprotein, α_2-macroglobulin and prealbumin in the human conceptus. *Journal of Clinical Investigation*, 48, 1433–46.

Gitlin, J.D. & Gitlin, D. (1976). Protein binding by cell membranes and the selective transfer of proteins from mother to young across tissue barriers. In *Maternofoetal Transmission of Immunoglobulins*, ed. W.A. Hemmings, *Clinical and Experimental Immunoreproduction 2*, pp. 113—21. Cambridge University Press.

Guyer, R.L., Koshland, M.E. & Knopf, P.M. (1976). Immunoglobulin binding by mouse intestinal epithelial cell receptors. *Journal of Immunology*, 117, 587—93.

Hemmings, W.A. (1975). The variation of protein degradation with administered protein concentration in young mice fed mouse IgG: a test of the Brambell hypothesis. *IRCS Medical Science*, 3, 249.

Hemmings, W.A. & Williams, E.W. (1976). The attachment of IgG to cell components of transporting membranes. In *Maternofoetal transmission of Immunoglobulins*, ed. W.A. Hemmings, *Clinical and Experimental Immunoreproduction 2*, pp. 91—111. Cambridge University Press.

Hillman, K., Schlamowitz, M. & Shaw, A.R. (1977). Characterisation of IgG receptors of the foetal rabbit yolk sac membrane. *Journal of Immunology*, 118, 782—8.

Jacobs, S., Chang, K. & Cuatrecasas, P. (1975). Estimation of hormone-receptor affinity by competitive displacement of labelled ligand. *Biochemical and Biophysical Research Communications*, 66, 687—92.

Jenkinson, E.T., Billingham, W.G. & Elson, J. (1976). Detection of receptors for immunoglobulin on human placenta by EA rosette formation. *Journal of Clinical and Experimental Immunology*, 23, 456.

Johnson, P.M., Page-Faulk, W. & Wang, A.-C. (1976). Immunological studies of human placentae: subclass and fragment specificity of binding of aggregated IgG by placental endothelial cells. *Immunology*, 31, 659—64.

Jones, E.A. & Waldmann, T.A. (1972). The mechanism of intestinal uptake and transcellular transport of IgG in the neonatal rat. *Journal of Clinical Investigation*, 51, 2916—27.

Kanaseki, T. & Kadota, K. (1969). The 'vesicle in a basket'. *Journal of Cell Biology*, 42, 202—20.

Langley, J.N. (1878). On the physiology of the salivary secretion. *Journal of Physiology*, 1, 339—69.

Lloyd, J.B. (1977). Cellular transport of lysosomal enzymes — an alternative hypothesis. *Biochemical Journal*, 164, 281—2.

McNabb, T., Koh, T.Y., Dorrington, K.J. & Painter, R.H. (1976). Structure and function of immunoglobulin domains. *Journal of Immunology*, 117, 882—8.

Mellbye, O.J., Natvig, J.B. & Kranstein, B. (1970). Presence of IgG subclasses and Clq in human cord sera. In *Protides of the Biological Fluids*, vol. 18, ed. H. Peeters, pp. 127—31. Oxford: Pergamon Press.

Mehrishi, J.N. (1976). Cell surface membrane interactions in reproductive biology. In *Maternofoetal Transmission of Immunoglobulins*, ed. W.A. Hemmings, *Clinical and Experimental Immunoreproduction 2*, pp. 25—46. Cambridge University Press.

Morris, B. & Morris, R. (1974). The absorption of ^{125}I-labelled IgG by different regions of the gut in young rats. *Journal of Physiology*, 241, 761—70.

Morris, B. & Morris, R. (1976). Quantitative assessment of the transmission of labelled protein by the proximal and distal regions of the small intestine of young rats. *Journal of Physiology,* 255, 619–34.

Morris, I.G. (1964). The transmission of antibodies and normal γ-globulins across the young mouse gut. *Proceedings of the Royal Society of London, Series B,* 160, 276–92.

Nagasawa, J., Douglas, W.W. & Schulz, R.A. (1971). Coated microvesicles in neurosecretory terminals of posterior pituitary glands shed their coats to become smooth synaptic vesicles. *Nature, London,* 232, 340–2.

Nagura, H. & Asai, J. (1976). Pinocytosis by macrophages: kinetics and morphology. *Journal of Cell Biology,* 70, 265A.

Novikoff, A.B. (1964). GERL, its form and function in neurons of rat spinal ganglia. *Biological Bulletin,* 127, 358A.

Ockleford, C.D. (1977). Antibody clearance by micropinocytosis: a possible role in foetal immunoprotection. *Lancet,* i, 310.

Ockleford, C.D. & Whyte, A. (1977). Differentiated regions of human placental cell surface associated with exchange of materials between maternal and foetal blood: coated vesicles. *Journal of Cell Science,* 25, 293–312.

Ockleford, C.D., Whyte, A. & Bowyer, D.E. (1977). Variation in the volume of coated vesicles isolated from human placenta. *Cell Biology International Reports,* 1, 137–46.

Olsen, R.W., Mennier, J.C. & Changeux, J.P. (1972). Progress in the purification of the cholinergic receptor protein from *Electrophorus electricus* by affinity chromatography. *FEBS Letters,* 28, 96–100.

Paton, W.D.M. (1961). A theory of drug action based on the rate of drug–receptor combination. *Proceedings of the Royal Society of London, Series B,* 154, 21–69.

Pearse, B.M.F. (1975). Coated vesicles from pig brain: purification and biochemical characterisation. *Journal of Molecular Biology,* 97, 93–8.

Rask, L., Klareskog, L., Ostberg, L. & Peterson, P.A. (1975). Isolation and properties of a murine spleen cell Fc receptor. *Nature, London,* 257, 231–3.

Rees, A.R. & Wallace, K. (1979). Approaches to structural studies on Fc receptors. In *Protein Transmission through Living Membranes,* ed. W.A. Hemmings, pp. 119-27. Amsterdam: Elsevier/North Holland.

Rees, A.R. & Wallace, K. (in press). Studies on the IgG-Fc receptor from neonatal rat small intestine. *Biochemical Journal.*

Rodewald, R. (1973). Intestinal transport of antibodies in the newborn rat. *Journal of Cell Biology,* 58, 189–211.

Roth, T.F. & Porter, K.R. (1964). Yolk protein uptake in the oocyte of the mosquito *Aedes aegypti. Journal of Cell Biology,* 20, 313–31.

Roth, T.F., Cutting, J.A. & Atlas, S.B. (1976). Protein transport: a selective membrane mechanism. *Journal of Supramolecular Structure,* 4, 527–48.

Scatchard, G. (1949). The attractions of proteins for small molecules and ions. *Annals of the New York Academy of Sciences,* 51, 660.

Schlamowitz, M. (1976). Membrane receptors in the specific transfer of immunoglobulins from mother to young. *Immunological Communications,* 5, 481–500.

Sonoda, S. & Schlamowitz, M. (1972). Kinetics and specificity of transfer of IgG and serum albumin across rabbit yolk sac in utero. *Journal of Immunology*, **108**, 807.

Stay, B. (1965). Protein uptake in the oocytes of the *Cecropia* moth. *Journal of Cell Biology*, **26**, 49–62.

Steinman, R.M. & Cohn, Z.A. (1972). The interaction of particulate horse-radish peroxidase (HRP)–anti-HRP immune complexes with mouse peritoneal macrophages *in vitro*. *Journal of Cell Biology*, **55**, 616–34.

Tsay, D.D. & Schlamowitz, M. (1975). Comparison of the binding affinities of rabbit IgG fractions to the rabbit foetal yolk sac membrane. *Journal of Immunology*, **115**, 939–42.

Unkeless, J.C. & Eisen, H.N. (1976). Binding of monomeric immunoglobulins to Fc receptors of mouse macrophages. *Journal of Experimental Medicine*, **142**, 1520–33.

Waldmann, T.A. & Jones, E.A. (1976). The role of IgG-specific cell surface receptors in IgG transport and catabolism. In *Maternofoetal Transmission of Immunoglobulins*, ed. W.A. Hemmings, *Clinical and Experimental Immunoreproduction 2*, pp. 123–36. Cambridge University Press.

Wallace, R.A. & Jared, D.W. (1976). Protein incorporation by isolated amphibian oocytes. V. Specificity for vitellogenin incorporation. *Journal of Cell Biology*, **69**, 345–51.

Wernet, P. & Kunkel, H.B. (1975). Immunochemical aspects of Ia type B cell membrane specificities as candidates for a human MLC stimulator antigen system: molecular heterogeneity and distinction from the Fc receptors. In *Histocompatibility Testing*, ed. F. Kissmeyer-Nielson, pp. 731–4. Copenhagen: Munksgard.

Wild, A.E. (1970). Protein transmission across the rabbit foetal membranes. *Journal of Embryology and Experimental Morphology*, **24**, 313–30.

Wild, A.E. (1975). Role of the cell surface in selection during transport of proteins from mother to foetus and newly born. *Philosophical Transactions of the Royal Society, Series B*, **271**, 395–407.

Wild, A.E. (1976). Mechanism of protein transport across the rabbit yolk sac endoderm. In *Maternofoetal Transmission of Immunoglobulins*, ed. W.A. Hemmings, *Clinical and Experimental Immunoreproduction 2*, pp. 155–65. Cambridge University Press.

Wild, A.E. & Dawson, P. (1977). Evidence for Fc receptors on rabbit yolk sac endoderm. *Nature, London*, **268**, 443–5.

Woods, J.W., Woodward, M.P. & Roth, T.F. (1978). Common features of coated vesicles from dissimilar tissues: composition and structure. *Journal of Cell Science*, **30**, 87–97.

9

Coated secretory vesicles

JÜRGEN KARTENBECK

Studies of the secretory process during the past decades have shown that materials destined for secretion, such as protein, glycoproteins, polysaccharides, and associated minerals are synthesised, transported and released in membrane-bound compartments. In the case of protein and glycoprotein secretion for example, such compartments include rough and smooth endoplasmic reticulum (ER), the Golgi apparatus, the condensing vacuoles, and the secretory vesicles (for review see Palade, 1975). The fusion of the secretory vesicle membrane with the plasma membrane leads to the discharge of the secretory product into the extracellular lumen and to the incorporation of the vesicle membrane components into the plasma membrane, be it transitory or for extended periods of time (for discussion see Franke & Kartenbeck, 1976). The transport of the secretion product along the discontinuous pathway through the cell occurs by fusions and fissions of the membranes involved. The recognition and interaction between these membranes is limited to definite partners and apparently takes place with a high degree of specificity (Palade, 1975). Examples include the fusion of ER-derived vesicles with the Golgi apparatus and the fusion of ER-derived or Golgi-apparatus-detached secretory vesicles with the plasma membrane. Gradual changes with respect to enzyme activities, lipid, carbohydrate and polypeptide composition, and the morphology of the specific membranes or secretory products, as well as the different turnover rates of the membrane polypeptides, might be explained by the existence of local membrane heterogeneities and the selection of membrane domains that are involved in the vesicle flow processes, as opposed to other membrane domains that do not participate in secretory membrane translocations (Morré, Keenan & Huang, 1974; Franke & Kartenbeck, 1976). These local membrane differentiations might be induced by intracisternal components, e.g. by the secretory products themselves, or by specific membrane-associated cytoplasmic elements.

In special situations local membrane differentiation is obviously correlated with the association of characteristic coat structures with cytoplasmic membrane surfaces. The appearance of a special form of fuzzy coat structure (Roth & Porter, 1964) can be correlated with the involvement in various membrane translocation processes, including endocytic phenomena such as uptake of extracellular macromolecules and membrane resorption as well as membrane transport phenomena between specific intracellular compartments such as ER, Golgi apparatus, lysosomes, and multivesicular bodies (for references see Roth & Porter, 1964; Fawcett, 1965; Friend & Farquhar, 1967; Palade & Bruns, 1968; Lagunoff & Curran, 1972; Heuser & Reese, 1973; Geuze & Kramer, 1974; Anderson, Brown & Goldstein, 1977; Herzog & Farquhar, 1977; Farquhar, 1978).

Recent observations in various animal and plant cells have also demonstrated the appearance of such coat structures on vesicles which appear to be involved in secretion and exocytosis (Franke & Herth, 1974; Franke *et al.*, 1976*a, b*, 1977; Kartenbeck, Franke & Morré, 1977; Herth, 1979). The possible participation of coated vesicles in secretory processes has been repeatedly discussed by several authors (e.g. Sheffield, 1970; Bonneville, Weinstock & Wilgram, 1968; Droz, 1969) but in view of the lack of morphologically identifiable secretion products in most situations examined a clear interpretation has not been possible. The recent descriptions of morphologically identified secretory contents in coated vesicles and the appearance of coated areas on membranes engaged in membrane renewal and growth have provided evidence for the existence of coated exocytic vesicles. Some of these examples of the occurrence of projecting coat structures on secretory vesicles are described in the following section.

Observations
Lactating mammary gland epithelial cells
The characteristic ultrastructure of the lactating mammary epithelial cell has been described in a variety of species (for references see Bargmann & Knoop, 1959; Pitelka *et al.*, 1973; Hollmann, 1974; Franke *et al.*, 1976*b*). Large vesicles ranging from 0.6 to 1.5 μm in diameter and containing one or several casein micelles are closely associated with the Golgi apparatus and apparently represent forming secretory vesicles, equivalent to the 'condensing vacuoles' (Plate 1*a*) of Jamieson & Palade (1967) and Palade (1975). In addition, smaller vesicles (0.2 to 0.4 μm in diameter) containing one or two casein aggregates occur in the apical cytoplasm and represent mature secretory vesicles (Plate 1*b*). These casein-containing vesicles show on their cytoplasmic side an association with regularly spaced projections or ridges (Plates 1 and 2*a*), similar to that of

the well-known smaller coated vesicles (diameters 50 to 90 nm) that are often
recognised in tangential sections in the form of a typical hexagonal–pentagonal
pattern (Plate 2c). For references on the architecture of these small coated
vesicles see Kanaseki & Kadota (1969), Crowther, Finch & Pearse (1976),
Ockleford (1976) and other chapters in this book. These coat structures cover
parts of the larger casein-containing vesicles ('condensing vesicles') in the
vicinity of the dictyosomes (Plate 1a). By contrast, most of the smaller casein-
micelle-containing vesicles, which are predominant in the cell apex, are completely covered by such a coat (Plates 1b and 2a). The casein micelles often
reveal fine fibrillar connections with the inner aspect of the vesicle membrane
which are often preserved in the apical plasma membrane upon the exocytic
release of the vesicle content. Thus the observed association of both casein
micelles and coat structures allows the identification of coated vesicle membrane as the origin of newly inserted apical surface membrane (Plate 2b).

Hepatocytes
Another example of the involvement of coat structures in a secretory
process has been described in hepatocytes (Kartenbeck *et al.*, 1977; see
also Plates 3 and 4). In isolated rat hepatocytes for example (Plates 3 and 4a,
b), we observed peripheral sacs of the dictyosomes containing aggregates of
densely stained material similar in structure to serum lipoproteins. Some of
these vesicles, which are filled with lipoprotein granules, are associated with
coat structures, usually covering only a part of their surface (Plates 3 and 4a).
Frequently, individual lipoprotein granules appeared in bulbous outpocketings
of these vesicles, and it is these regions which are characterised by a conspicuous
covering with coat material, a situation which suggests either fusion or budding
of smaller coated vesicles (Plate 3). Such smaller (about 0.15–0.25 μm in
diameter) and usually completely coated vesicles contain single lipoprotein
particles and are frequently observed in the cell periphery (Plate 4b) in what
appears to be a phase of exocytic fusion with the plasmalemma. As shown in
Plate 3 the coat structures can be recognised in association with the plasma
membrane during the exocytic release of the lipoprotein granules.

Seminal vesicles
The glandular epithelium cells of the seminal vesicle are characterised
by the appearance of large secretory vacuoles (up to 2 μm in diameter) which
are especially prominent in the apical cytoplasm. The densely stained, relatively
large aggregates contained in these vesicles probably are carbohydrate-associated
proteinaceous material because of the occurrence of some unusual basic proteins in such secretions (cf. Higgins, Burchell & Mainwaring, 1976). Projecting

coat structures can be identified which partly cover these larger vacuoles (Plate 5a, b). In addition, coat structures are seen over the entire surface of smaller secretory vesicles (0.1–0.2 μm in diameter). Upon discharge of the secretory material the coat structures seem to remain attached to the post-exocytic membrane caveolae of the plasma membrane (Plate 5c). The identification of the coat structures on these secretory vesicles seems, as in other cell systems, to depend on the specific conditions of the fixation.

Happlopappus gracilis *cells grown in culture*

Dividing cells among exponentially growing cells of the plant *Happlopappus gracilis* show an equatorial aggregation of numerous vesicles, which fuse with each other and form larger vesicles, thus contributing to the growing cell plate and the new plasma membrane (Plate 6a). The content of these vesicles represents finely textured primary cell wall material (Plate 6c). During this process of cell plate formation coated vesicles are detected close to the growing cell plate. Bleb-like coated protrusions as well as extended coat-covered regions are seen on these cell-plate-forming vesicles, indicative of the fusion of coated vesicles and their involvement in the formation of the new plasma membrane and the primary wall (Plate 6). At early stages of cell plate maturation up to 60 % of the total surface of this de-novo-formed membrane can be seen to be covered with projections. At later stages of cell plate growth the coat-bearing proportion of the plasmalemma of the equatorial plate is clearly reduced.

Concluding remarks

The occurrence of morphologically identified intracellular markers, be they endogenously formed or exogenously applied, helps in the analysis of cellular translocation processes. The involvement of the typical coat structures described above in the endocytic uptake of extracellularly added molecules and particles and in the intracellular transport of such endocytic coated vesicles has been established in a variety of experiments, mainly using externally added marker substances (Roth & Porter, 1964; Friend & Farquhar, 1967; Lagunoff & Curran, 1972; Heuser & Reese, 1973; Geuze & Kramer, 1974; Roth, Cutting & Atlas, 1976; Anderson *et al.*, 1977; Herzog & Farquhar, 1977; Farquhar 1978). The examples of the occurrence of projecting coat structures on certain secretory vesicles summarised above, all reveal a morphologically identifiable secretory content and thus suggest a participation of coated vesicles in the reverse process of membrane translocation, i.e. in the exocytic pathway.

Some related findings have been described by other authors. In hepatocytes of ethanol-treated rats Ehrenreich, Bergeron, Siekewitz & Palade (1973) have noted coated vesicles containing single lipoprotein particles in the vicinity of the plasma membrane as well as partially coated vesicles that are in the process of exocytically discharging their content into the space of Disse. In pancreatic exocrine cells, coat structures covering small sectors of the limiting membrane of the 'condensing vacuoles' have been noted by Kramer & Geuze (1974), and Jamieson & Palade (1971) have described coated vesicles adjacent to the distal face of the Golgi apparatus or directly on Golgi apparatus cisternae. The appearance of secretory-vesicle-associated coat structures has also been shown in rat anterior pituitary gland (Pelletier, Peillon & Vila-Porcile, 1971; see their plates 1 and 4) and on secretory-granule-containing vesicles on Clara cells of rat bronchioles (Kuhn, Callaway & Askin, 1974; see their figure 11) during exocytic release. In spermiogenesis of mammals Susi, Leblond & Clermont (1971) and Mollenhauer, Hass & Morré (1976) have described the budding of coated vesicles from distal cisternae of Golgi apparatus as well as the incorporation of such vesicles into the growing acrosomal vesicle, a process which is partly related to secretory phenomena. The fusion of coated vesicles with the growing cell plate during plant mitosis has been observed in bean root tips by Hepler & Newcomb (1967) and these authors already have discussed that 'it is conceivable that the coats of these vesicles play a role in effecting fusion with the plasmalemma of the young plate'. Another example of the involvement of coat structures in membrane growth and exocytosis occurs during cyst formation in the caps of the green alga *Acetabularia mediterranea* (Plates 7 and 8). In these stages, the frequent occurrence of coated outpocketings on free edges of large vesicles suggests (Plate 8*a–c*) that these mostly elongated vesicles grow by fusion of their edges with smaller coated vesicles. These large vesicles finally fuse with each other as well as with the plasma membrane and the tonoplast membrane and thereby dissect the cytoplasm into cyst-equivalent portions. Because of the morphological similarity of the hyaline cell wall material on the one hand, and the contents of the Golgi-apparatus-derived vesicles and the large vesicles on the other hand, this fusion with the plasma membrane might be regarded as a special form of exocytosis combined with secretion, in the case of slime material of the cell wall.

The typical coat structures may be regarded as membrane markers. The fusion of the coat-bearing secretory vesicle membrane with the plasma membrane in rapidly dividing cells (cf. the examples of *Happlopappus* and *Acetabularia* presented above) or in cells, in which a continuous loss of apical plasma membrane occurs, such as in the lactating mammary gland cell, illustrates how endomembranous material might contribute to the biogenesis of surface membrane.

Association of coated secretory vesicles with microfilaments and microtubules is commonly observed and seems to be mediated by fine thread-like cross-bridge elements. It remains to be shown, however, whether such, or other, associations of vesicles with cytoskeletal elements (e.g. Plate 9a) really reflect the interaction of these components in secretion or membrane translocation processes. In addition, we have observed, in regions of close apposition of the coated secretory vesicles with the apical plasma membrane, coat-free surfaces of otherwise coated vesicles, presumably reflecting the partial removal of the coat elements prior to the exocytic process (Plate 9b). Another form of close apposition of the secretory vesicle to the cell membrane was formed by an intermembrane 'attachment plaque' constituted by regularly spaced conspicuous cross-bridge elements (Plate 9c) (for details see Satir, 1974; Franke *et al.*, 1976a, b, 1977; Davidson, 1976). Whether these 'pre-exocytic attachment plaques' are formed with the involvement of coat material or whether they represent a membrane transition stage of the exocytic process cannot be decided.

Origin and fate as well as turnover and function of the coat material are so far unknown. The association of the coat material with the secretory vesicles appears to be a transient one and is possibly characteristic of phases of membrane translocation such as vesicle budding from ER membranes, Golgi apparatus, or the 'condensing vacuoles'. Although the morphological similarity of the polygonal coat structures on the small 'endocytic' or 'intermembranous' coated vesicles and the larger exocytic (secretory) coated vesicles is obvious, it remains to be clarified whether in both types of vesicles the coat is made up of the same major protein 'clathrin' (Pearse, 1975). In addition, one should be able to explain the apparent absence of coat structures on some types of secretory vesicles in relation to the question of the functional significance of the coat structures in the secretory process.

I wish to thank Dr W.W. Franke for helpful discussion and for reading and correcting the manuscript. I further thank my colleagues Drs W. Herth and H. Spring for their valuable co-operation.

References
Anderson, R.G.W., Brown, M.S. & Goldstein, J.L. (1977). Role of the coated endocytotic vesicle in the uptake of receptor bound low density lipoprotein in human fibroblasts. *Cell*, 10, 351—64.
Bargmann, W. & Knoop, A. (1959). Über die Morphologie der Milchsekretion. Licht- und elektronenmikroskopische Studien an der Milchdrüse der Ratte. *Zeitschrift für Zellforschung und Mikroskopische Anatomie*, 49, 344—88.
Bonneville, M.A., Weinstock, M. & Wilgram, F.F. (1968). An electronmicroscope

study of cell adhesion in psoriatic epidermis. *Journal of Ultrastructure Research*, 23, 15–43.

Crowther, R.A., Finch, J.T. & Pearse, B.M.F. (1976). On the structure of coated vesicles. *Journal of Molecular Biology*, 103, 785–98.

Davidson, L.A. (1976). Ultrastructure of the membrane attachment sites of the extrusomes of *Ciliophrys marina* and *Heterophrys marina* (Actinopoda). *Cell and Tissue Research*, 170, 353–65.

Droz, D. (1969). Protein metabolism in nerve cells. In *International Review of Cytology*, 25, 363–90.

Ehrenreich, J.H., Bergeron, J.J.M., Siekevitz, P. & Palade, G.E. (1973). Golgi fractions prepared from rat liver homogenates. I. Isolation procedure and morphological characterisation. *Journal of Cell Biology*, 59, 45–72.

Farquhar, M.G. (1978). Recovery of surface membrane in anterior pituitary cells. Variations in traffic detected with anionic and cationic ferritin. *Journal of Cell Biology*, 77, R35–R42.

Fawcett, D.W. (1965). Surface specialisations of absorbing cells. *Journal of Histochemistry and Cytochemistry*, 13, 75–91.

Franke, W.W. & Herth, W. (1974). Morphological evidence for de novo formation of plasma membrane from coated vesicles in exponentially growing cultured plant cells. *Experimental Cell Research*, 89, 447–51.

Franke, W.W. & Kartenbeck, J. (1976). Some principles of membrane differentiation. In *Progress in Differentiation Research*, ed. N. Müller-Bérat, pp. 213–43. Amsterdam: North-Holland.

Franke, W.W., Kartenbeck, J. & Spring, H. (1976a). Involvement of bristle coat structures in surface membrane formation and membrane interactions during coenocytotomic cleavage in caps of *Acetabularia mediterranea*. *Journal of Cell Biology*, 71, 196–206.

Franke, W.W., Lüder, M.R., Kartenbeck, J., Zerban, H. & Keenan, T.W. (1976b). Involvement of vesicle coat material in casein secretion and surface regeneration. *Journal of Cell Biology*, 69, 173–95.

Franke, W.W., Spring, H., Kartenbeck, J. & Falk, H. (1977). Cyst formation in some Dasycladacean green algae. I. Vesicle formations during coenocytotomy in *Acetabularia mediterranea*. *Cytobiologie*, 14, 229–52.

Friend, D.S. & Farquhar, M.G. (1967). Functions of coated vesicles during protein absorption in the rat vas deferens. *Journal of Cell Biology*, 35, 357–76.

Geuze, J.J. & Kramer, M.F. (1974). Function of coated membranes and multivesicular bodies during membrane regulation in stimulated exocrine pancreas cells. *Cell and Tissue Research*, 156, 1–20.

Hepler, P.K. & Newcomb, E.H. (1967). Fine structure of cell plate formation in the apical meristem of *Phaseolus* roots. *Journal of Ultrastructure Research*, 19, 498–513.

Herth, W. (1979). The site of β-chitin formation in centric diatoms. II. The chitin-forming cytoplasmic structures. *Journal of Ultrastructure Research*, 68, 16–27.

Herzog, V. & Farquhar, M.G. (1977). Luminal membrane retrieved after exocytosis reaches most Golgi cisternae in secretory cells. *Proceedings of the National Academy of Sciences, USA*, 74, 5073–7.

Heuser, J.E. & Reese, T.S. (1973). Evidence for recycling of synaptic vesicle membrane during transmitter release at the frog neuromuscular junction. *Journal of Cell Biology*, 57, 315–44.

Higgins, S.J., Burchell, J.M. & Mainwaring, W.I.P. (1976). Androgen dependent synthesis of basic secretory proteins by the rat seminal vesicle. *Biochemical Journal*, 158, 271–32.

Hollmann, K.H. (1974). Cytology and fine structure of the mammary gland. In *Lactation*, ed. B.L. Larson & V.R. Smith, vol. 1, pp. 3–95. New York & London: Academic Press.

Jamieson, J.D. & Palade, G.E. (1967). Intracellular transport of secretory proteins in the pancreatic exocrine cell. I. Role of the peripheral elements of the Golgi complex. *Journal of Cell Biology*, 34, 577–96.

Jamieson, J.D. & Palade, G.E. (1971). Synthesis, intracellular transport, and discharge of secretory proteins in stimulated pancreatic exocrine cells. *Journal of Cell Biology*, 50, 135–58.

Kanaseki, T. & Kadota, K. (1969). The 'vesicle in a basket'. A morphological study of the coated vesicle isolated from the nerve endings of the guinea pig brain, with special reference to the mechanism of membrane movements. *Journal of Cell Biology*, 42, 202–20.

Kartenbeck, J., Franke, W.W. & Morré, D.J. (1977). Polygonal structures on secretory vesicles of rat hepatocytes. *Cytobiologie*, 14, 284–91.

Kramer, M.F. & Geuze, J.J. (1974). Redundant cell-membrane regulation in the exocrine pancreas cells after pilocarpine stimulation of secretion. In *Advances in Cytopharmacology*, ed. B. Cecarelli, F. Clementi & J. Meldolesi, vol. 2, pp. 87–97. New York: Raven Press.

Kuhn, C., Callaway, L.A. & Askin, F.B. (1974). The formation of granules in the bronchiolar Clara cells of the rat. I. Electron microscopy. *Journal of Ultrastructure Research*, 49, 387–400.

Lagunoff, D. & Curran, D.E. (1972). Role of bristle-coated membrane in the uptake of ferritin by rat macrophages. *Experimental Cell Research*, 75, 337–46.

Mollenhauer, H.H., Hass, B.S. & Morré, D.J. (1976). Membrane transformation in Golgi apparatus of rat spermatids. A role for thick cisternae and two classes of coated vesicles in acrosome formation. *Journal de Microscopie et de Biologie Cellulaire*, 27, 33–6.

Morré, D.J., Keenan, T.W. & Huang, C.M. (1974). Membrane flow and differentiation: origin of Golgi apparatus membranes from endoplasmic reticulum. In *Advances in Cytopharmacology*, ed. B. Cecarelli, F. Clementi & J. Meldolesi, vol. 2, pp. 107–25. New York: Raven Press.

Ockleford, C.D. (1976). A three-dimensional reconstruction of the polygonal pattern on placental coated vesicle membranes. *Journal of Cell Science*, 21, 83–91.

Palade, G.E. (1975). Intracellular aspects of the process of protein synthesis. *Science*, 189, 347–58.

Palade, G.E. & Bruns, R.R. (1968). Structural modulations of plasmalemmal vesicles. *Journal of Cell Biology*, 37, 633–49.

Pearse, B.M.F. (1975). Coated vesicles from pig brain: purification and biochemical characterisation. *Journal of Molecular Biology*, 97, 93–8.

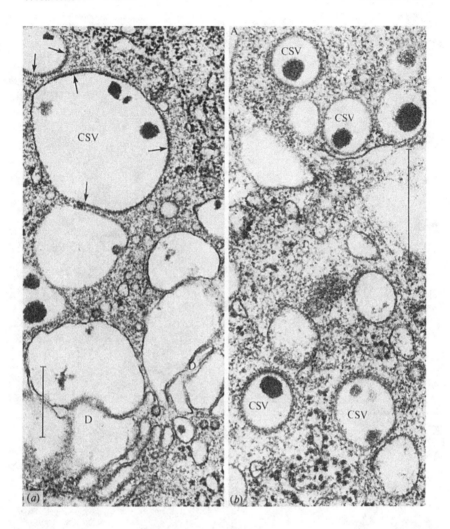

For explanation of plates see p. 251

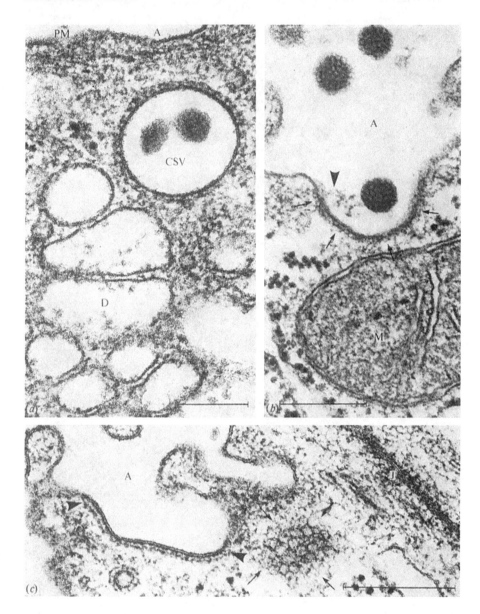

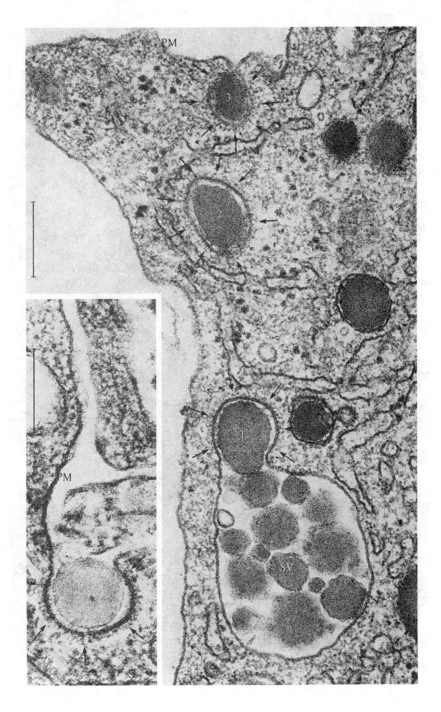

J. Kartenbeck   Plate 4

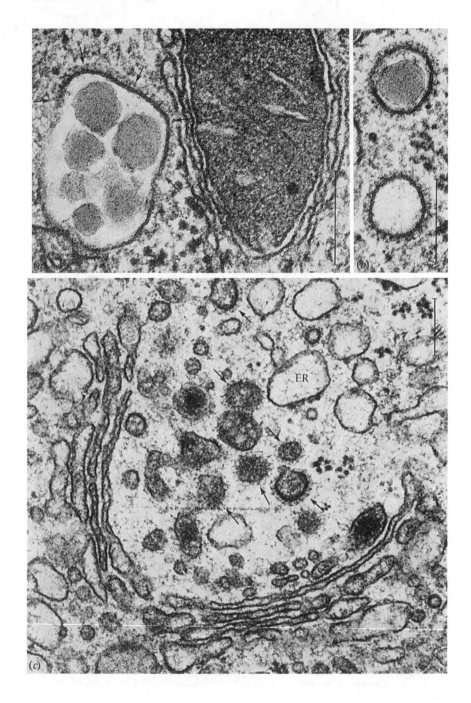

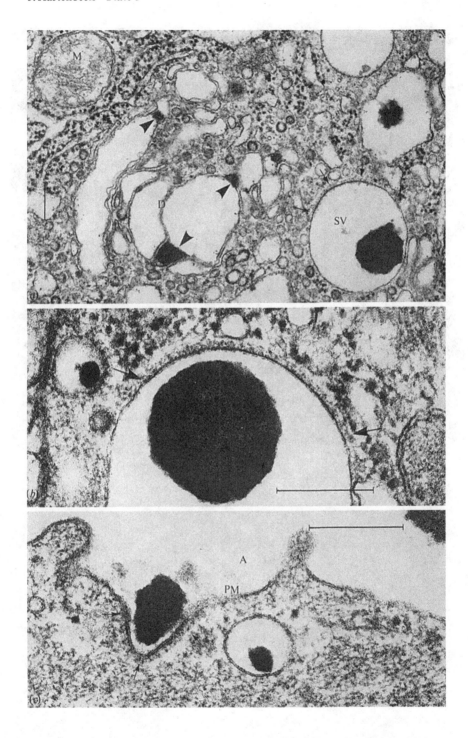

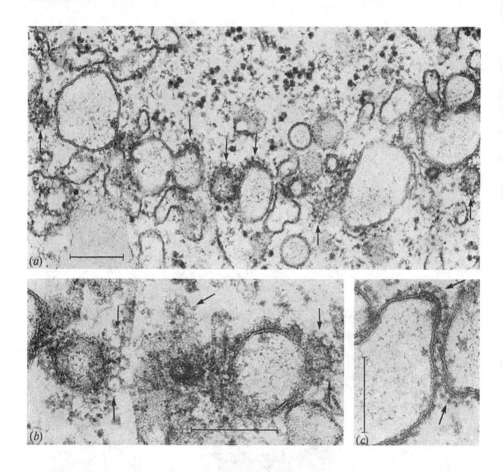

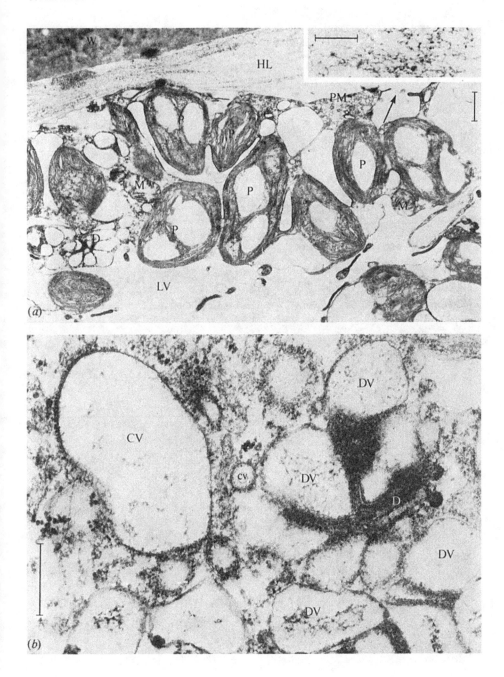

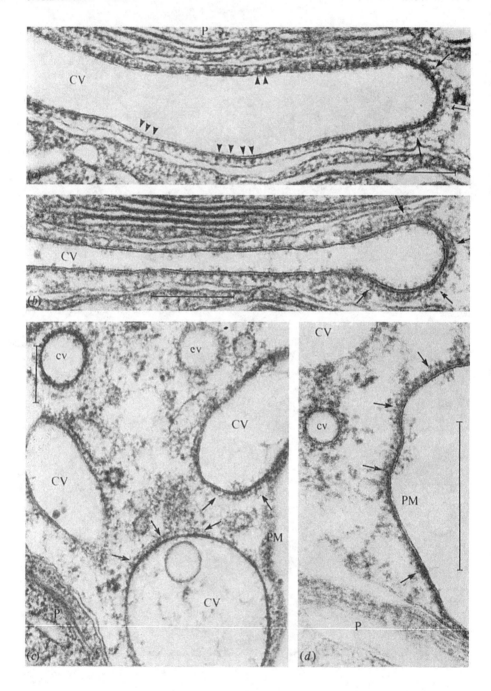

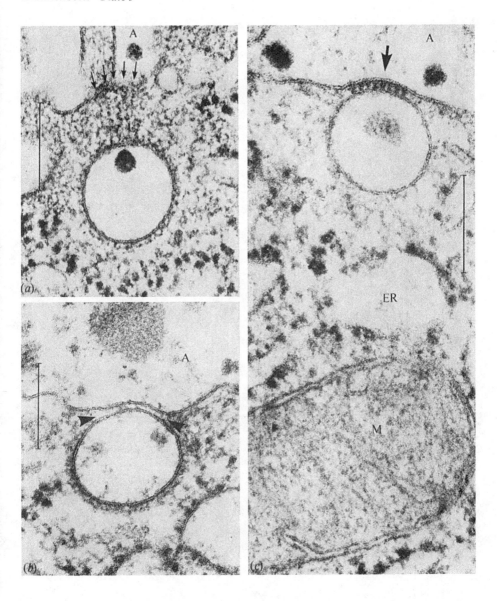

Pelletier, G., Peillon, F. & Vila-Porcile, E. (1971). An ultrastructural study of sites of granule extrusion in the anterior pituitary of the rat. *Zeitschrift für Zellforschung und Mikroskopische Anatomie*, 115, 501–7.

Pitelka, D.R., Hamamoto, S.T., Duafala, J.G. & Nemanic, M.K. (1973). Cell contacts in the mouse mammary gland. I. Normal gland in postnatal development and the secretory cycle. *Journal of Cell Biology*, 56, 797–818.

Roth, T.F. & Porter, K.R. (1964). Yolk protein uptake in the oocyte of mosquito *Aedes aegyptii* L. *Journal of Cell Biology*, 20, 313–32.

Roth, T.F., Cutting, J.A. & Atlas, S.B. (1976). Protein transport: a selective membrane mechanism. *Journal of Supramolecular Structure*, 4, 527–48.

Satir, B. (1974). Membrane events during the secretory process. In *Transport at the Cellular Level*, ed. M.A. Sleigh & D.H. Jennings, *Society for Experimental Biology Symposium 28*, pp. 399–418. Cambridge University Press.

Sheffield, J.B. (1970). Studies on aggregation of embryonic cells: initial cell adhesion and the formation of intercellular junctions. *Journal of Morphology*, 132, 245–64.

Susi, F.R., Leblond, C.P. & Clermont, Y. (1971). Changes in the Golgi apparatus during spermiogenesis in the rat. *American Journal of Anatomy*, 130, 251–68.

Plates

Plate 1. Electron micrographs showing apical portions of epithelial cells of lactating rat mammary gland. (*a*) Large secretory vesicles (CSV) containing one of several casein micelles are situated next to a dictyosome (D). Frequently, parts of the surface of such vesicles are covered with coat structures (arrows). (*b*) Smaller secretory vesicles most of which contain only one casein micelle are found concentrated in the cell apex. Most of these vesicles have projections on their cytoplasmic surfaces. A, alveolar space. Scale bars represent 0.5 μm.

Plate 2. Details of the cell apex in the lactating rat mammary gland epithelial cell. (*a*) An entirely coated secretory vesicle close to the plasma membrane (PM) and a dictyosome (D). (*b*) During the process of exocytosis fibrillar connections (arrowhead) between casein micelles and the plasma-membrane-integrated coated secretory vesicle membrane (arrows) are frequently maintained. (*c*) Typical aspect of the apical plasma membrane showing fusion with secretory vesicles and incorporation of coated vesicle membrane (arrowheads). The arrows point to a section tangential to a larger coated vesicle, most probably a secretory vesicle, which demonstrates the characteristic polygonal arrangement of the coat material. A, alveolar space with some free casein micelles (*b*); J, junctional complex; M, mitochondrion. Scale bars represent 0.3 μm.

Plate 3. Low-power electron micrograph of the cell periphery of an isolated rat hepatocyte. A secretory vesicle (SV) with several lipo-protein aggregates shows a coated protrusion (1). Smaller coated secretory vesicles with single lipoprotein granules are seen in the vicinity of the plasma membrane (2) or during fusion with the plasma membrane (3). The release of a lipoprotein granule is demonstrated in the insert. The arrows point to the membrane-associated coat structures. PM, plasma membrane. Scale bars represent 0.2 μm.

Plate 4. (*a*) and (*b*) Secretory vesicles of isolated rat hepatocytes filled with lipoprotein aggregates are partly (arrows in *a*) or completely (*b*) covered by coated membrane. Scale bars represent 0.2 μm. (*c*) Dictyo-some of a mouse hepatocyte showing the accumulation of several secretory vesicles filled with lipoprotein granules. Coat structures are identified in association with most of these vesicles (arrows). ER, rough endoplasmic reticulum. Scale bar represents 0.2 μm.

Plate 5. Electron micrographs of secretory epithelial cells of the rat seminal vesicle. (*a*) Survey picture of the peridictyosomal region. Distinct aggregates of electron-dense secretory material are recognised in the peripheral portions of cisternae (arrowheads) of the dictyosomes (D) and in secretory vesicles (SV). Some of the secretory vesicles have coat structures on their cytoplasmic surface (arrows). (*b*) Coated cap of a secretory vesicle at higher magnification. (*c*) Release of the secretion into the alveolar lumen (A). The arrows denote a coated region of the apical plasma membrane (PM). M, mitochondrion. Scale bars represent 0.2 μm.

Plate 6. (*a*)–(*c*) Electron micrographs of *Happlopappus gracilis* cells cultured *in vitro*, showing the equatorial accumulation of vesicles revealed in early stages of cell plate formation. A high percentage of the nascent plasma membrane is covered with coating structures (arrows) which, in oblique section, show the typical polygonal pattern (arrows in *b*). The finely textured material of the primary cell wall is visible within these vesicles. Scale bars represent 0.2 μm.

Plate 7. Electron micrographs of caps of *Acetabularia mediterranea* during early stages of cyst formation. (*a*) Low-power micrograph showing the fibrillar material of the 'hyaline layer' (HL) (see insert for higher magnification) interspersed between the cap wall (W) and the plasma membrane (PM). The multinucleate cytoplasm contains numerous plastids (P), dictyosomes (D), and mitochondria (M) and is limited by the tonoplast membrane of the large central vacuole (LV). Abundant large vesicular profiles (putative 'cytotomic vesicles') are closely associated with the plastids (P). Invaginations of the plasma membrane (arrow at the right) most probably represent fusion with a cytotomic vesicle. Scale bars represent 1 μm and 0.3 μm (insert). (*b*) Higher magnification of a dictyosomal area. Small coated vesicles (cv) are frequently located between the dictyosomes (D) and the 'cytotomic vesicles' (CV). Note the morphological similarity of the fibrillar material contained in the dictyosomal vesicles (DV) with the contents of the cytotomic vesicles (CV). Scale bar denotes 0.5 μm.

Plate 8. Electron micrographs of the cap periphery of *Acetabularia mediterranea* at higher magnification. (*a*) and (*b*) Detail of the 'growing' tips of the large cytoplasmic vesicles (CV). The close and parallel association of the cytotomic vesicles (CV) with the outer membranes of plastids (P) and mitochondria (not shown here; cf. Franke *et al.*, 1977) is usually mediated by small-columnar cross-bridge structures (arrowheads), whereas the free edges of these vesicles are covered by typical coat structures (arrows). (*c*) and (*d*) The accumulation of smaller coated vesicles (cv) and larger cytotomic vesicles (CV) showing coat structures in parts of their perimeters (arrows in *c*) are seen in the vicinity of the plasma membrane (PM). Extended regions of the plasma membrane with a coat on its cyto-plasmic surface (in *d*) suggest an origin by local fusion with cytotomic vesicles. Scale bars represent 0.2 μm.

Plate 9. (*a*)–(*c*) Various forms of association of casein-containing secretory vesicles with the apical surface membrane of lactating rat mammary gland epithelial cells. (*a*) Parallel finely filamentous (3–5 nm thick) connections between coat structures of the secretory vesicle membrane and the plasma membrane are indicated by arrows. (*b*) Where secretory vesicles are in close apposition to the plasma membrane, coat structures are frequently absent in the narrow intermembranous space (arrowheads). (*c*) Secretory vesicles intimately associated with the apical plasma membrane often form char-acteristic 'attachment plaques' with regularly spaced membrane-to-membrane cross-bridges (arrow). A, alveolar lumen; ER, rough endoplasmic reticulum; M, mitochondrion. Scale bars represent 0.2 μm.

10

Dynamic aspects of coated vesicle function

C.D. OCKLEFORD & E.A. MUNN

One of the continuing intellectual challenges in the field of coated vesicle
research is extrapolation from morphological and experimental data, obtained
from observations at single points in time, to a concept of vesicle behaviour as
a continuum. We would like to know more of the dynamics of coated vesicles
expressed in supramolecular and, hopefully, molecular terms. Although little
of the data presently available is conclusive there is sufficient circumstantial
evidence to enable us to attempt to set up hypotheses about how coated
vesicles are formed, move and are discharged. Specifically consideration will
be given to the following problems:

(1) What biochemical pathways and structures are essential to coated
vesicle function? We can expect the results of investigations using stimulators
and inhibitors of endocytosis to eliminate some of the mechanisms put forward
to account for coated vesicle behaviour.

(2) How does the membrane invaginate into the configuration of a coated
vesicle? The induction of curvature entails changes in the local molecular
arrangement or composition of the membrane or associated structures probably brought about by movement of constituent molecules. We will define
this as *membrane motility*.

(3) Once a coated vesicle has formed it travels to one of a number of
specified destinations (see Chapter 2). This process might or might not require
a structured motile system. We will call this dynamic process *vesicle motility*
to distinguish it from the membrane motility just described. We will consider
the evidence relevant to it.

(4) The general problem of how two membranes fuse has special cases in
the fusion of the coated vesicle membrane with that of the target organelle to
which it is directed, and in the self fusion of the membrane of the neck of the
caveola when a forming vesicle finally closes off from a phospholipid bilayer.
This problem will only be considered briefly.

Endocytosis: pharmacological and other factors
Inhibitors of endocytosis

In Chapter 1 Wild has reviewed the evidence for the separate status of coated vesicles in the classification of organelles of endocytosis. Several types of endocytic vesicle are distinguishable (Simson & Spicer, 1973; Allison & Davies, 1974). Some of these are illustrated in Plate 1. In cells where more than one class of endocytosis occurs it is sometimes possible using particular inhibitors to interfere with just one (Nagura & Asai, 1976).

Table 1 is a summary of the effects of various substances on different types of endocytosis in macrophages. These results have been selected because they are as comparable as any group available. Even here, however, caution should be exercised before undertaking comparison between different sets of data because groups of the same cell type from the same animal but obtained from different sites may be affected differently in the same experiment. This point is illustrated most clearly in the demonstration by Cohn (1970) that mouse peritoneal and alveolar macrophages exhibit different sensitivity to the same metabolic inhibitors.

The metabolic inhibitors used for studies of this kind include iodoacetate, sodium fluoride, 2, 4-dinitrophenol and potassium cyanide. All of the metabolic inhibitors employed in these experiments have been defined in Table 1 not by name but according to their inhibitory effect. From this table it can be seen that two distinct kinds of experimental approach underly the research. Some experimenters have used ultrastructural monitoring and have defined the morphological type of vesicle involved, using either ferritin or horseradish peroxidase as a label. Both of these labelling molecules are in certain circumstances endocytosed in the unconjugated form (e.g. Van Deurs, Møller & Amtorp, 1978). They therefore fall into a class of experiments employing 'suspect labels' with the consequent necessity for applying rigorously controlled experimentation, but give the possibility of studying the uptake of more than one molecule by a particular form of endocytosis. Other experimenters have studied uptake of radiolabelled particulates and solutes using wide-aperture counting systems. These methods alone do not have the resolution to discriminate between individual cells let alone organelles but can be used to study very accurately in quantitative terms the net effect of possibly more than one form of endocytosis. Care has been taken over the terminology used in Table 1 because where experiments of a particular type are referred to, only the appropriate conclusion is listed. This explains why the effects of some inhibitors are described in terms of their action on coated vesicles while the effects of others are given in more general terms as effects on micropinocytosis.

There are two possible types of experiment directly relevant to the study

Table 1. *Effects of inhibitors and stimulators of endocytosis in macrophages*

Agent	Effect				
	Phagocytic vacuoles	Macropinocytic vacuoles	Smooth surface pinocytic vesicles (> 0.15 μm diam.)	Micropinocytosis	Coated micropinocytic vesicles
Glycolysis inhibitors	Inhibition (1, 2)	Inhibition (1, 2)	Inhibition (5)	No effect (1); slight inhibition (3); inhibition (4)	No effect (5)
Oxidative phosphorylation inhibitors	Inhibition (1, 2)	Inhibition (1, 2)	Inhibition (5)	No effect (1); slight inhibition (4); inhibition (3)	No effect (5)
Cytochalasin B	Inhibition (6, 7)	Inhibition (8)	No effect (5)	No effect (9); slight inhibition (4)	No effect (5)
Colchicine				No effect (4, 13)	–
Low temperature (4 °C)	Inhibition (10)	Inhibition (10)	Inhibition (5)	No effect (10, 11); inhibition (3)	No effect (5)
Serum		Stimulation (4, 12)	Slight inhibition (4)		
Phospholipase C			Inhibition (5)		Slight stimulation (5)
Triton X-100			Inhibition (5)		Slight stimulation (5)

References: 1, Casley-Smith (1969a); 2, Cohn (1970); 3, Allison & Davies (1974); 4, Munthe-Kaas (1977); 5, Nagura & Asai (1976); 6, Malawista et al. (1971); 7, Allison (1973); 8, Allison et al. (1971); 9, Wills et al. (1972); 10, Casley-Smith (1966); 11, Casley-Smith (1969b); 12, Cohn (1966); 13, Bhisey & Freed (1971).

of the energy requirement for the formation of coated vesicles: those in which inhibitors cause all uptake of material to cease (therefore of necessity coated vesicles if present must be inhibited) and those where ultrastructural monitoring at high resolution indicates coated vesicle function or lack of it in the presence of an inhibitor.

The only work relating specifically to coated vesicles in macrophages (i.e. in the latter category) is that of Nagura & Asai (1976), which indicates that uptake by these organelles does not require any metabolic energy supply. (In mosquito oocytes, however, work of the former type (Roth, Cutting & Atlas, 1976) indicates that metabolic energy may be required in some circumstances.) Nagura & Asai (1976) also maintain that low-temperature treatment (4 °C) does not abolish micropinocytosis by coated vesicles, although the rate was not determined. This is an interesting conclusion because it sets coated vesicle micropinocytosis apart from several other life processes. Possibly the only other type of event involving microscopically detectable and organised movement of molecules in the cell surface which may be similarly unaffected by metabolic inhibitors and low temperature (4 °C) is the patching of multivalent cell surface ligands (de Petris & Raff, 1972, 1973). The literature on the effects of low temperature on endocytosis is united in finding that large-scale endocytosis is inhibited at 4 °C (Table 1). However there is need for caution because the effects of low temperature on micropinocytosis are apparently variable. Allison & Davies (1974), for example, were able to abolish the uptake of a radioisotope of colloidal gold by macrophages at 4 °C. Similarly several other authors have noted near-total inhibition of uptake of radiolabelled molecules at 4 °C (Ryser, 1968; Cohn, 1970; Davies, Allison & Haswell, 1973; Steinman, Silver & Cohn, 1974). On the other hand Contractor & Krakauer (1976) have concluded that some form of pinocytosis (presumably micropinocytosis) may continue at 4 °C. These authors' evidence that radioiodinated haemoglobin is accumulated by cultured trophoblastic cells in the period 14–16 hours after commencement of incubation at 4 °C is important because of two features. First, the temperature-control method was particularly well specified. Secondly, residual pinocytosis which might occur during the initial cooling period and adsorption to the cell surface are certain to have reached maximum levels and stabilised well before the 14-hour time point. Similar evidence of uptake continuing at 4 °C has been obtained by Casley-Smith (1969a), Schmidtke & Unanue (1971) and Munthe-Kaas (1977). The uptake of [^{3}H] IgG into chorionic villi of human placenta in vitro (Fig. 1) also continues at 4 °C.

The temperature at which a membrane's phospholipid bilayer becomes fluid will vary with its molecular composition. This is, however, usually at around 0 °C (Blasie & Worthington, 1969) but may be even lower (Ladbrooke

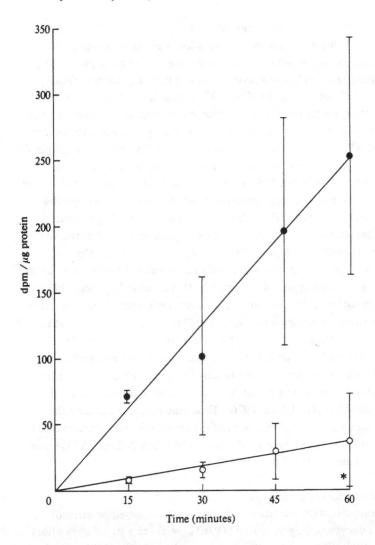

Fig. 1. This illustration shows the change in rate of uptake with time of [³H]IgG into human chorionic villi *in vitro*. The uptake is shown at 37 °C (solid circles) and 4 °C (open circles). Temperature control at 37 °C was carried out in a water bath and at 4 °C in a constant temperature room. Tissues and solutions were maintained at incubation temperature prior to the experiments. Background radiation level is indicated (*). Each point is averaged from three experiments and the error bars show the standard error. (Unpublished experiments of J.M. Clint & C.D. Ockleford.)

& Chapman, 1969). Precise regulation of the temperature at 4 °C may therefore permit patching or coated vesicle formation whereas performing an
experiment on ice may cause more extensive inhibition of these events. The
potential importance of low temperature causing loss of membrane fluidity
has been pointed out by Munthe-Kaas (1977). Overall the data are consistent
with there being a basic separation of endocytic mechanisms between, on the
one hand, large-scale mechanisms such as macropinocytosis and phagocytosis,
and, on the other hand, micropinocytosis (Allison, 1973). Micropinocytosis
(and hence coated vesicle formation) emerges as being less dependent on
metabolic energy supply and high temperature than other forms of endocytosis.
The fact that the fungal metabolite cytochalasin B, a drug which depresses
actin-based motile processes (Wessels *et al.*, 1971), inhibits phagocytosis and
macropinocytosis (Allison, 1973) but not micropinocytosis by either smooth
or coated vesicles (Nagura & Asai, 1976) is important. It implies that an
invagination mechanism based on interaction of membrane regions with an
underlying ATP—actomyosin system, such as that envisaged by Bray (1973)
for phagocytosis, probably will not explain micropinocytosis as well. Because
micropinocytosis is insensitive to colchicine (Bhisey & Freed, 1971) and to
low temperature (Nagura & Asai, 1976), an important role for cytoplasmic
microtubules (Behnke & Forer, 1967) in coated membrane invagination is
unlikely. This is corroborated by the finding that microtubules do not occur
in close apposition to coated vesicle *caveolae*, at least in human placental
syncytiotrophoblast (Ockleford, 1976). These findings, together with the
obvious lack of sensitivity of coated vesicles to metabolic inhibitors, indicate
that a microtubule/dynein type motile system (Burns & Pollard, 1974) is not
involved in membrane motility of this type.

Stimulators of endocytosis

Compared with inhibition little research has centred on attempts to
stimulate endocytosis. Nagura & Asai (1976) have noted a stimulatory effect
on coated vesicle pinosome formation in macrophages treated with phospholipase C and Triton X-100. Both these substances have a destructive effect on
the cell surface, presumably by removing phospholipid molecules from the
plasma membrane. Thus the effect may be exerted through weakening of the
membrane making it more flexible and hence facilitating pinocytosis. It has
been suggested that selective removal of phospholipids from the outer leaflet
of the membrane will destabilise it, actually causing small intuckings of the
cell surface. Whilst this is an interesting possibility it is unlikely that phospholi-
pases in the extracellular spaces of tissues cause micropinocytosis *in vivo*
because there is no obvious means by which such a system would be able to

provide the necessary control and spatial discrimination required for pinosome formation.

Compounds such as vitamin A and primaquine applied at high concentrations stimulate endocytosis in erythrocytes to such an extent that the cells eventually haemolyse (Glauert, Daniel, Lucy & Dingle, 1963; Ginn, Hochstein & Trump, 1969). This result is interpreted as destabilisation resulting from incorporation of these molecules into the structure of the erythrocyte plasma membrane. The vesicles so formed are usually rather larger than micropinocytic vesicles and it seems unlikely that these molecules would trigger coated vesicle formation. Conceivably other molecules found in serum might produce smaller vesicles in a similar manner, but there is no evidence for the occurrence of any such molecules.

Stossel (1977) lists a number of pinocytosis stimulators including lectins, lipoproteins, lipid-soluble drugs and many charged soluble substances. He makes the interesting suggestion that the degree of stimulation is proportional to the number of receptors that a molecule can bind. The observed higher rate of pinocytosis induced by aggregated as compared to soluble substances, and by multivalent as compared to monomeric molecules, supports this view. Unfortunately it is not known whether the pinocytic mechanisms stimulated include the formation of coated vesicles.

Membrane motility

The events which lead to the invagination of the membrane, forming first a pit (caveola) and then a mature coated vesicle, are remarkable. Any one of a known number of mechanisms may potentially be involved. These are briefly considered here.

Actomyosins

From the evidence given above that coated vesicle activity continues at low temperature and in the presence of metabolic and other inhibitors, it may be concluded that membrane motility does not fall into any currently accepted pattern of motile activity. Mechanisms in which filamentous or tubular protein polymers transmit forces developed by the release of chemical energy from the splitting of ATP by myosin or dynein ATPases (Huxley, 1973; Burns & Pollard, 1974) are therefore inappropriate.

Spasmin

There is at least one motile mechanism which appears to be independent of metabolic energy in the short term. The protein spasmin forms an organelle, the spasmoneme, which undertakes cycles of contraction and relaxation caused apparently only by variation in the local concentration of Ca^{2+} (Amos, 1971). This type of system is rare. It is unlikely to be involved in coated vesiculation for a number of reasons, of which the most obvious is that motility of this type depends upon organised structure of greater permanence than is typical of the ephemeral structures seen in pinocytosis.

Brownian motion

It might be suggested that random impact of particles in Brownian motion upon the cell surface effectively causes micropinocytosis. Such collisions have been viewed as deforming and occasionally splitting off portions of the cell surface as a membrane-enclosed structure. If the particles were directed centripetally they might produce a pinocytic vesicle. However on this basis some numbers of impacts would be expected from particles within the cell in Brownian motion moving outward, producing buds and in extreme cases pinched off portions of cytoplasm. Such structures are rarely observed associated with normal healthy cells. If this objection did not hold, Brownian motion might cause smooth micropinocytic vesicle formation, but it is incapable of providing an explanation for the formation of the clathrin and surface coats. Casley-Smith (1969a) has suggested that the final stage of the formation of a micropinocytic vesicle — breaking away from the cell surface — might be the result of Brownian impacts.

Cations and potential gradients

Douglas (1974) has considered the possibility that coated vesicle formation is a calcium-dependent process. He wished to test whether the transmembrane potentials induced by the electrochemical properties of ions were of importance to uptake or whether the chemical properties alone were of greatest importance.

He incubated hamster adrenal glands in a Ca^{2+}-free environment, exposed them to elevated potassium concentrations to depolarise the cells and then added Ca^{2+}. Addition of Ca^{2+} evoked secretion and coated vesicle formation, processes which are coupled in this situation. The membrane potential (30 mV) in these cells might, he postulated, have caused the orientation of molecules in the cell membrane. These were molecules which had, prior to exocytic fusion, been disordered in the membranes of secretory vesicles. Ordered orientation of membrane proteins might have been the cue for the binding of clathrin and the initiation of endocytosis. Experimental results, however, showed that electrical potential change alone was not the cause of vesiculation and focussed attention on the cations themselves. Other experiments indicated that sodium was not essential for vesicle formation and the proposal of a direct role for calcium as a 'membrane vesiculating' agent was put forward.

ATP binding

Penniston & Green (1968) induced pinocytosis-like vesicles to form in the cell surfaces of erythrocyte ghosts. They proposed that these were produced as a consequence of conformational changes in membrane molecules triggered by the binding of ATP. Various factors indicate that these events are irrelevant to the process of coated vesicle formation. These include the large sizes of vesicles so formed and the obvious requirement for ATP.

Particle-free membrane zones

Clustering of intramembranous particles (IMPs) in erythrocyte ghosts is induced when the underlying filamentous protein spectrin is caused to aggregate. Presumably the two components are linked in some way. When this occurs particle-free membrane areas form and then invaginate to become vesicles (Elgsaeter, Shotton & Branton, 1976). This, it is proposed, comes about because the minimum energy state for the phospholipid bilayer is a small sphere, and the membrane is only prevented from forming these *in vivo* by the membrane-associated particles in the form of IMPs. As will be explained, this is an almost directly contrary sequence of events to that which occurs when a coated vesicle forms. The mechanism proposed for erythrocyte ghost vesiculation does not therefore appear to be applicable to coated vesicle formation.

Clathrin

Kanaseki & Kadota (1969) proposed an important model of membrane motility for coated vesicle formation. They suggested that curvature developed in the membrane when particular hexagons (components of extended lattices of hexagons underlying the cell surface) became converted into pentagons with the same length of side. The evidence for specifity in uptake by coated vesicles reviewed in Chapter 8 is easily accommodated by this theory if the polygonal lattice is coextensive with groups of receptors on the external surface of the membrane. Binding of ligand might form a suitable trigger for polygon transformation and the resultant induction of curvature. The energy for membrane deformation would presumably be derived from the alteration in binding of the molecules of clathrin in their lattice, from the receptor—ligand interaction, or a similar event.

The precise mechanism by which the lattice transforms has not been specified but would probably involve the ejection of clathrin molecules. Since the clathrin molecules in the original lattice are presumably equivalent, the transformation of any polygon must be specifically triggered. This event might be a consequence of the binding of ligand to receptor, in which case there is no obvious means by which the transformation of *particular* polygons in the lattice might be accomplished. Random transformation of polygons in lattices of a fixed size is expected to give rise to the formation of different structures. For example where the concentration of ligand is high, receptors would all rapidly saturate producing a large proportion of pentagons, a smaller radius of curvature in the lattice and thus smaller vesicles. The excess lattice of this model must become detached, because coated vesicles with attached extra lattice are not observed. We can see from this model how the actual variation in vesicle volume (Ockleford, Whyte & Bowyer, 1977) might be achieved through randomness in the processes of ligand binding or variations in the concentrations of ligand. There is however a built-in lack of economy because wherever the random transformation of the polygons in a lattice induces vesicles with small radii of curvature there is the possibility of the need to discard extra lattice. Flat sheets of hexagonally packed clathrin lattice have not been described on the cytoplasmic surface of plasma membranes. In addition, assuming ligand binding is the trigger for polygon transformation, the uptake would be inefficient. This follows because even after vesiculation was complete most of the polygons would remain untransformed, and a similarly high proportion of receptors would remain unoccupied. These and other objections (discussed later) to the fixed differentiated area model are probably not fatal. They have, however, prompted a proposed modification (Ockleford, 1976; Ockleford & Whyte, 1977) which is expanded in this chapter. We suggest that coated vesicles are formed by a process analogous to patching. This is illustrated in Fig. 2 and can be outlined as follows. Specific receptors are exposed at the external surface of the phospholipid bilayer forming the cell surface. In the particular case of the human placental IgG receptor, the sterically important part of the receptor molecule is proteinaceous (Schlamowitz, 1979). This finding is consistent with known properties of other receptors (Changeux, 1972) and does not preclude the possibility that other receptor molecules have sugar moieties which may in some cases be important for ligand binding.

Either a portion of this receptor molecule, or of the clathrin molecule to which it may become connected, spans the bilayer of the membrane, or, alternatively, these two molecules may be linked together by a third molecule which spans the phospholipid bilayer. The receptor or linker becomes capable of binding clathrin under unspecified circumstances. The trigger to this latter event could be a variety of factors but can be envisaged as the binding of a

molecule such as the ligand itself. In Fig. 2 the site at which these events
(1–3) take place is the membrane surface of a microvillus. The model is flexible to the extent that only stages 1 and 2 need occur at this site. The clathrin
molecules shown here as freely diffusing within the microvillus might, for
example, be excluded by the axial core composed of microfilaments (Mooseker
& Tilney, 1976), and so bind later. A crucial part of the suggestion is the

Fig. 2. This scheme illustrates the dynamic model of coated vesicle
formation proposed in the text. The first event (1) is binding of a
trigger (e.g. a specific macromolecule which is to be taken up). This
leads to a change in a linker (2) which permits binding of clathrin (3).
Clathrin is in turn altered (4) so that it is in a condition suitable for
polymerisation. For some reason (e.g. excessive 'rigidity' of the
system) polymerisation only occurs to a limited extent (5) or not at
all within the microvillus. When molecules, possibly by random
motion, reach the base of the microvillus, however, there are no restraints on formation of a curved surface so polymerisation proceeds
by successive addition of clathrin–linker–receptor units, singly or
in small groups, and caveolae are formed (6 and 7).

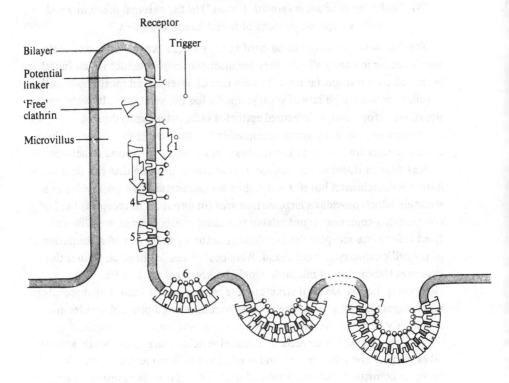

movement centripetally of the receptor and ligand in the plane of the phospholipid bilayer and their aggregation into a complex. Clathrin attachment could
take place as late as this stage. The result of the lateral diffusion of the ligand-
loaded receptors in the membrane and their aggregation is to concentrate the
ligand. This concentration is of importance to the selective nature of the
transport mechanism thought to be mediated by coated vesicles.

Clathrin is altered (step 4 in Fig. 2) by binding to the triggered linker or
receptor molecule so that it is able to participate in lattice formation. Growth
of the lattice, which comprises both hexagonal and pentagonal facets, would
induce curvature in the membrane (step 6 in Fig. 2). This process then continues until the vesicle separates from the membrane continuum and the
curved lattice closes thus forming a complete coated vesicle.

At this stage we are not able to specify the initial cross-linking of components which gives rise to the formation of lattice in the membrane. It could
result from:

(1)   clathrin cross-linking,
(2)   multivalent ligand attaching to more than one receptor, or
(3)   inducing substances (growth factors?) in the external medium which
       cross-link exposed portions of ligand-bound receptors.

The following evidence can be cited in the model's favour. Although coated
vesicles occur in nearly all cells they are most common in tissues which function
in the selective transportation of protein (see Chapter 1) and most importantly,
in cells in which a high rate of protein uptake has been initiated. In such
situations — for example intestinal epithelial cells, osteoclasts, thyroid
epithelium and placental syncytiotrophoblast — the sites of formation of the
coated vesicles are at or between the bases of microvilli or related structures
such as flaps or ridges of cell surface. It is of course possible that this distribution is coincidental but if it is not then we can envisage the microvillus as a
structure which provides a large surface area for exposure of receptors (and of
course, other components not related to coated vesicle formation). Given a
fixed cell-surface receptor density these structures can be viewed as amplifiers
of the cell's capacity to bind ligand. Were coated vesicles fixed structures the
extensive elaboration of microvilli would be irrelevant to their function.
However if the two kinds of structures are related then it follows that receptors
must migrate laterally in the membranes because coated pits and vesicles are
found only between the *bases* of the microvilli, not along their length.

The evidence which suggests that coated-vesicle-mediated pinocytic activity
takes place at low temperatures and is relatively independent of metabolic
energy is consistent with the proposed model. On this basis pinosome formation should cease only when the temperature is reduced to such an extent that

the phospholipid bilayer loses its fluidity. In many membranes this will not occur until temperatures below 0 °C are reached. Certainly in-vitro uptake of IgG into human syncytiotrophoblast is not abolished at 4 °C (Fig. 1). Admittedly it is still not proven that human placentally transported IgG is taken up by coated vesicles, but comparative arguments (see Chapter 5) suggest that it is. Many treatments which might be expected to inhibit function based on other models are completely ineffective in preventing coated vesicle formation (Table 1). The general lack of sensitivity to various potentially inhibitory agents is quite remarkable. In this regard coated vesicle formation is similar to patching, also a process of molecular aggregation within or on membrane.

Examination of extensive numbers of preparations (e.g. figure 5 in Ockleford, 1976) reveals what may best be described as multiple vesicles. These are dumbbell shaped or multilobed. At the cell surface multiple caveolae may also be found; presumably these give rise to multiple vesicles. Consideration of how these form is important in distinguishing between the present modification and Kanaseki & Kadota's original model. Assuming two curved aggregates nucleate nearer to each other than the diameter of the smallest coated vesicle then on the dynamic model the curved lattices will grow outwards until they make contact and fuse in a zone of irregular packing. Had two complete flat lattices of hexagons become attached to the cytoplasmic surface of the phospholipid bilayer side by side there would be two possible consequences: either the hexagonal lattices would not link up, in which case two normal coated vesicles would be formed side by side, or, if they linked together to form a larger plaque, a double-sized vesicle would result. This type of effect might be expected to show itself in graphs of the distribution of vesicle volumes (see Chapter 11). This is not the case.

Both the original and the modified model require that eventually receptors and clathrin lattice become co-extensive (see also Roth, Cutting & Atlas, 1976). We have proposed that this occurs via linkage molecules binding receptors to clathrin through the membrane. We have made it clear that the linker may be a separate molecule, an arm of the clathrin molecule or equally possibly a portion of the receptor molecule. There is evidence for the occurrence of structures which could perform such a linking role. A deep-etch freeze-fracture ultrastructural study of caveolae in cultured human fibroblasts which could be positively designated as forming coated rather than smooth micropinocytic vesicles revealed a population of particles (mean diameter 11 nm including platinum—carbon coating) between the two leaflets of the plasma membrane (Orci *et al.*, 1978). These particles (Plate 2) are rather larger than other IMPs in this membrane and may form a distinct population. They occur both within caveolae and in other portions of the cell surface membrane. Plate 3 shows some of our own preliminary data suggesting the presence of IMPs in the microvillar membrane of human syncytiotrophoblast. In fibroblasts the number of large IMPs per unit area is greater within caveolae (mean $1434/\mu m^2$) than it is in the surrounding membrane (mean $851/\mu m^2$) (Orci et al., 1978). Using the model presented here these large IMPs in the surrounding membrane are easily explicable as being linker molecules which will eventually diffuse to the site of a forming membrane complex. The observed closer relationship of large IMPs in caveolae is also predicted.

Components of the glycocalyx of the cell surface are seen within coated vesicles, in coated pits, and on other parts of the cell surface. The material is apparently more densely packed in pits and vesicles than on other parts of the cell surface (Jollie & Triche, 1971; Ockleford & Whyte, 1977). There is evidence that these molecules contain sialic acid residues: polygonal arrangements of stains specific for sialic acid have been detected on the cell surfaces of tissues at sites rich in coated vesicle caveolae (Tighe, Garrod & Curran, 1967; Marx, Graf & Wesemann, 1973). These may represent sugar moieties of receptor glycomolecules forming the membrane complexes. It is probably these molecules which are represented in polyacrylamide gels of isolated coated vesicles as a high molecular weight band stainable with periodic acid–Schiff (Whyte, 1978). Although there is no conclusive evidence for sugar moieties being implicated in the process of ligand binding this does not exclude their forming other portions of receptor molecules. Possibly, as has been suggested previously for endocytosis generally, their function is as a space filler (Stockem, 1977). By this means a coated vesicle might reduce the amount of material passively acquired within the lumen when it forms. This has obvious advantages for a selective uptake mechanism.

Further evidence for the fluid dynamic nature of coated vesicle formation comes from the observations of Heuser & Reese (1975) reviewed in Chapter 5. Their evidence for a class of neuronal IMPs called vesicle particles indicates that they migrate into forming coated pits. In the case of coated vesicles from brain, which are presumably predominantly neuronal, there is little evidence for the presence of an externally exposed receptor (see Chapter 11).

Adding to the weight of evidence that clathrin, IMPs and glycocalyx material cluster together in caveolae, there is good evidence to suggest that ligand bound to cell surface receptors also clusters. Ferritin-labelled low-density lipoprotein becomes specifically bound to two groups of receptors (Orci et al., 1978). One group is clustered at high density in caveolae of coated vesicles whilst the other is scattered at lower density over other parts of the cell surface. The function of these scattered receptors is very difficult to explain on all but the present model, which predicts that they are destined to move laterally in the plane of the membrane eventually entering membrane complexes which effect their internalisation.

One final item of support can be derived from the conclusions of Stossel (1977), provided that they are applicable to coated vesicles and not restricted to other forms of pinocytosis. He points out that, in general, multivalent molecules are potent stimulators of endocytosis. This effect is consistent with the dynamic model, which provides a rational explanation. Cross-linking of identical receptors by multivalent ligands at the external surface of the cell would tend to increase the rate of curved complex formation by producing locally elevated concentrations of receptors. Once one of the receptors bound to the multivalent ligand becomes engaged in the membrane complex the others will have their diffusion restricted to the immediate region of the forming vesicle. Because they are unable to move away, they are likely to enter the complex rapidly, speeding the formation of a vesicle.

The lipid composition of the coated vesicle membrane is not typical of plasma membrane. Pure preparations of coated vesicles isolated from porcine brain contain some 25 % lipid by weight (Pearse, 1975). A large proportion of the actual membrane is therefore probably in the form of proteins. Membrane proteins of at least one type appear to command a collar of phospholipids with which they are contiguous (Metcalfe & Warren, 1977). These collar molecules are not necessarily typical of the phospholipids in other parts of the membrane. If as we suggest, receptors or linker protein molecules are free to diffuse laterally through the membrane they may well carry their collar of atypical phospholipids with them. The proteins forming the membrane complexes may in this way actually function as segregators of phospholipids. A lipid composition within the membrane of coated vesicles which is different from that of the plasma membrane is not necessarily an indication that there are permanently differentiated patches of cell surface which alone have the capacity for coated vesicle formation.

The fact that coated vesicles do form in cell surfaces largely devoid of microvilli has been dealt with briefly earlier (p. 266). We see microvilli not as vital to membrane motility of this type but as structures providing greater cell surface area in which more receptors can be accommodated. Amplification of the number of receptors exposed for ligand binding in this way is only expected in the tissues most highly adapted for specific uptake of protein at a high rate. Microvilli or functionally equivalent structures are found in most if not all such tissues.

Wild (Chapter 1) points out that smooth micropinocytic vesicles appear to carry out endocytosis without the aid of clathrin and concludes that clathrin is irrelevant to membrane motility. In fact this argument serves to sharpen our appreciation of the importance of different components of the membrane complex.

First consider the similarities between micropinocytosis by smooth and coated vesicles. Both may be processes which are relatively insensitive to low temperature and inhibitors. They also have similarity in that, as is the case for coated caveolae, IMPs appear to cluster in the membrane of probably smooth caveolae (Orci & Perrelet, 1973). Then consider the differences. Smooth micropinosomes do not exhibit a denser glycocalyx than other parts of the membrane. They also appear not to have a greatly specific adsorptive action in uptake. Presumably the luminal surface of a smooth micropinocytic vesicle will not be loaded with ligand. The smooth caveolar phospholipid bilayer is probably more flexible than the coated caveola in this respect. Nevorotin introduces this argument in Chapter 2. The essential differences between coated and smooth forms of micropinocytosis may lie in the need for a more massive curved membrane complex, including clathrin on/in the cytoplasmic surface, to deform the membrane plus receptor-bound ligand.

Coated vesicle motility

Once a vesicle has pinched off from a phospholipid bilayer it apparently travels through the cell prior to its fusion with a phospholipid bilayer at another site. The route taken is variable and depends among other things on the particular tissue and its physiological state (see Chapter 2). There are various possible methods by which vesicle motility may be accomplished. They may be permutated by assuming the presence or absence of three factors. These are recognition, direction and propulsion.

Recognition

To establish the role of recognition we may use as an example the diacytic (transepithelial vacuolar) transport of intact IgG across the yolk sac placenta (Wild, 1976; Moxon & Wild, 1976). Here, coated vesicles apparently form at the apical surface of a cell, carry IgG across the cell and then fuse with the basal or lateral cell surface. This process involves discrimination of apical from other cell surface by coated vesicles. Although we cannot be sure that in this system nascent coated vesicles are prevented from immediately fusing with the cell surface at which they have just formed, this would be a desirable state of affairs. We can be sure, however, that coated vesicles do not fuse with lysosomes (see Chapter 1) and we could probably add a list of other membrane-bounded structures, such as nuclei and mitochondria, which are not subject to fusion with coated vesicles. The recognition or discrimination implied by these observations might operate in a positive or in a negative sense:

in a positive way by providing specific structures in the lateral and basal cell surface which permit coated vesicle fusion or in a negative way by providing all other membranes with structures which resist fusion. Seen in this light the clathrin lattice alone cannot be viewed as a mechanism for controlling the fusion or absence of fusion of a coated vesicle with a particular membrane site. Rather because coated vesicles do fuse with target membranes, particular sites must have the power to recognise, bind and break open the clathrin lattice prior to fusion of the adjacent phospholipid bilayers. These recognition sites would represent the other half of the control system if positive discrimination were operative. The less economical negative discrimination would of course be partly dependent on the fusion-resisting structures being sited where fusion is undesirable.

In mosquito and chicken oocytes which are taking up egg proteins the coated vesicles apparently lose their clathrin prior to the vesicles fusing with larger protein-containing vacuoles (Roth *et al.*, 1976).

Direction

Random diffusion of coated vesicles coupled with a means of basal or lateral cell-surface recognition could possibly account for the observed features of transepithelial IgG transport. However the transport time could be reduced and efficiency of the process could be improved by directing the motion of the vesicles. By analogy with other forms of intracellular transport of membrane-bounded organelles we could perhaps suggest the involvement of cytoskeletal elements in directing vesicle transport. It has been shown that membrane-bounded structures which move through cytoplasm are observed in close physical and sometimes chemically specific contact with cytoskeletal structures such as microtubules and actin filaments (Allen, 1975; Burridge & Phillips, 1975).

There is evidence that inhibitors of motile processes based on actin or tubulins cause abnormal motile behaviour of vesicles. This in some situations leads to a reduction in the efficiency of fusion with target organelles (Bhisey & Freed, 1971; Wagner & Rosenberg, 1973; Ginsel, Debets & Daems, 1975; Neve, Rocmans & Ketelbant-Balasse, 1975).

Evidence that coated vesicle transport may be directional in a similar manner is restricted to a few observations. Coated vesicles occur in close proximity to microtubules (Ockleford, 1976). Coated vesicles co-purify with a protein of the same molecular weight as actin (Ockleford & Whyte, 1977). Under this heading can also be included the evidence given in Chapter 5 that there is a specific association between neurofilaments and coated vesicles. This area obviously requires further investigation.

Propulsion

To examine the possibility that micropinocytic coated vesicles are actively translocated through the cytoplasm of the cell by a force-producing system, we have to know rates of vesicle movement. We should then compare these rates with rates of movement found in other motile systems, and with the calculated expected rate of diffusion of a vesicle if it were free to diffuse in conditions similar to those which obtain in the cytoplasm of a cell. In fact not all these data are available and we are not at the stage where we can decide this point one way or the other.

Some labelled uptake studies are relevant in that they fix a lower limit for the rate of transit. Halliday (1955) and Rodewald (1976) studied coated vesicle transport of IgG in the neonatal rat gut and concluded that IgG was completely transported within 30 minutes. Obviously vesicle pinching off and fusion, which take an unknown length of time, must occur within this period, as well as transit. Knowing the dimensions of the epithelium, we can conclude that transit must take place at some speed faster than 10 μm/hour. Similar calculations based on Bruns & Palade's (1968) study of smooth micropinocytic transport of ferritin across muscle capillary endothelium indicate a lower limit of about 2 μm/hour. Present evidence therefore does not rule out the possibility that smooth and coated vesicles translocate across cells in a similar manner.

If pinching off and fusion do not take an excessively large proportion of the total transport time, then diffusion alone can probably account for the vesicles' movement. Green & Casley-Smith (1972) have used Chandrasekhar's modification of the Fokker–Planck equation to calculate rates of diffusion of 0.3 to 0.7 mm/hour for vesicles with diameter 70 nm moving through a cell of viscosity 0.2 poise.

Examination of the measured rates of a number of different motile systems (Table 2) indicates that various structured motile systems give rise to transport speeds slightly faster than Green & Casley-Smith (1972) have calculated for diffusion. This is the case for fast neuronal transport, which moves vesicles and materials of a range of sizes, probably including particles of 70 nm diameter, along axons at a rate of 14 mm/hour.

We think it unlikely that a cell would expend chemical energy in the form of high-energy phosphates to translocate material at the same rates as it would diffuse, although the expenditure of energy by the cell on a system to guide the process of diffusion is not an improbable explanation. A Brownian-motion-powered ratchet mechanism permitting movement in favourable directions, for example alongside microtubules or microfilaments, is one possibility. An alternative possibility is that the assumptions underlying Green & Casley-Smith's calculation are not appropriate in this context.

Table 2. *Rates of movement within cells*

System studied	Approximate transport rate	Reference
Ferritin transport across muscle capillary endothelium	> 2 μm/hour	Bruns & Palade (1968)
Neonatal rat gut antibody (IgG) transport	> 10 μm/hour	Halliday (1955); Rodewald (1976)
Extension of fibroblast margins	30–60 μm/hour	Harris (1973)
Glass particle transport on fibroblast cell surface	180–240 μm/hour	Harris & Dunn (1972)
Trypsin-digested sea urchin ciliary microtubule doublet sliding[a]	1 mm/hour	Summers & Gibbons (1971)
Fast neuronal transport	14 mm/hour	Heslop (1974)
Oyster smooth muscle myofilament sliding	45–72 mm/hour	Ruegg (1968)

[a] At room temperature.

The obvious need exposed by this discussion is for an accurate determination of the time taken for a coated vesicle to travel a known distance within a cell. This may be difficult to accomplish because the rate should be measured in a period starting after pinching off from the membrane in which it forms and before fusion with its target membrane commences.

Membrane fusion

The membrane fusion which occurs in the phospholipid bilayer when a micropinocytic vesicle pinches from the cell surface and seals up and when it eventually fuses with its target organelle is probably very similar to the events occurring in a number of biological processes — for example when mitochondria fuse and separate, when a cell surface wound heals, when two daughter cells finally separate in the process of cytokinesis, or when a lysosome fuses with a phagocytic vesicle. These are just a few examples of what is an extremely important fundamental and commonly occurring event in cell dynamics.

The events as they affect phospholipid bilayers in general are the subject of other reviews (Lucy, 1975, 1978; Edelson & Cohn, 1979) and will not be dealt with here.

The coated vesicle, however, is enclosed in a clathrin lattice which is partially or totally shed prior to the fusion of its contained phospholipid bilayer with a target membrane (Douglas, 1974). Since the target membrane may, in the case of transepithelial transport, be a portion of cell surface, exocytosis by coated vesicles must also be preceded by partial or total lattice loss. Shedding of a portion of lattice prior to membrane fusion might explain why empty clathrin lattices have a smaller mean diameter than do clathrin lattices enclosing phospholipid bilayers (Ockleford *et al.*, 1977).

If, as seems probable, energy is required to break open clathrin lattices, then the presence of a Ca^{2+}-dependent ATPase within vesicles (Blitz, Fine & Toselli, 1977) may be of importance to the process.

Conclusion

Whereas there is still a need for much direct experimentation in the area of coated vesicle dynamics, a considerable amount of circumstantial evidence is relevant and this has been reviewed. In general it supports a fluid component model of coated vesicle membrane motility in which ligands, receptors, IMPs and clathrin become clustered into a curved membrane complex. Such a model accounts simply for the motile process of membrane invagination and also provides a mechanism for the concentration of bound ligand into the coated vesicle lumen. This latter property is of course a useful facet of a *selective* transport system. Studies of vesicle motility and coated vesicle membrane fusion are at an early stage.

Part of this work was funded by the Medical Research Council.

Dynamic aspects of coated vesicle function 275

References

Allen, R.D. (1975). Evidence for firm linkages between microtubules and membrane bound vesicles. *Journal of Cell Biology*, **64**, 497–503.

Allison, A.C. (1973). The role of microfilaments and microtubules in cell movement, endocytosis and exocytosis. In *Locomotion of Tissue Cells, Ciba Foundation Symposium 14*, pp. 110–48. Amsterdam: Elsevier.

Allison, A.C. & Davies, P. (1974). Mechanisms of endocytosis and exocytosis. In *Transport at the Cellular Level*, ed. M.A. Sleigh & D.H. Jennings, *Society for Experimental Biology Symposium 28*, pp. 521–46 Cambridge University Press.

Allison, A.C., Davies, P. & de Petris, S. (1971). Role of contractile microfilaments in movement and endocytosis. *Nature New Biology*, **232**, 153–5.

Amos, W.B. (1971). A reversible mechanochemical cycle in the contraction of *Vorticella*. *Nature, London*, **299**, 127–8.

Behnke, O. & Forer, A. (1967). Evidence for four classes of microtubules in individual cells. *Journal of Cell Science*, **2**, 169–92.

Bhisey, A.N. & Freed, J.J. (1971). Altered movement of endosomes in colchicine-treated cultured macrophages. *Experimental Cell Research*, **64**, 430–8.

Blasie, J.K. & Worthington, C.R. (1969). Planar liquid-like arrangement of photopigment molecules in frog retinal receptor disc membranes. *Journal of Molecular Biology*, **39**, 417–39.

Blitz, A.L., Fine, R.E. & Toselli, P.A. (1977). Evidence that coated vesicles isolated from brain are calcium-sequestering organelles resembling sarcoplasmic reticulum. *Journal of Cell Biology*, **75**, 135–47.

Bray, D. (1973). Model for membrane movements in the neural growth cone. *Nature, London*, **244**, 93–5.

Bruns, R.R. & Palade, G.E. (1968). Studies on blood capillaries. II. Transport of ferritin molecules across the wall of muscle capillaries. *Journal of Cell Biology*, **37**, 277–99.

Burns, R.G. & Pollard, T.D. (1974). A dynein-like protein from brain. *FEBS Letters*, **40**, 274–80.

Burridge, K. & Phillips, J.H. (1975). Association of actin and myosin with secretory granule membranes. *Nature, London*, **254**, 526–9.

Casley-Smith, J.R. (1969a). Endocytosis: the different energy requirements for the uptake of particles by small and large vesicles into peritoneal macrophages. *Journal of Microscopy*, **90**, 15–30.

Casley-Smith, J.R. (1969b). The dimensions and numbers of small vesicles in cells endothelial and mesothelial and the significance of these for endothelial permeability. *Journal of Microscopy*, **90**, 251–68.

Casley-Smith, J.R. & Day, A.J. (1966). The uptake of particulate lipid preparations by macrophages *in vitro*: an electron microscopical study. *Quarterly Journal of Experimental Physiology*, **51**, 1–10.

Changeux, J.P. (1972). Receptor proteins. In *Polymerisation in Biological Systems, Ciba Foundation Symposium 7*, pp. 289–90. Amsterdam: Elsevier.

Cohn, Z.A. (1966). The regulation of pinocytosis in mouse macrophages. I. metabolic requirements as defined by the use of inhibitors. *Journal of Experimental Medicine*, 124, 557–71.

Cohn, Z.A. (1970). Endocytes and intracellular digestion. In *Mononuclear Phagocytes*, ed. R. van Furth, pp. 121–32. Oxford: Blackwell.

Contractor, S.F. & Krakauer, K. (1976). Pinocytosis and intracellular digestion of ^{125}I-labelled haemoglobin by trophoblastic cells in tissue culture in the presence and absence of serum. *Journal of Cell Science*, 21, 595–607.

Davies, P.F., Allison, A.C. & Haswell, A.D. (1973). The quantitative estimation of pinocytosis using radioactive colloidal gold. *Biochemical and Biophysical Research Communications*, 52, 627–34.

de Petris, S. & Raff, M.C. (1972). Distribution of immunoglobulin on the surface of mouse lymphoid cells as determined by immunoferritin electron microscopy. Antibody induced, temperature-dependent redistribution and its implications for membrane structure. *European Journal of Immunology*, 2, 523–35.

de Petris, S. & Raff, M.C. (1973). Fluidity of the plasma membrane and its implications for cell movement. In *Locomotion of Tissue Cells, Ciba Foundation Symposium 14*, pp. 27–52. Amsterdam: Elsevier.

Douglas, W.W. (1974). Involvement of calcium in exocytosis and the exocytosis vesiculation sequence. *Biochemical Society Symposia*, 39, 1–28.

Edelson, P. & Cohn, Z. (1979). Membrane interactions in endocytosis. In *Membrane Fusion*, vol. 5, *Cell Surface Reviews*, ed. G. Poste & G.L. Nicolson, pp. 388–402. Amsterdam: Elsevier.

Elgsaeter, A., Shotton, D.M. & Branton, D. (1976). Intramembrane particle aggregation in erythrocyte ghosts. II. The influence of spectrin aggregation. *Biochimica et Biophysica Acta*, 426, 101–22.

Ginn, F.L., Hochstein, P. & Trump, B. (1969). Membrane alterations in hemolysis: internalisation of plasmalemma induced by primaquine. *Science*, 164, 843–5.

Ginsel, L.A., Debets, W.F. & Daems, W.T. (1975). The effect of colchicine on the distribution of apical vesicles and tubules in absorptive cells of cultured human small intestine. In *Microtubules and Microtubule Inhibitors*, ed. M. Borgers & M. de Brabander, pp. 187–98. Amsterdam: Elsevier.

Gitlin, D. & Biasucci, A. (1969). Development of γG, γA, γM, B_{1C}, B_{1A}, $C'1$ esterase inhibitor, ceruloplasmin, transferrin, hemopexin, haptoglobin, fibrinogen, plasminogen, α_1-antitrypsin, orosomucoid, β-lipoprotein, α_2-macroglobulin and prealbumin in the human conceptus. *Journal of Clinical Investigation*, 48, 1433–46.

Gitlin, J.D. & Gitlin, D. (1976). Protein binding by cell membranes and the selective transfer of proteins from mother to young across tissue barriers. In *Maternofoetal Transmission of Immunoglobulins*, ed. W.A. Hemmings, *Clinical and Experimental Immunoreproduction 2*, pp. 113–19. Cambridge University Press.

Glauert, A.M., Daniel, M.R., Lucy, J.A. & Dingle, J.T. (1963). Studies on the mode of action of excess of vitamin A. VII. Changes in fine structure of erythrocytes during haemolysis by vitamin A. *Journal of Cell Biology,* **17**, 111–21.

Green, H.S. & Casley-Smith, J.R. (1972). Calculation on the passage of small molecules across endothelial cells by Brownian motion. *Journal of Theoretical Biology,* **35**, 103–11.

Halliday, R. (1955). The absorption of antibodies from immune sera by the gut of the young rat. *Proceedings of the Royal Society of London, Series B,* **143**, 408–13.

Harris, A.K. (1973). Cell surface movements related to cell locomotion. In *Locomotion of Tissue Cells, Ciba Foundation Symposium 14,* pp. 3–20. Amsterdam: Elsevier.

Harris, A.K. & Dunn, G. (1972). Centripetal transport of attached particles on both surfaces of moving fibroblasts. *Experimental Cell Research,* **73**, 519–23.

Heslop, J.P. (1974). Fast transport along nerves. In *Transport at the Cellular Level,* ed. M.A. Sleigh & D.H. Jennings, *Society for Experimental Biology Symposium 28,* pp. 209–27. Cambridge University Press.

Heuser, J.E. & Reese, T.S. (1975). Redistribution of intramembranous particles from synaptic vesicles: direct evidence for vesicle recycling. *Anatomical Record,* **181**, 374.

Huxley, H.E. (1973). Muscular contraction and cell motility. *Nature, London,* **243**, 445–9.

Jollie, W.P. & Triche, T.J. (1971). Ruthenium labelling of micropinocytotic activity in the rat visceral yolk sac placenta. *Journal of Ultrastructure Research,* **35**, 541–53.

Kanaseki, T. & Kadota, K. (1969). The 'vesicle in a basket'. A morphological study of the coated vesicle isolated from the nerve endings of the guinea pig brain with special reference to the mechanism of membrain movement. *Journal of Cell Biology,* **42**, 202–20.

Ladbrooke, B.D. & Chapman, D. (1969). Thermal analysis of lipids, proteins and biological membranes. Review and summary of some recent studies. *Chemistry and Physics of Lipids,* **3**, 304–56.

Lucy, J.A. (1975). The fusion of cell membranes. In *Cell Membranes: Biochemistry, Cell Biology and Pathology,* ed. G. Weissmann & R. Claiborne, pp. 79–83. New York: HP Publishing Co.

Lucy, J.A. (1978). Mechanisms of chemically induced cell fusion. In *Membrane Fusion,* ed. G. Poste & G.L. Nicolson, pp. 267–304. Amsterdam: Elsevier.

Malawista, S.E. (1975). Effects of colchicine and vinblastine on the mobilisation of lysosomes in phagocytising human leukocytes. In *Microtubules and Microtubule Inhibitors,* ed. M. Borgers & M. de Brabander, pp. 199–206. Amsterdam: Elsevier.

Malawista, S.E. Gee, J.B.L. & Bensch, K.G. (1971). Cytochalasin B reversibly inhibits phagocytosis: functional, metabolic and ultrastructural effects in human blood leukocytes and rabbit alveolar macrophages. *Yale Journal of Biology and Medicine,* **44**, 286–300.

Marx, R., Graf, E., & Wesemann, W. (1973). Histochemical and biochemical demonstration of sialic acid and sulphate in vesicles and membranes isolated from nerve endings of rat brain. *Journal of Cell Science*, 13, 237–55.

Metcalfe, J.C. & Warren, G.B. (1977). Lipid–protein interactions in a reconstituted calcium pump. In *International Cell Biology*, ed. B.R. Brinkley & K.R. Porter, pp. 15–23. New York: Rockefeller University Press.

Mooseker, M.S. & Tilney, L.G. (1976). Organisation of an actin filament–membrane complex: filament polarity and membrane attachment in the microvilli of intestinal epithelial cells. *Journal of Cell Biology*, 67, 724–43.

Morphis, L.G. & Gitlin, D. (1970). Maturation of the maternofoetal transport system for human γG in the mouse. *Nature, London*, 228, 573.

Moxon, L.A. & Wild, A.E. (1976). Localisation of proteins in coated micropinocytotic vesicles during transport across rabbit yolk sac endoderm. *Cell and Tissue Research*, 171, 175–93.

Munthe Kaas, A.C. (1977). Uptake of macromolecules by rat Kupffer cells in vitro. *Experimental Cell Research*, 107, 55–62.

Nagura, H. & Asai, J. (1976). Pinocytosis by macrophages: kinetics and morphology. *Journal of Cell Biology*, 70, 89A.

Neve, P., Rocmans, P. & Ketelbant-Balasse, P. (1975). Microtubules and microtubule inhibitors and cytochalasin in the thyroid gland. In *Microtubules and Microtubule Inhibitors*, ed. M. Borgers & M. de Brabander, pp. 177–86. Amsterdam: Elsevier.

Ockleford, C.D. (1976). A three dimensional reconstruction of the polygonal pattern on placental coated-vesicle membranes. *Journal of Cell Science*, 21, 83–91.

Ockleford, C.D. & Whyte, A. (1977). Differentiated regions of human placental cell surface associated with exchange of materials between maternal and foetal blood: coated vesicles. *Journal of Cell Science*, 25, 293–312.

Ockleford, C.D., Whyte, A. & Bowyer, D.E. (1977). Variation in the volume of coated vesicles isolated from human placenta. *Cell Biology International Reports*, 1, 137–46.

Orci, L., Carpentier, J.-L., Perrelet, A., Anderson, R.G.W., Goldstein, J.L. & Brown, M.S. (1978). Occurrence of low density lipoprotein receptors within large pits on the surface of human fibroblasts as demonstrated by freeze-etching. *Experimental Cell Research*, 113, 1–13.

Orci, L. & Perrelet, A. (1973). Membrane-associated particles: increase at sites of pinocytosis demonstrated by freeze-etching. *Science*, 181, 868–9.

Pearse, B.M.F. (1975). Coated vesicles from pig brain: purification and biochemical characterisation. *Journal of Molecular Biology*, 97, 93–8.

Penniston, J.T. & Green, D.E. (1968). The conformational basis of energy transformations in membrane systems. IV. Energised states and pinocytosis in erythrocyte ghosts. *Archives of Biochemistry and Biophysics*, 128, 339–50.

Rodewald, R. (1976). Intestinal transport of peroxidase-conjugated IgG fragments in the neonatal rat. In *Maternofoetal Transmission of Immuno-globulins*, ed. W.A. Hemmings, *Clinical and Experimental Immuno-reproduction 2*, pp. 137–49. Cambridge University Press.

Röhlich, P. & Allison, A.C. (1976). Oriented pattern of membrane associated vesicles in fibroblasts. *Journal of Ultrastructure Research*, 57, 94–103.

Roth, T.F., Cutting, J.A. & Atlas, S.B. (1976). Protein transport: a selective membrane mechanism. *Journal of Supramolecular Structure*, 4, 527–48.

Ruegg, J.C. (1968). Contractile mechanisms of smooth muscles. In *Aspects of Cell Motility*, ed. P.L. Miller, *Society for Experimental Biology Symposium 22*, pp. 45–67. Cambridge University Press.

Ryser, H.J.P. (1968). Uptake of protein by mammalian cells: an undeveloped area; the penetration of foreign proteins into mammalian cells can be measured and their functions explored. *Science,* 159, 390–6.

Schlamowitz, M. (1979). Fc specificity and properties of IgG receptors of fetal rabbit yolk sac membrane. In *Transport of Proteins Across Cell Membranes*, ed. W.A. Hemmings, pp. 13–26. Amsterdam: North-Holland.

Schmidtke, J.R. & Unanue, E.R. (1971). Macrophage–antigen interaction, uptake, metabolism and immunogenicity of foreign albumin. *Journal of Immunology*, 107, 331–8.

Sharpey-Shafer, J.M. & Ockleford, C.D. (1978). Uptake of protein by cultured human trophoblast cells. *Cell Biology International Reports*, 2, 579–89.

Simson, J.V. & Spicer, S.S. (1973). Activities of specific cell constituents in phagocytosis (endocytosis). In *International Review of Experimental Pathology*, vol. 12, ed. G.W. Richter & M.A. Epstein, pp. 79–118. New York & London: Academic Press.

Steinman, R.M., Silver, J.M. & Cohn, Z.A. (1974). Pinocytosis in fibroblasts; quantitative studies *in vitro*. *Journal of Cell Biology*, 63, 949–69.

Stockem, W. (1977). Endocytosis. In *Mammalian Cell Membranes*, vol. 5, ed. G.A. Jamieson & D.M. Robinson, pp. 151–95. London: Butterworth.

Stossel, T.P. (1977). Endocytosis. In *Receptors and Recognition*, Series A, vol. 4, ed. P. Cuatrecasas & M.F. Greaves, pp. 105–41. London: Chapman & Hall.

Summers, K.E. & Gibbons, I.R. (1971). Adenosine triphosphate-induced sliding of tubules in trypsin treated flagella of sea-urchin sperm. *Proceedings of the National Academy of Sciences, USA*, 68, 3092–6.

Tighe, J.R., Garrod, P.R. & Curran, R.C. (1967). The trophoblast of the human chorionic villus. *Journal of Pathology and Bacteriology*, 93, 559–67.

Van Deurs, B., Møller, M. & Amtorp, O. (1978). Uptake of horseradish per-oxidase from CSF into the choroid plexus of the rat, with special reference to transepithelial transport. *Cell and Tissue Research*, 187, 215–34.

Wagner, R.C. & Rosenberg, M.D. (1973). Endocytosis in Chang liver cells: the role of microtubules in vacuole orientation and movement. *Cytobiologie*, 7, 20–7.

Wessels, N.K., Spooner, B.S., Ash, J.F., Bradley, M.O., Luduena, M.A., Taylor,
    E.L., Wrenn, J.T. & Yamada, K.M. (1971). Microfilaments in cellular
    and developmental processes. *Science*, 171, 135–43.
Whyte, A. (1978). Proteins of coated micropinocytic vesicles isolated from
    human placentae. *Biochemical Society Transactions*, 6, 299–301.
Wild, A.E. (1976). Mechanisms of protein transport across the rabbit yolk sac
    endoderm. In *Maternofoetal Transmission of Immunoglobulins*, ed.
    W.A. Hemmings, *Clinical and Experimental Immunoreproduction 2*,
    pp. 155–67. Cambridge University Press.
Wills, E.J., Davies, P., Allison, A.C. & Haswell, A.D. (1972). The failure of
    cytochalasin B to inhibit pinocytosis by macrophages. *Nature New
    Biology*, 240, 58–60.

Plates

Plate 1. Micrographs showing the appearance of various types of endocytic vesicle. (*a*) Coated micropinocytic vesicle (*) in a cultured human trophoblastic cell. Scale bar represents 0.1 μm. (*b*) Isolated negatively stained coated vesicle. Scale bar represents 0.1 μm. (*c*) Smooth-walled microvesicle formation and pinocytic channel (*) in a cultured human trophoblast cell. The electron-dense reaction product indicates peroxidase adsorbed to the luminal surfaces of caveolar and vesicle membranes (Sharpey-Shafer & Ockleford, 1978). Scale bar represents 0.2 μm. (*d*) A phagocytic vacuole encloses this Sephadex bead (*) which was preadsorbed with haemoglobin. Scale bar represents 50 μm. (*e*) Phagocytosis is effected by this protozoan despite the presence of radially projecting axopodia (arrow). Scale bar represents 50 μm.

Plate 2. (*a*) The P fracture face (cytoplasmic leaflet) of a frozen-etched fibroblast plasma membrane. Small depressions (i) are the necks of flask-shaped membrane-associated vesicles (Röhlich & Allison, 1976). These can be clearly distinguished from the large pits (*) which are coated vesicle caveolae. Scale bar represents 0.1 μm. (*b*) The P fracture face running through several coated vesicle caveolae (*). The pits contain a population of large intramembranous particles (arrow heads). These also occur in surrounding membrane (arrows) but are more densely packed in caveolae. Scale bar represents 0.1 μm. (These micrographs were kindly supplied by Dr L. Orci.)

Plate 3. Electron micrographs of replicas of frozen-etched human placental tissue obtained after therapeutic terminations of pregnancy performed in the 10th to 14th week of gestation. Small groups of chorionic villi were isolated by dissection and fixed in 6 % glutaraldehyde in M/15 Sorenson phosphate buffer (pH 7.2). About 12 hours prior to freezing this was replaced with 20 % glycerol in the fixation medium. Small portions of trophoblast were mounted on gold support discs and frozen in liquid Freon 22 at −155 °C prior to storage in liquid nitrogen. Chorionic villi were fractured and etched for one minute in a Balzers Freeze Etcher. Fracture faces were replicated using platinum and carbon. Replicas were cleaned using solutions of sodium hypochlorite and hydrochloric acid, and examined in a Philips 300 electron microscope. (*a*) A secantial fracture of the irregular syncytiotrophoblast cell surface showing a polygonally shaped invagination which may be a coated vesicle caveola (arrow). Scale bar represents 0.1 μm. (*b*) The fracture plane through this microvillus (*) is transverse to the phospholipid bilayer until it reaches the point arrowed. Above this level the fracture plane enters the bilayer revealing intramembranous particles. Scale bar represents 0.1 μm. (*c*) Numerous microvilli (*) protrude from the cell surface. These bear intramembranous particles where the fracture plane reveals the interior of the phospholipid bilayer. Scale bar represents 0.3 μm.

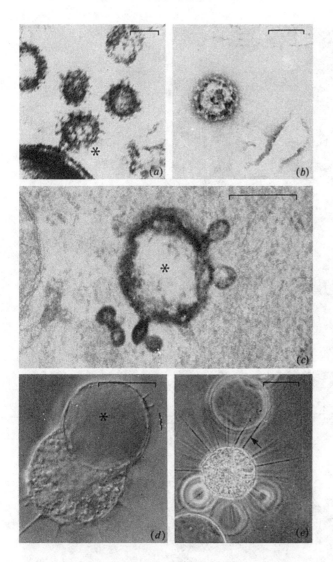

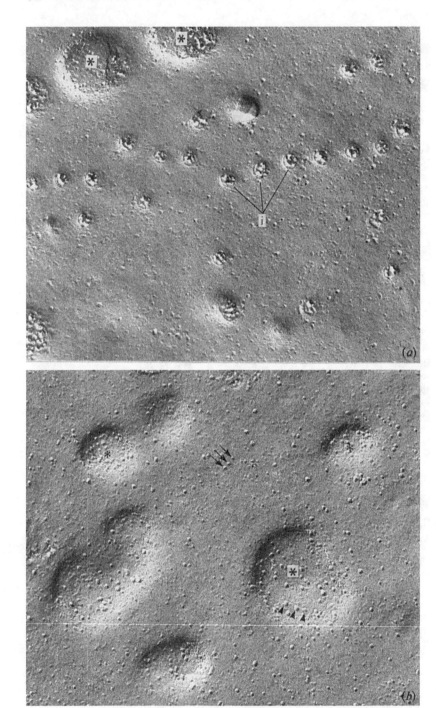

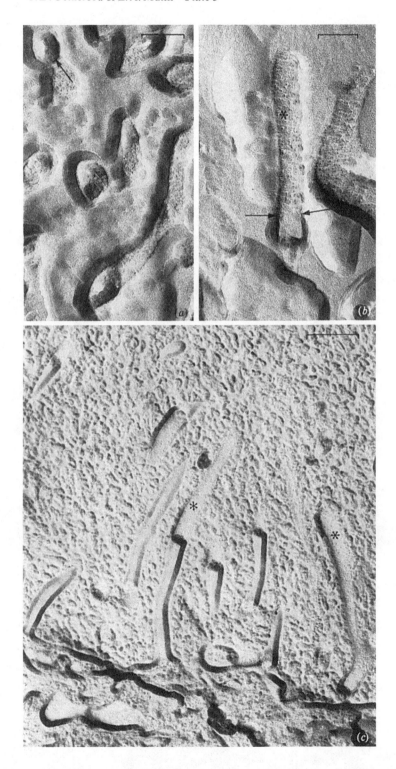

11

Structural aspects of coated vesicles at the molecular level

A. WHYTE & C.D. OCKLEFORD

The growth in our understanding in recent years of the class of pinocytic vesicles known as 'coated vesicles' has been substantial, as evidenced by the contributions of others to this volume. It is arguably in the sphere of biochemical and structural aspects of these vesicles, however, that this growth has been most rapid. It was only a few years ago, for example, that the cytoplasmic network on coated vesicles was thought to be merely protruding 'bristles' or 'hairs', or even that this structure observed in thin-sectioned material was an artifact of the tissue-processing procedure. Although some of the myths surrounding these ephemeral cytoplasmic organelles have been dispelled, there is still a long way to go before we can be fully confident of our interpretations of their structure and function.

Isolation of coated vesicles

Unlike the isolation of smooth pinosomes, the isolation of coated micropinocytic vesicles can be easily monitored by morphological methods, requiring only electron-microscopical observation for estimation of both purity and yield (Plate 1). In spite of this, it was not until 1969 that two groups of workers independently made attempts to isolate these subcellular organelles. The first group, that of Schjeide et al. (1969), isolated coated vesicles from oocytes by use of a simple pressing device, separating the vesicles from other components by density gradient centrifugation. The second group, that of Kanaseki & Kadota (1969), isolated vesicles from guinea pig brains. These latter authors countered the misnomers of 'bristle-coat', 'hair-like coat', etc., and replaced them with their own phrase — 'vesicle in a basket'. These coated vesicle fractions were, however, only partially purified and it was Barbara Pearse who produced the first technique for the purification of these vesicles to near homogeneity (Pearse, 1975). She isolated coated vesicles from pig brains by density gradient centrifugations. The homogenisation medium used,

which was similar to those sometimes employed for the isolation of microtubules, was probably of some importance in maintaining the stability of the
vesicles. The guanosine triphosphate used in this method, however, was
apparently omitted from a later application of the same technique to the
isolation of coated vesicles from brain and adrenal medulla of bullock and
a lymphoma of mouse (Pearse, 1976). Partial purification was also obtained
from pig liver and mouse Krebs ascites tumour cells (Pearse, 1976). The applicability of the method of Pearse (1975) to different tissues has been demonstrated by other authors. Blitz, Fine & Toselli (1977) used it to purify coated
vesicles from brain, and also to partially purify them from pancreas, adrenal
medulla and parotid gland. Ockleford & Whyte (1977) and Whyte (1978) have
used the basic technique of Pearse to isolate coated vesicles from first-trimester
and full-term human placentae. Woods, Woodward & Roth (1978) have also
used this method to prepare vesicles from porcine brain and chicken oocytes.
It appears probable that Pearse's (1975) method will prove applicable to the
isolation of vesicles from most tissues which contain reasonable quantities of
these components. Many authors have, however, adapted the isolation method
of Pearse (1975). Whether these modifications have any significant effect on
the yield or purity of vesicles is unclear. Woods *et al.* (1978) modified it by
omitting the first and third density gradient centrifugation steps of Pearse
(1975), and inserting one or two differential-centrifugation steps. In certain
circumstances, it would appear that a low-speed spin (300*g*, 15 minutes) may
increase the yield of coated vesicles (Whyte, 1978). Because this increased
yield can also be achieved by washing the pellet resulting from a medium-speed
spin (5000*g*, five minutes) (Woods *et al.*, 1978), this effect may be due to
trapping of coated vesicles by cellular debris in the initial centrifugation used
by Pearse (1975) – 20 000*g*, 30 minutes. However, it is the last density
gradient centrifugation step of Pearse (1975) which would appear to be the
most contentious. Woods *et al.* (1978), Ockleford & Whyte (1977) and
Whyte (1978) omitted this step as, at least in one case, it was found not to
increase the purity of the vesicle preparation, but rather to markedly reduce
the yield (unpublished work associated with Whyte, 1978). Blitz *et al.* (1977)
also modified the last sucrose-gradient step of Pearse (1975). Even if the
method eventually becomes modified beyond recognition, however, Pearse's
development of a successful isolation technique for coated vesicles will be
remembered for the surge of interest it generated in these organelles.

Blitz *et al.* (1977) suggested that coated vesicles were unstable upon
storage and that centrifugation was necessary to remove smooth-membraned
vesicle aggregates which formed. It is not clear, however, whether these arose
from coated vesicles which had lost their coats prior to fusion. We have found

that coated vesicles will retain their characteristic morphological appearance
after storage in MES (2-(*N*-morpholino)-ethanesulphonic acid) buffer (see
Pearse, 1975) for 1–2 weeks at 4 °C (unpublished work). It has also been
observed that lattices of coated vesicles retain their integrity after a period of
at least 24 hours in distilled water (Ockleford, unpublished work).

Biochemistry of coated vesicles
Protein composition
'*Clathrin*'. Pearse (1975) showed that the final density gradient pellet from her
isolation technique gave rise to one band of molecular weight 180 000 on
electrophoresis in sodium dodecyl sulphate (SDS)-containing polyacrylamide
gels. Treatment of coated vesicles with trypsin or pronase removed the outer
network of the vesicles when examined by electron microscopy, and also
deleted the 180 000 mol. wt component on SDS-electrophoresis. She therefore
concluded that the 180 000 mol. wt protein subunit was responsible for the
lattice network on the cytoplasmic aspects of the vesicles, and further proposed that it be termed clathrin (Pearse, 1975). Clathrin, or at least a protein
of the same molecular weight, has been observed on SDS-electrophoresis of
isolated vesicles from several sources: bullock brain and mouse lymphoma
(Pearse, 1976); brain (Blitz *et al.*, 1977); porcine brain and chicken oocyte
(Woods *et al.*, 1978); and human placentae (Ockleford & Whyte, 1977;
Whyte, 1978). The molecule thus appears to be highly conserved across both
tissue and species barriers. Indeed, one-dimensional peptide mapping of
clathrin derived from cattle brain and mouse lymphoma coated vesicles
following cleavage at methionine and cysteine residues did reveal conservation
of the protein component (Pearse, 1976).

The amino acid compositions of clathrin from various sources are shown
in Table 1. There is a close similarity between the different tissues, again
indicating conservation of the protein moiety of clathrin. Schjeide *et al.* (1969)
published an amino acid analysis of oocyte coated vesicles. The composition
was similar to that of Pearse (1975) (Table 1) except that there was more
alanine, glycine and serine, and less phenylalanine and tyrosine.

Clathrin has an isoelectric point of 5.8, and carbohydrate does not appear
to be associated with the molecule as it does not stain with the periodic acid–
Schiff reagent for 1,2-glycol groups (Whyte, 1978). Low levels of hexosamine
have been observed by Pearse (1976) although, as this author pointed out, if
clathrin is to freely associate with and disassociate from membranes then it
may be expected to be non-glycosylated (Pearse, 1976).

Kadota, Kadota & Gray (1976) showed that the clathrin cytoplasmic
lattice was labile to the actions of trypsin, pronase and 'nagase', and was also

Table 1.   *Amino acid compositions of clathrin of coated vesicles isolated from different sources (expressed as moles %)*

	Pig brain	Bullock brain	Adrenal medulla
Aspartic acid	10.20	10.39	10.52
Threonine	4.22	4.28	4.39
Serine	5.23	5.33	5.64
Glutamic acid	15.36	15.49	15.22
Proline	5.24	5.13	4.69
Glycine	3.29	3.22	3.89
Alanine	6.02	5.93	5.76
Cysteine	0.93	0.81	ND[a]
Valine	5.95	5.89	6.23
Methionine	2.72	2.71	2.72
Isoleucine	5.03	5.02	5.21
Leucine	9.36	9.42	10.05
Tyrosine	5.21	5.20	3.64
Phenylalanine	6.09	5.77	5.70
Histidine	2.67	2.82	2.60
Lysine	6.58	6.58	6.65
Arginine	5.92	6.00	7.09

After Pearse (1976).
[a]ND, not determined.

destroyed by alkaline pH (9.6) treatment. The clathrin vesicle material was, however, different from the 'particle/chain' material also derived from the synaptosomes. They further concluded that the vesicle coat material did not appear to resemble tubulin (Kadota *et al.*, 1976).

Other protein components. Woods *et al.* (1978) suggested that protein subunits of 120 000 and 55 000 mol. wt, in addition to clathrin, may contribute to the vesicle structure. Neither of these proteins, however, has been observed in isolated placental coated vesicles (Ockleford & Whyte, 1977; Whyte, 1978). At this stage it would appear that the only protein found in all isolated vesicle preparations is clathrin. Several protein bands are usually observed on SDS-electrophoresis of a coated vesicle isolate (Fig. 1). A protein of 100 000 mol. wt has been observed in brain coated vesicles (Blitz *et al.*, 1977) and placental coated vesicles (Ockleford & Whyte, 1977). A minor protein band of 97 000 mol. wt was observed in brain and oocyte coated vesicles (Woods *et al.*, 1978). It has been suggested that the protein of molecular weight 100 000, at least in brain coated vesicles, may be a calcium-stimulated adenosine triphosphatase (ATPase) (Blitz *et al.*, 1977). The vesicles resembled sarcoplasmic reticulum in terms of their SDS-polyacrylamide gel pattern. The vesicles also contained an ATPase which was stimulated by calcium in the presence of Triton X-100 and which displayed maximal activity at 8×10^{-7} M of Ca^{2+}. The vesicles sequestered calcium ions from the medium, and this uptake was stimulated by ATP and by potassium oxalate which is a calcium-trapping agent. In addition, it was found that the 100 000 mol. wt protein of the brain coated vesicles displayed immunological cross-reactivity with an antiserum directed against the calcium-stimulated ATPase of sarcoplasmic reticulum (Blitz *et al.*, 1977). The 100 000 mol. wt component observed in other situations may be an ATPase. For example, in placental coated vesicles, a 100 000 mol. wt band is observed (Ockleford & Whyte, 1977). The histochemical cation stain, potassium pyroantimonate, revealed localisation of cations to the luminal surface of the vesicle walls of coated vesicles (Ockleford & Whyte, 1977). The calcium-chelating agent ethylenediaminetetraacetic acid (EDTA) did not affect the distribution of coated vesicles in the syncytial trophoblast of the placenta (Ockleford & Whyte, 1977), which might suggest that a Ca^{2+}-dependent ATPase is not involved in the formation of coated vesicles.

Fig. 1. Separation of the proteins of isolated placental coated vesicles by SDS-electrophoresis. The gel was stained with Coomassie Blue and scanned at 540 nm. The apparent molecular weights of the protein bands are indicated, determined relative to standard protein mobilities. (From Ockleford & Whyte, 1977.)

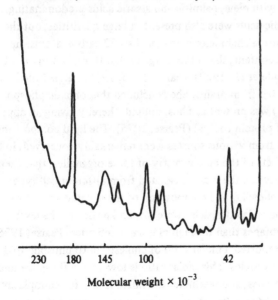

Molecular weight $\times 10^{-3}$

Kadota & Kadota (1973) described a nucleoside diphosphate phosphohydrolase enzyme in coated vesicles isolated from guinea pig brains. These
authors suggested that such an enzyme may accelerate the phosphoenolpyruvate
(PEP)-kinase reaction in the presence of high levels of ATP, and other purine
nucleoside triphosphates, to favour gluconeogenesis. They further postulated
that, in view of the regulatory roles of adenine nucleotides or nicotinamide
adenine dinucleotide in brain glycolytic pathways or acetylcholine synthesis,
coated vesicles may exercise some control over energy metabolism and subsequent acetylcholine synthesis in brain nerve-endings (Kadota & Kadota,
1973). In this connection, it is of interest that the uptake of protein by
oocytes, presumably a function of coated vesicle activity, was prevented by
glycolytic inhibitors (e.g. indoacetic acid, fluoride and 2-deoxyglucose) but
not by inhibitors of respiration or oxidative phosphorylation (e.g. cyanide,
fluorocitrate and antimycin A) (Roth, Cutting & Atlas, 1976).

Lipid composition

Schjeide *et al.* (1969) made a detailed study of the lipids of coated
vesicles isolated from oocytes. They found that the vesicles were 48 % lipid,
predominantly phospholipid. The phospholipid was mainly phosphatidyl
choline, although sphingomyelin and phosphatidyl ethanolamine were well
represented. About 10–15 % sterol, 3 % sterol esters, 7–10 % triglyceride and
1.5 % non-esterified fatty acid were present. The fatty acid component of the
total lipid was simple, with oleic, palmitic and stearic acids predominating.
Linoleic and arachidonic acids were also present in large quantities, but there
were few fatty acids with a chain length greater than 22 carbon atoms. In
spite of this large lipid content, the authors argued that it was not involved in
a transport role (Schjeide *et al.*, 1969). Pearse (1975, 1976) studied the lipids
of coated vesicles isolated from brains. She concluded that organic phosphate
(410 nmol/mg protein) was present as phospholipid, thereby giving an approximate 3 : 1 w/w ratio of protein to lipid (Pearse, 1975). The lipid compositions
of the vesicles isolated from various sources were remarkably conserved, indicating, as do other aspects of the biochemistry of these organelles, that there
are close similarities between the vesicles isolated from different cell types.
Table 2 illustrates the phospholipid composition of the isolated coated vesicles.
Cholesterol levels in the isolated vesicles were low and nearer to the levels
found in internal membranes than those in plasma membranes (Pearse, 1975).
Schjeide *et al.* (1969) reported a molar ratio of cholesterol to phospholipid
of 0.4 in oocyte coated vesicles. This molar ratio is lower than those commonly
found in plasma membranes, and resembles the ratios found for endoplasmic
reticulum. Because coated vesicles are surface-membrane-derived structures,

Table 2. *Phospholipid composition of coated vesicles.*

Phospholipid	%
Phosphatidyl choline	42–43
Phosphatidyl ethanolamine	30–33
Phosphatidyl inositides	10–11
Phosphatidyl serine	4
Sphingomyelin	8–12

After Pearse (1975, 1976).

the observation of such a low cholesterol content is perhaps a little unexpected. It could be explained, however, by an extension of the hypothesis that coated vesicles function in the selective uptake of protein and their receptors (Roth & Porter, 1964; Ockleford & Whyte, 1977). Coated vesicles may also selectively internalise plasma membrane lipids, avoiding those cholesterol-rich domains characteristic of plasma membranes (Bretscher, 1976).

Nucleic acid contents
Both DNA and RNA appear to be integral components of oocyte coated vesicles (Schjeide *et al.*, 1969). The RNAs (about 4 % of the dry weight) had an average composition of G 32, C 24, A 22, U 22, which appears to be markedly different from that of microsomes in terms of the GC:AU ratio (Schjeide *et al.*, 1969). The RNA could be rapidly labelled with [^{3}H] uridine and appeared to be a structural component as it resisted salt extraction. Coated vesicle DNA from the oocytes was similarly unusual. The DNAs were of unique density and rapidly incorporated radiolabelled thymidine. Schjeide *et al.* (1969) speculated that the RNA was transcribed by the coated vesicle DNA. Coated vesicle nucleic acid may play a morphic role as a nucleating factor in vesicle morphogenesis, similar to the role played by the 5′-end RNA bases in tobacco mosaic virus assembly (Klug, 1972).

'Receptors' inside vesicles
Isolated coated vesicles which contain phospholipid bilayers (Plate 1) will also presumably contain the contents of the vesicle. The hypotheses that coated vesicles can function in the selective uptake of protein and the like (Rodewald, 1973; Lloyd & Beck, 1974) would imply that these molecules and their receptors are sequestered into the vesicles. SDS-polyacrylamide gel scans of isolated vesicles must therefore also reveal components which represent these 'loaded' receptors (Fig. 1). It has been suggested that high molecular weight Schiff-positive glycoproteins in isolated placental coated

vesicles may represent glycocalyx enclosed within these vesicles (Whyte, 1978). This postulate was based upon the observation that these high molecular weight glycomolecules remained in a Triton-insoluble coated vesicle fraction. When examined by electron microscopy these vesicles were found to have lost their clathrin cytoplasmic lattices while leaving the phospholipid-bilayer vesicles apparently intact. In view of the observation that some of the clathrin also remained in the detergent-insoluble fraction, it was suggested that clathrin might span the phospholipid bilayer and be attached to the receptors of the glycocalyx inside the vesicle (Whyte, 1978). A morphological association between the clathrin ridges in thin-sectioned vesicles and the vesicle contents has been noted in placental coated vesicles (Ockleford, 1976). Blitz *et al.* (1977) found that urea treatment also removed the clathrin cytoplasmic lattice of isolated vesicles, and some clathrin remained in the pellet (see figure 3 in Blitz *et al.*, 1977). In addition, these authors suggested that the 100 000 and 55 000 mol. wt components which were also present were intrinsic membrane-associated proteins. In coated vesicles isolated from first-trimester human placentae clathrin may be associated with a protein of 100 000 mol. wt (Whyte, 1978). Because a protein of the same molecular weight has been observed in plasma membranes isolated from equivalent human placentae (Whyte, unpublished work) it may be that this protein represents a subunit of a surface-membrane-located receptor in the placenta.

Receptors for low-density lipoprotein (LDL), the major transport material for cholesterol in serum, have been shown to occur within coated vesicle pits of cultured human fibroblasts (Anderson, Brown & Goldstein, 1977a). Study of fibroblasts cultured from a patient (J.D.) with the clinical phenotype of homozygous familial hypercholesterolaemia or hyper-β-lipoproteinaemia, an hereditary disorder characterised by a marked increase in plasma LDL levels, revealed that although these fibroblasts had membrane receptors for LDL, internalisation of LDL into coated vesicles did not occur (Anderson, Goldstein & Brown, 1977b). Most patients with the homozygous form of hyper-β-lipoproteinaemia fail to bind LDL (Anderson *et al.*, 1977b). It was postulated that J.D. possessed two mutant alleles at the LDL receptor locus: the first, maternally inherited, is the R^{b^0} allele which specifies a receptor molecule unable to bind LDL and is biochemically 'silent'; and the second the R^{b+,i^0} allele, paternally inherited, specifies a receptor that can bind LDL normally but cannot internalise it (Anderson *et al.*, 1977b). In view of the similarities in the processes of uptake of LDL and other proteins, such as epidermal growth factor (Carpenter & Cohen, 1976) and transcobalamin (Youngdahl-Turner, Allen & Rosenberg, 1977), further studies of the R^{b+,i^0} mutant allele should prove interesting (Anderson *et al.*, 1977*b*). Freeze-fracture studies indicate that the LDL receptor may be a transmembrane protein (Orci *et al.*, 1978) and that the R^{b+,i^0} allele may be deficient in the cytoplasmic protein (clathrin?) which internalises these receptors into coated vesicles (Anderson *et al.*, 1977*b*).

The extent of the involvement of coated vesicles in processes other than endocytosis is apparent from the contributions of others to this volume (see, in particular, Chapters 1 and 9). Coated vesicles may therefore be expected to contain not only receptors for uptake, but may also contain receptor molecules involved in the recognition of materials for export or digestion within the cell.

Morphological analyses of coated vesicles

Isolation of coated vesicles permitted detailed observations on their morphology (Ockleford, Whyte & Bowyer, 1977*a*), an advantage over measurements made on vesicles in ultrathin sections, which might be expected to give rise to misleading results. Kanaseki & Kadota (1969) made the first study of a coated vesicle isolate. They suggested that it was a transformation of hexagonal facets on the cytoplasmic aspect of the plasma membrane into pentagons that induced curvature and the subsequent formation of a pit (caveola) and finally a vesicle.

Friend & Farquhar (1967) suggested that coated vesicles in rat vas deferens occurred in two size classes with discrete functions: the smaller vesicles (< 75 nm in diameter), which performed Golgi-associated or lysosomal functions, and the larger vesicles (> 100 nm), which are endocytic. They suggested that no vesicles had diameters between these two size ranges. Such morphological dichotomy, however, is not found in all tissues. Ockleford *et al.* (1977*a, b*) have made the most extensive study to date on the measurements of coated vesicles isolated from first-trimester human placenta. The vesicles were classed in two groups according to their shape: spherical and prolate spheroidal (i.e. a sphere extended in its polar axis). A further subdivision was made according to whether or not the vesicles possessed a central lipid bilayer. When the volumes of the isolated vesicles are plotted graphically (Fig. 2), unimodal distributions are apparent. A similar distribution was obtained when the volume of all vesicles was studied (Fig. 3). Unimodal distributions were also observed when histograms of the number of vertices on the circumference, number of polygons per hemisphere, and longest and shortest vesicle axes were constructed (Ockleford *et al.*, 1977*a*). Woods *et al.*

Fig. 2. The volume distributions of isolated placental coated vesicles.
(a) Spherical vesicles; (b) prolate spheroidal vesicles; (c) vesicles containing phospholipid bilayers; (d) vesicles without phospholipid
bilayers. All distributions are unimodal. (Fig. 2 (c) is from Ockleford,
Whyte & Bowyer, 1977.)

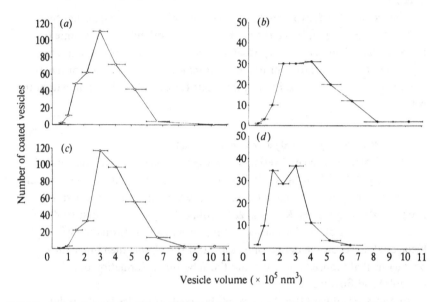

Vesicle volume ($\times 10^5$ nm^3)

(1978) observed that coated vesicles isolated from porcine brain and chicken
oocytes also showed unimodal distributions as assessed by their external
diameters. The association of a particular size of coated vesicles with a specific
cellular function, such as suggested for vas deferens (Friend & Farquhar,
1967), therefore does not appear to hold true for some other tissues. The sum
of the published observations on isolated coated vesicles indicates they occur
in a normally distributed fashion, often with positive skew in the histograms,
and that the lowest value for the diameter (50–60 nm) corresponds with the
lowest theoretical vesicle volume (Ockleford *et al.*, 1977*a*). In addition, the
mean diameter appears to be about 75–80 nm and the upper limit about 110
nm, although vesicles up to 130 nm in diameter have been observed (see Table
4). In thin-sectioned material coated vesicle lattices up to 540 nm in diameter
have been observed (Ockleford & Whyte, 1977). There also appears to be a
correlation between vesicle size and the presence of a phospholipid bilayer –
those vesicles containing a bilayer being significantly larger than those which
do not (Ockleford *et al.*, 1977*a*) – and the smaller vesicles without bilayers
may be interpreted as representing parts of larger vesicles which have lost their
coats prior to fusion with other membranous organelles (Ockleford *et al.*,
1977*b*). Finally, those vesicles classed as prolate spheroidal are significantly

larger than the spherical vesicles (Table 3), which may be due to a change in ratio of the polygons in the lattice wall (Ockleford *et al.*, 1977*b*).

Molecular models of structure

Although the structure of coated vesicles shows considerably more variability in assembly than other similar structures, e.g. certain viruses (Ockleford *et al.*, 1977*b*), various attempts have nevertheless been made to rationalise their configuration. The most extensive has been that of Crowther, Finch & Pearse (1976). These authors analysed the structure of vesicles isolated from pig brain by use of a technique similar to that which Finch & Klug (1967) used to establish the structure of spherical viruses. They based their analysis on the assumption that twelve pentagons were the minimum necessary in order for a hexagon/pentagon configuration to have sufficient curvature to close. The number of hexagons associated with the twelve pentagons was variable. They identified three structures which we will consider in more detail, although other structures were observed, one of which, containing four hexagons, accounted for a proportion of the empty coats (Crowther *et al.*, 1976).

Fig. 3. Frequency distribution of the total population of coated vesicles shown in Fig. 2. Again a unimodal distribution is apparent.

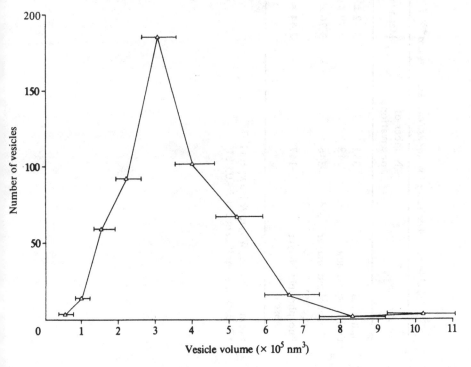

Table 3. *Analysis of volumes (expressed as nm^3 × 10^{-5}) of coated vesicles isolated from human placenta*

Vesicle class	Number of observations	Mean volume ± S.D.	95 % confidence limits
Spherical	352	3.12 ± 1.29	2.99–3.25 ⎱ *
Prolate spheroidal	142	3.99 ± 1.97	3.66–4.32 ⎰
Phospholipid bilayer present	346	3.70 ± 1.58	3.54–3.87 ⎱ *
Phospholipid bilayer absent	147	2.44 ± 1.12	2.25–2.64 ⎰

From Ockleford, Whyte & Bowyer (1977).
*Represents significance at $P<0.001$.

Structure A. This contains eight hexagons and twelve pentagons (Fig. 4*a*), and has top- and bottom-located hexagons with rows of six pentagons, six hexagons and six pentagons respectively in between. Crowther *et al.* (1976) noted that a similar structure was seen in other biological situations though on a larger scale. The particle has six-fold symmetry and a diameter of about 54 nm.

Structure B. This consists of four hexagons and twelve pentagons (Fig. 4*b*). The hexagons lie at the vertices of a tetrahedron. The diameter of structure B is about 47 nm.

Structure C. This consists of eight hexagons and twelve pentagons, and is similar to structure A except that two of the equatorial hexagons and the two pentagons in the occluded angle between them have been rotated through 90° (Fig. 4*c*). The result is two arcs of four hexagons joined together by twelve pentagons. The apparent diameter of structure C is 54 nm.

As Crowther *et al.* (1976) pointed out, an increase in the number of hexagons allows increasing numbers of arrangements for vesicle structure. With eight hexagons there are three possible structures, two of which have been observed in coated vesicle populations (structures A and C). Crowther *et al.* (1976) did not observe any images corresponding to the structure suggested by Kanaseki & Kadota (1969), an icosahedron consisting of twenty hexagons and twelve pentagons, although such a structure is consistent with their theoretical models.

Woods *et al.* (1978) observed four classes of coat structure in vesicles isolated from pig brain and chicken oocyte. Their observations were made on Markham rotations performed on negatively stained vesicles. At least two of their structures, however, might be accounted for by the models of Crowther *et al.* (1976). The first, four pentagons and two hexagons surrounding a central hexagon (Woods *et al.*, 1978), may arise from structure A viewed at (90° 30°) (figures 3 and 4*b* of Crowther *et al.*, 1976). The second, consisting of six pentagons surrounding a central hexagon, might represent the (0° 0°) view of structure A (Crowther *et al.*, 1976). The other two structures described by Woods *et al.* (1978) are eight small polygons surrounding a central larger polygon, and ten small polygons surrounding a central larger polygon. Because these conclusions were based upon Markham rotations of the isolated vesicles, however, they might be expected to contain information derived from two opposing sides of the vesicle superimposed upon each other and therefore may be misleading. Because the number of possible structural permutations increases markedly with increasing vesicle size (Crowther *et al.*, 1976), it

(a)

(b)

(c)

Fig. 4. Stereopairs of the three structures for coated vesicles proposed by Crowther *et al.* (1976). (*a*) Structure A consists of eight hexagons and twelve pentagons. (*b*) Structure B comprises four hexagons and twelve pentagons. (*c*) Structure C contains eight hexagons and twelve pentagons. For details see text. (From Crowther, Finch & Pearse, 1976.)

appears uncertain whether we will be able to formulate guidelines upon which all vesicles can be constructed.

The most marked feature of these structural studies is the difference in vesicle dimensions reported by the authors. Thus, although both Woods *et al.* (1978) and Crowther *et al.* (1976) studied vesicles isolated from pig brains, their range of measurements are totally exclusive of each other (Table 4). Because Crowther *et al.* (1976) looked at vesicles suspended in stain over holes in the carbon support, their measurements may be expected to be nearer to the true situation, as vesicles in stain on a carbon film tend to show a degree of flattening. However, this alone would probably be insufficient to account for the difference shown in Table 4, and it is therefore possible that the different isolation methods of these two groups of authors have selected for different populations of coated vesicles. This point should be borne in mind when comparing results obtained by different authors.

In conclusion, it is pertinent to consider the molecular dimensions of the walls of the polygons and the possible role of clathrin in the structure of the walls. Kanaseki & Kadota (1969) measured the length of the sides of the polygons as 24–27 nm. Crowther *et al.* (1976) estimated it as 15 nm. In view of the preceding remarks, it is probable that the true length lies somewhere nearer the lower value. Based on the molecular weight of empty coats determined by ultracentrifugation as 22×10^6 (Crowther *et al.*, 1976), then one might expect about 100 to 120 molecules of clathrin per vesicle. Crowther *et al.* (1976) suggested that each vertex of a polygon is contributed to in part by three clathrin molecules so that each edge is formed by two clathrin molecules (Fig. 5). This arrangement would result in structures A and C each containing 108 molecules of clathrin, and structure B containing 84 (Fig. 4), predictions which are consistent with the above experimental estimates. The

Table 4. *The dimensions of coated vesicles isolated from different sources*

Source	Mean diameter (range) (nm)	Reference
Chicken oocyte	85 (60–105)	Woods *et al.* (1978)
Pig brain	75 (60–105)	
Pig brain	53 (45–60)	Crowther *et al.* (1976)
Bullock adrenal medulla	73 (55–90)	
Human placenta	80 (50–130)	Ockleford & Whyte (1977)

Note that although both Woods *et al.* (1978) and Crowther *et al.* (1976) isolated vesicles from the same material, their ranges of measurements were totally exclusive of each other, suggesting that they were looking at different populations of coated vesicles.

Fig. 5. A diagrammatic representation of a possible arrangement of individual clathrin molecules in the structure of hexagons and pentagons as suggested by Crowther et al. (1976).

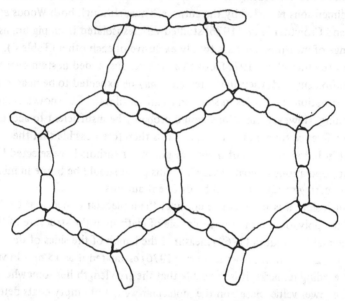

molecular weight of structures A and C would therefore be $108 \times 180\,000 = 19.44 \times 10^6$. The molecular weight of phospholipid-bilayer-containing vesicles determined by ultracentrifugation is about 28×10^6 (Crowther et al., 1976). The difference, i.e. the phospholipid bilayer and its contents, would therefore have a molecular weight of 8.56×10^6. If the vesicle contained 10 000 phospholipids (Pearse, 1975) each with an average molecular weight of 760 (estimated from the percentage composition illustrated in Table 2), then the contents of the vesicle, i.e. apart from clathrin and phospholipid, would have a molecular weight of $(8.56 \times 10^6) - (7.6 \times 10^6) = 9.6 \times 10^5$. This value is probably within the error limit of these calculations and therefore indicates that coated vesicles isolated from pig brains do not have significant levels of vesicle contents. Thus, they may be performing a lipid transfer role within this cell type (Pearse, 1976), avoiding uptake of plasma membrane proteins as suggested by Bretscher (1976) (but see pp. 112-15). Coated vesicles in other situations, however, do appear to endocytose proteins and a similar calculation performed on vesicles from these sources might give some idea of the quantities of protein taken up.

In coated vesicles which contain phospholipid bilayers the central vesicle appears to be smaller than the inside diameter of the coat (Plate 1). Whether the clathrin forms a rigid sphere somewhat larger than the membrane vesicle

(Kanseki & Kadota, 1969) or merely forms ridges of material along the vesicle (Ockleford, 1976), is presently unclear (Woods *et al.*, 1978). What is certain, however, is that clathrin is a sufficiently large molecule to form much more than the polyhedral vertices. Crowther *et al.* (1976) calculated that if each clathrin molecule contributed a double α-helical coiled-coil to each edge, of length 15 nm, this would require 200 amino acids or a molecular weight of about 22 000. Even if the sides of the polygons are 27 nm (Kanaseki & Kadota, 1969) a molecular weight of only 40 000 would be required. As Crowther *et al.* (1976) pointed out, this will leave plenty of the clathrin protein to form the projecting knobs seen at the vertices of the coat and also to provide connections to the inner vesicle, perhaps to protein receptors in the glycocalyx inside the vesicle (Whyte, 1978).

Conclusions

In view of the many roles performed by coated vesicles in different cell types, it may be naive to expect vesicles from different sources to be identical to each other both biochemically and structurally. Indeed, differences are found. But particularly striking are the homologies. All coated vesicles studied so far appear to be composed of a single protein, clathrin, which perhaps is solely responsible for the lattice network on the cytoplasmic aspects of these vesicles. In addition, there is a certain degree of similarity in the structure of vesicles isolated from different sources. A range of polymorphism is observed, however, and although there must be intermolecular specificity as only hexagons and pentagons occur, nevertheless the degree of flexibility is unlike that seen in spherical viruses or enzyme complexes (Crowther *et al.*, 1976; Ockleford *et al.*, 1977*b*).

Much benefit will be derived from further study of the proteins associated with vesicles and the receptors they internalise. One can envisage such knowledge eventually extending to include virtually all cell types, but in the perhaps not too distant future major rewards may come from the iatrogenic potentialities of selective uptake. It would undoubtedly be useful to physicians to know whether a drug tagged to a certain protein will, or will not, be taken up by a particular cell type.

We thank the Medical Research Council for its support.

References

Anderson, R.G.W., Brown, M.S. & Goldstein, J.L. (1977*a*). Role of the coated endocytic vesicle in the uptake of receptor-bound low density lipoprotein in human fibroblasts. *Cell*, 10, 351–64.
Anderson, R.G.W., Goldstein, J.L. & Brown, M.S. (1977*b*). A mutation that

impairs the ability of lipoprotein receptors to localise in coated pits on the cell surface of human fibroblasts. *Nature, London,* **270**, 695–9.

Blitz, A.L., Fine, R.E. & Toselli, P.A. (1977). Evidence that coated vesicles isolated from brain are calcium-sequestering organelles resembling sarcoplasmic reticulum. *Journal of Cell Biology,* **75**, 135–47.

Bretscher, M.S. (1976). Directed lipid flow in cell membranes. *Nature, London,* **260**, 21–3.

Carpenter, G. & Cohen, S. (1976). ^{125}I-labelled human epidermal growth factor. Binding, internalisation, and degradation in human fibroblasts. *Journal of Cell Biology,* **71**, 159–71.

Crowther, R.A., Finch, J.T. & Pearse, B.M.F. (1976). On the structure of coated vesicles. *Journal of Molecular Biology,* **103**, 785–98.

Finch, J.T. & Klug, A. (1967). Structure of broad bean mottle virus. I. Analysis of electron micrographs and comparison with turnip yellow mosaic virus and its top component. *Journal of Molecular Biology,* **24**, 289–302.

Friend, D.S. & Farquhar, M.G. (1967). Functions of coated vesicles during protein absorption in the rat vas deferens. *Journal of Cell Biology,* **35**, 357–76.

Kadota, K. & Kadota, T. (1973). A nucleoside diphosphate phosphohydrolase present in a coated-vesicle fraction from synaptosomes of guinea-pig whole brain. *Brain Research,* **56**, 371–6.

Kadota, T., Kadota, K. & Gray, E.G. (1976). Coated-vesicle shells, particle/chain material, and tubulin in brain synaptosomes. An electron microscope and biochemical study. *Journal of Cell Biology,* **69**, 608–21.

Kanaseki, T. & Kadota, K. (1969). The 'vesicle in a basket'. A morphological study of the coated vesicle isolated from the nerve endings of the guinea pig brain, with special reference to the mechanism of membrane movements. *Journal of Cell Biology,* **42**, 202–20.

Klug, A. (1972). The polymorphism of tobacco mosaic virus protein and its significance for the assembly of the virus. In *Polymerisation in Biological Systems, Ciba Foundation Symposium 7,* pp. 207–15. Amsterdam: North-Holland.

Lloyd, J.B. & Beck, F. (1974). Lysosomes. In *The Cell in Medical Science,* ed. F. Beck & J.B. Lloyd, vol. 1, pp. 272–313. New York & London: Academic Press.

Ockleford, C.D. (1976). A three-dimensional reconstruction of the polygonal pattern on placental coated-vesicle membranes. *Journal of Cell Science,* **21**, 83–91.

Ockleford, C.D. & Whyte, A. (1977). Differentiated regions of human placental cell surface associated with exchange of materials between maternal and foetal blood: coated vesicles. *Journal of Cell Science,* **25**, 293–312.

Ockleford, C.D., Whyte, A. & Bowyer, D.E. (1977*a*). Variation in the volume of coated vesicles isolated from human placenta. *Cell Biology International Reports,* **1**, 137–46.

Ockleford, C.D., Whyte, A. & Bowyer, D.E. (1977*b*). Negative staining of coated vesicles. *Micron*, 8, 233–5.

Orci, L., Carpentier, J., Perrelet, A., Anderson, R.G.W., Goldstein, J.L. & Brown, M.S. (1978). Occurrence of low density lipoprotein receptors within large pits on the surface of human fibroblasts as demonstrated by freeze-etching. *Experimental Cell Research*, 113, 1–13.

Pearse, B.M.F. (1975). Coated vesicles from pig brain: purification and biochemical characterisation. *Journal of Molecular Biology*, 97, 93–8.

Pearse, B.M.F. (1976). Clathrin: a unique protein associated with intracellular transfer of membrane by coated vesicles. *Proceedings of the National Academy of Sciences, USA*, 73, 1255–9.

Rodewald, R. (1973). Intestinal transport of antibodies in the newborn rat. *Journal of Cell Biology*, 58, 189–211.

Roth, T.F., Cutting, J.A. & Atlas, S.B. (1976). Protein transport: a selective uptake mechanism. *Journal of Supramolecular Structure*, 4, 527–48.

Roth, T.F. & Porter, K.R. (1964). Yolk protein uptake in the oocyte of the mosquito *Aedes aegypti* L. *Journal of Cell Biology*, 20, 313–31.

Schjeide, O.A., San Lin, R.I., Grellert, E.A., Galey, F.R. & Mead, J.F. (1969). Isolation and preliminary chemical analysis of coated vesicles from chicken oocytes. *Physiological Chemistry and Physics*, 1, 141–63.

Whyte, A. (1978). Proteins of coated micropinocytic vesicles isolated from human placentae. *Biochemical Society Transactions*, 6, 299–301.

Woods, J.W., Woodward, M.P. & Roth, T.F. (1978). Common features of coated vesicles from dissimilar tissues: composition and structure. *Journal of Cell Science*, 30, 87–97.

Youngdahl-Turner, P., Allen, R.H. & Rosenberg, L.E. (1977). Binding and uptake of transcobalamin II by human fibroblasts. *Clinical Research*, 25, 472A.

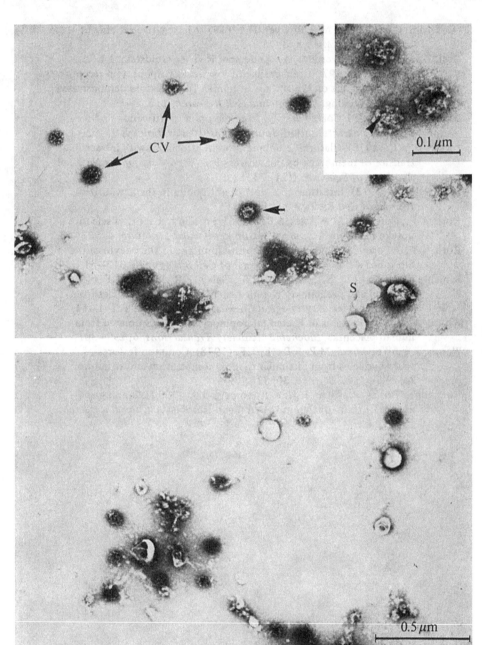

Plate

Plate 1. Electron micrograph of coated vesicles (CV) isolated from human placenta and negatively stained with uranyl acetate. The polygonally patterned network or 'coat' of clathrin is apparent (inset). There are some smooth-surfaced membrane structures (S) in the samples. Some of the coated vesicles (arrow) contain an electron-lucent region which may represent the phospholipid membrane bilayer of the vesicle. Occasionally the sides of the polygons show a beaded appearance (arrowhead in inset). (From Ockleford & Whyte, 1977.)

12

Coated vesicles in medical science

F. BECK

Coated vesicles varying considerably in morphology have been described in
various types of cells and a survey of the literature leads one to consider it
probable that more than one functional entity has been given this name.
Friend & Farquhar (1967) in a widely quoted paper present evidence for two
types of coated vesicles in the rat vas deferens: a large type (> 100 nm in
diameter) which they equate with heterophagosomes and to which they ascribe
the functions of protein uptake and membrane resorption, and a smaller
variety (< 75 nm in diameter) which they consider to be primary lysosomes.
Kallio, Garant & Minkin (1971) described the ruffled border of active osteoclasts in fish and rats. They drew attention to deep invaginations and villous
extensions of the plasma membrane which possess a coat of repeating units
along their cytoplasmic surface. The presence of numerous smooth-walled and
coated vesicles in the cytoplasm deep to the ruffled border was also noted.
The authors suggest that the particulate coat on the surface cytoplasmic membrane may be of a different nature to that present on the coated vesicles.
Tangential sections of the coating on the cytoplasmic aspect of the surface
membrane failed to reveal the polygonal pattern characteristic of the coated
vesicles in the apical cytoplasm and it was thought possible that while the
coated vesicles played a part in protein resorption, the particles on the ruffled
border may represent sites of enzymatic activity connected with bone demineralisation. Yet other appearances were described by Mollenhauer, Hass &
Morre (1976) in their description of the Golgi apparatus and surrounding
organelles of rat spermatids in which, once again, two classes of coated vesicles
were described. Small vesicles lie between the smooth endoplasmic reticulum
and the immature face of the Golgi, and apart from their cytoplasmic coating
have membrane characteristics morphologically similar to the endoplasmic
reticulum, while larger vesicles with bounding membranes that resemble the
acrosome membrane lie between the mature face of the Golgi and the acrosome.

The 'honeycomb' nature of the coat material was clearly evident for both large and small coated vesicles. In yet another situation coated vesicles ranging from 50 to 200 nm in diameter have been implicated in the transport of IgG across the rabbit yolk sac endoderm *in utero* (Moxon, Wild & Slade, 1976) and across the proximal small intestine in the neonatal rat (Rodewald, 1973: see Chapter 4 in this volume). In these situations the vesicles seem to pass across the cells without losing their cytoplasmic coating and to be unassociated with primary lysosomes or with the Golgi apparatus. In the human placenta, where coated vesicles are found principally in the region of the syncytiotrophoblastic surface, a unimodal distribution of vesicle size has been described (Ockleford, Whyte & Bowyer, 1977), a state of affairs which would be consistent with a functionally discrete population of organelles. Heuser & Reese (1973), using horseradish peroxidase as a marker, produced persuasive evidence suggesting that coated vesicles in frog sartorius neuromuscular junctions retrieved synaptic vesicle membrane which had been added to the surface by exocytosis consequent upon stimulation. They advanced a working hypothesis that synaptic vesicles coalesce with the surface membrane at specific regions adjacent to the muscle and that equal amounts of surface membrane are then retrieved by coated vesicles arising from plasma membrane adjacent to the Schwann sheath. Subsequently the vesicles lose their coating and coalesce to form cisternae which accumulate in the region of the nerve ending where synaptic vesicle depletion has occurred. The cisternae then divide to form new synaptic vesicles. The authors support the idea that the coat apparatus is part of a mechanism for deforming the plasma membrane in the process of vesicle formation (see Chapter 10). They calculate that a resting end plate could contain about 10^4 free coats and suggest that during continuous stimulation, membrane retrieval would balance exocytosis of transmitter if each coat took about one minute to attach to the plasma membrane, pinch off a vesicle and detach from the vesicle to reorganise on the plasma membrane again. Heuser & Rees's (1973) theory concerning the function of coated vesicles in frog neuromuscular junctions must be reconciled with Clark, Hurlbut & Mauro's (1972) observations on changes in these structures caused by black widow spider venom. Application of this material depletes the neuromuscular junction of synaptic vesicles with a concomitant increase in the frequency of *miniature* end plate potentials. At the same time, while there appears to be a net increase in the area of presynaptic surface membrane *there is no loss of coated vesicles* in the nerve terminal. In fact these structures are much easier to locate in terminals depleted of synaptic vesicles. The authors also draw attention to Nickel & Potter's (1970) observation that the electric organ of *Torpedo* apparently contains few, if any, coated vesicles. Clark and his co-workers believe that the coated vesicles at the frog neuromuscular junction are, as elsewhere, concerned with protein uptake. Clearly many other factors, such as the rapid assembly of fresh membrane from soluble macromolecular pools and its destruction in lysosomes, are operative in membrane dynamics. The question is comprehensively reviewed by Holtzman, Schacher, Evans & Teichberg (1977) who conclude that 'The most prudent interpretation of available information probably is that fusion of coated vesicles to form larger sacs which then break-up into smaller vesicles has yet to be demonstrated decisively'.

The role of coated vesicles

We are faced with a variety of suggestions concerning the function of coated vesicles, all of which may be true depending upon the location and type of vesicle under discussion. It is in the context of these functions that suggestions concerning their role in medical science must be seen and the fragmentary reports of disturbed activity evaluated.

Protein transport

In 1892 Ehrlich immunised mice with plant toxins and observed the pre- and postnatal transmission of passive immunity to their offspring. He was of the opinion that the greater part of this transmission occurred after birth by way of the milk, and his views are now generally accepted. Culbertson (1939) was the first to show that young rats absorb antibodies from immune serum administered orally as readily as they did from their mothers milk and Halliday in 1955, using high-titre rat *Salmonella pullorum* antiserum, showed that the efficiency of the intestine *as a whole* to absorb antibody from a constant dose given between birth until day 18 is constant. After day 18 its efficiency declines rapidly and after day 21 no transmission was observed. Rodewald (1970) demonstrated that the proximal one-third of the rat intestine was concerned with antibody transport across the gut epithelium, while Williams & Beck (1969) showed that the distal part of the gut was principally concerned with the uptake and intracellular digestion — supposedly for nutritional purposes — of macromolecules present in the milk. In 1973 Rodewald, using ferritin and peroxidase-labelled IgG, presented evidence that uptake in the proximal gut was highly species selective. He indicated that species-specific antibodies are selectively bound at the apical cell surface of the enterocyte and transferred to the intercellular spaces between these cells (i.e. deep to the junctional complex) by coated vesicles.

Brambell (1970) has comprehensively reviewed the transmission of passive immunity in man and monkey and concluded that the principal route is by way of the chorioallantoic placenta. His conclusion rests heavily upon the

work of Bangham, Hobbs & Terry (1958) and Bangham (1960). Gitlin &
Gitlin (1974) have shown that the human placenta binds IgG and Ockleford &
Whyte (1977) have suggested that the coated vesicles, which they have convincingly shown near the surface of the syncytiotrophoblast, are formed in
regions of the surface membrane where specific surface receptor sites are
and which additionally are characterised by a very pronounced glycocalyx.
Besides transferring passive immunity from the mother to the foetus the
passage of IgG across the placenta can also result in the production of embryo-
pathies. The well-known condition of erythroblastosis foetalis results from
maternal immunisation against her own foetus. It occurs in its most common
form when a foetus carrying the antigen Rh(D) on its erythrocytes is implanted
in an Rh(d) mother. Probably chiefly at parturition, but perhaps to a smaller
extent during pregnancy also, leakage of foetal red cells across the placenta
occurs. This sometimes results in sensitisation and the production of anti-D
IgG by the mother. In subsequent pregnancies anti-D antibodies pass across
the placenta causing haemolysis of foetal red cells bearing the D antigen, and
this may lead to foetal death or to congenital jaundice, brain defect and other
abnormalities. Similar conditions occur more rarely with other 'Rhesus' antibodies and with blood group factors other than 'Rhesus'. Recently Rhesus-
negative mothers carrying Rhesus-positive babies have been given anti-Rh IgG
immediately after the birth of their first Rhesus-positive baby (which is usually
normal because sensitisation of the mother does not take place until leakage
of foetal blood at parturition). Apparently the anti-Rh antibody combines
with the Rh antigen on the foetal cells that leak at parturition into the
maternal circulation and destroys them. This suppresses subsequent immunisation of the mother and therefore the production of such Rhesus antibody
as would interfere with the succeeding pregnancies. This treatment has resulted
in a dramatic reduction in the incidence of what was previously the commonest
form of prenatal disease! (For review see Clarke, 1975.)

An intriguing question concerns the apparent loss of coating of the vesicles
in the deeper regions of the syncytiotrophoblast. Rodewald (1973) observed
that in the rat gut the coating is still present when the vesicles deliver their
contents in the intercellular space by exocytosis, and Wild (1976) has suggested
that the clathrin coating may act as a physical barrier to lysosomal attack.
There is evidence that other proteins are specifically taken up by cells at
receptor sites present on plasma membrane in regions of coated vesicle formation and are subsequently demonstrable within coated vesicles inside the cells.
An excellent example is the coated vesicles which form morphologically distinct 'fuzzy' regions of the plasma membrane and may be responsible for the
transport of ferritin into mammalian erythroblasts for the purpose of haemoglobin synthesis (Ghadially, 1975).

Anderson, Brown & Goldstein (1977) have shown that a proportion of the specific receptors for low-density lipoproteins are present on circumscribed coated segments of the plasma membrane of human fibroblasts. They estimated that at 37 °C a morphologically demonstrable population of coated segments (constituting about 2 % of the cell surface) was internalised within 10 minutes. Upon internalisation the vesicles lost their coats within two minutes and seemed to have fused with lysosomes within six minutes. The authors conceive of this mechanism as being a quick turnover system, which results in the ingested lipoprotein particle being degraded within lysosomes with the release of free cholesterol which then becomes available for cell membrane synthesis and for other aspects of cholesterol metabolism. It was demonstrated that exposure to substrate (i.e. low-density lipoprotein) was not required to induce movement of the specific receptors into the coated region. In addition to sites for immunoglobulins and lipoproteins, coated regions may also have specific receptors for polypeptide hormones destined to be taken up and degraded by cells. Carpenter & Cohen (1976) have demonstrated binding of epidermal growth factor to specific areas on human fibroblast membranes and Terris & Steiner (1975) reported insulin receptor sites on rat hepatocytes.

Clearly there exists a distinction between mechanisms which transport material *across* cells in coated vesicles (the system described by Rodewald as operative for homologous IgG in the rat gut) and protein *uptake* into a cell by coated organelles capable of binding and rapid interiorisation of a variety of specific macromolecules. Anderson, Brown & Goldstein (1977) tentatively suggest that this latter category of molecules might exert both nutritional and regulatory functions *within* the cell after losing their coating and possibly fusing with lysosomes. A dramatic extension of our knowledge of the factors which might be operative in coated vesicle formation has come from the description by Anderson, Goldstein & Brown (1977) of a rare mutant that impairs the capacity of low-density lipoprotein receptors to localise in coated pits. It will be remembered that Anderson, Brown & Goldstein (1977) had suggested that in normal human fibroblasts the specific receptors were localised in patches occupying some 2 % of the plasma membrane surface. While Ockleford & Whyte (1977) postulated that such localisation might result from the binding of the receptors to multivalent macromolecules, Anderson, Brown & Goldstein showed that the aggregation occurred independently of binding to low-density lipoprotein and was therefore not akin to ligand-induced patching or capping (see de Petris & Raff, 1972) unless it were the case that other materials present cross-linked the same receptors. Now Anderson, Goldstein & Brown (1977) describe an allele $(R^{b+,i^{0}})$ in which altered lipoprotein receptors, though still able to bind with their substrate, lack the

ability to become incorporated into coated pits. Because of their mislocation
the receptors cannot carry bound lipoprotein into the cell in significant
quantities and this results in a rare familial hypercholesterolaemia (different
from the commoner form resulting from the gene R^{b^0} which, if present in the
homozygous form, prevents the binding of lipoprotein to receptors). Anderson,
Goldstein & Brown (1977) have illustrated a working model for the formation
of coated vesicles (Fig. 1). It is suggested that the low-density lipoprotein
receptor is a transmembrane protein inserted at random into the plasma membrane (probably after glycosylation) and bears its specific lipoprotein binding
site on its external surface. The cytoplasmically located pole of the protein
contains a sequence (possibly clathrin) which is necessary for recognition of

Fig. 1. Schematic illustration of the proposed pathway by which low-
density lipoprotein (LDL) receptors become localised to coated pits on
the plasma membrane of human fibroblasts. The sequential steps in
this process are as follows: (1) synthesis of LDL receptors on poly-
ribosomes; (2) insertion of LDL receptors at random sites along
non-coated segments of plasma membrane; (3) clustering together
of LDL receptors in coated pits; (4) internalisation of LDL receptors
as coated pits invaginate to form coated endocytic vesicles; and (5)
recycling of internalised LDL receptors back to the plasma membrane. (From Anderson, Goldstein & Brown (1977), with permission.)

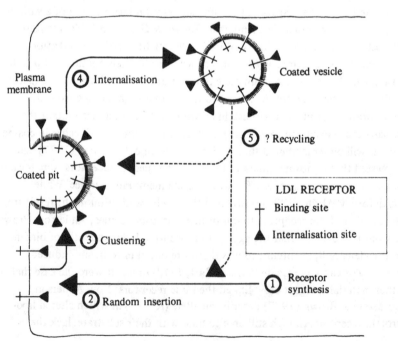

the receptor as a component of coated pits; this sequence causes the receptors to gather together into discrete areas which will form coated vesicles. Furthermore kinetic data suggest that recycling of internalised lipoprotein receptors back to the cell surface occurs even if protein synthesis is experimentally inhibited by cyclohexamide.

Membrane turnover

Franke *et al.* (1976) have made a careful study of vesicle coat material in lactating rat mammary epithelium. They found 'classical' coated vesicles (40–60 nm in internal diameter) in the secretory cells; these were often Golgi-associated and the authors describe their origin, fate and function as unknown. In addition, however, many casein-containing vesicles had, either in part or *in toto,* a morphologically similar coat showing classical polygonal patterns in appropriately sectioned material. It was suggested that the 'fuzzy' coat of these vesicles might be similar to that seen in association with regions of the plasma membrane; also that both are proteinaceous in nature and might contain both actin and myosin. Because of the close association of the polygonal coat with microtubules and microfilaments it was proposed that a basis exists for guiding secretory vesicles to their region of discharge as well as providing the necessary motility forces for their passage to the surface. The suppressive effect of colchicine on lactation (Patton, 1974) supports the hypothesis. Franke and his co-workers produce persuasive morphological evidence suggesting addition of secretory vesicle membrane to the apical plasma membrane of the epithelial cells. This provides a simple method for the replenishment of plasma membrane lost because many secreted fat globules are enveloped by plasma membrane.

Many other examples of membrane turnover involving coated vesicles have been described. The work of Heuser & Rees has already been referred to and may in some ways be a special case pertaining to neurons and synapses. In the general context of secretory cells the simplest models of membrane flow postulate lipoprotein synthesis in the rough endoplasmic recticulum. The rough endoplasmic reticulum is continuous with smooth endoplasmic reticulum from which transitional vesicles bud off to fuse with the Golgi apparatus. Then condensing vacuoles appear on the mature face of this organ and, becoming smaller in size, form secretory granules. The latter discharge their contents at the surface of the cell and their membrane is intercalated into the plasma membrane of the cell. There are, however, difficulties in accepting this sequence without some modification; many secretory proteins are greatly condensed in the Golgi apparatus or condensing vacuole and redundant membrane must therefore be produced. It has been suggested that the coated vesicles seen near or actually fused with mature secretion granules might be *removing* excess membrane rather than performing the function usually ascribed to them, namely the contribution of Golgi or endoplasmic reticulum products to membrane-bounded material (Kramer & Geuze, 1974).

Even in actively secreting glandular cells changes in surface area are transient, so that there is presumably a compensatory mechanism for removing secretion granule membrane or an equivalent area of plasma membrane from the apical cell surface. The endocytic budding of short tubules, vesicles and other structures is now a well-documented phenomenon in this context (Amsterdam *et al.*, 1971; Abrahams & Holtzman 1973; Kalina and Rabino-witch, 1975), but of course this does not exclude the possibility that a molecular dismantling of the surface membrane also occurs. Grynszpan-Winograd (1971) and Benedeczy & Smith (1972) have described coated endocytic vesicles forming from the plasma membrane present at the site of active exocytosis, and suggest that the membrane retrieved by endocytosis is the same as that added by exocytosis. Kramer & Geuze (1974), however, have shown that this may not always be so and that 'surface membrane circulation' may be an important phenomenon. Whether *all* vesicles that retrieve surface membrane are coated is not proven and may in fact turn out to be very difficult to investigate, since in many situations coated vesicles lose their coating soon after formation (Roth & Porter, 1964; Douglas, Nagasawa & Schulz, 1971). One may merely be dealing here with a question of fixation (Gray, 1975), but it is easy to believe that the coating itself provides the energy for the membrane reorganisation required in endocytosis (see above). The fact that such coated vesicles may be enriched on their luminal aspect by specific membrane proteins which act as receptors for individual substrate species could conceivably be related to the cytoplasmic coating only in certain situations. This would mean that only some coated vesicles are responsible for the endocytosis of specific macromolecules while others might be merely concerned with (say) membrane turnover (Ockleford & Whyte 1977).

Relationship of coated vesicles to the Golgi apparatus

Coated vesicles are frequently observed in the region of the Golgi apparatus. Their significance here is a subject of debate and may differ in various cell types. It seems likely that a proportion at least of these profiles are budded off from the Golgi apparatus or from GERL and contain specific secretion products. The most widely accepted suggestion in this context is that many are primary lysosomes transporting acid hydrolases to hetero-phagosomes containing endocytosed material (Holtzman, Novikoff & Villaverde, 1967; see also Friend & Farquhar, 1967). It is interesting to speculate, as have Ockleford & Whyte (1977), whether their coat is associated with specific receptors in their interior which imparts to them the capacity of segregating particular enzymes (i.e. acid hydrolases) from the Golgi secretions.

Another source of coated vesicles in the Golgi region has been postulated to be plasma membrane which has been interiorised and may be in the process of incorporation into the Golgi apparatus (Novikoff, Novikoff, Quintana & Hann, 1971). This may be of importance in membrane turnover kinetics but its true significance has not been evaluated.

Ion exchange

Studies by Kallio *et al.* (1971) of the osteoclasts from mouse, rat and fish have already been mentioned (see p. 303). While these workers believe that the 'conventional' coated vesicles found in active osteoclasts are connected with protein resorption, they suggest that the particular coat present on the cytoplasmic aspect of the plasma membrane is perhaps similar to that in the blowfly *Calliphora erythrocephala*. Here the membrane may be a site of hydrogen ion secretion and/or ion pumping (Berridge & Gupta, 1967) and thus the particles comprising the coat may represent the loci of enzymes related to bone demineralisation. Indeed, osteopetrosis — a rare condition in man — is probably largely due to failure of osteoclast function (Loutit & Sansom, 1976).

The calcium-sequestering action of coated vesicles isolated from the brain has been stressed by Blitz, Fine & Toselli (1977). They suggest that like the sarcoplasmic reticulum the coated vesicles contain Ca-ATPase in their coats. It is therefore possible that coated vesicles, by sequestering calcium, reduce the cytoplasmic calcium level after neurotransmitter release triggered by calcium influx, and thereby control the function of the presynaptic nerve terminals. In this context the observations of Kendrick, Blaustein, Friend & Ratzlaff (1977) are interesting. They have shown that the presynaptic terminals contain an ATP-dependent non-mitochondrial calcium storage system stimulated by oxalate and that an increase occurs in the number of coated vesicles present in frog motor end plates after lengthy stimulation (Heuser & Rees, 1973). As well as being concerned with membrane recycling, this phenomenon may therefore also lower the elevated cytoplasmic calcium concentration consequent upon nerve stimulation (Llinas, Blinks & Nicholson, 1972).

Conclusion

Coated vesicles have been considered in this chapter as an entity principally by virtue of the electron microscopic demonstration of a coating

on the cytoplasmic face of their bounding membrane. In some cells this coating has been characterised biochemically while in others identification has
been on purely morphological grounds. Although it will be apparent to the
reader of this volume that characteristics common to coated vesicles in a
variety of situations are in the process of being defined, it cannot be said with
any certainty that all the morphological forms described in the literature as
coated vesicles constitute a discrete structural and functional entity in the sense
of (say) the mitochondrion or the Golgi apparatus. For this reason it is pre-
mature to speak of the medical importance of 'coated vesicles' if by this one
necessarily implies common functional characteristics. It must be understood
therefore that a review of this nature is based upon an incomplete catalogue
of sometimes unconnected experimental findings in many fields, with perhaps
no underlying theme to unify all of them. Nevertheless the widespread presence
of coated vesicles in eukaryotic cells, and the functions that many of them
seem to share, makes speculation in this field useful, if only to direct attention
to the critical questions which still require answers.

   There is little doubt concerning the prime importance of the uptake of
specific proteins by various cells of the body. Equally it now seems likely that
this is largely if not completely a function of coated vesicles. An excellent
example is provided by the uptake of low-density lipoprotein by human fibroblasts and the hypercholesterolaemia which results from a disorder of this
function (p. 307). Other receptor-mediated protein uptake in the human is
well documented and may repay investigation from a number of points of
view. First, the protein may pass across the cell in an unchanged form, protected from lysosomal catabolism as described by Rodewald in the neonate
rat gut. This is obviously important in such situations as the prenatal transfer
of IgG from mother to young, which seems in man to occur across the chorioallantoic placenta. Gitlin & Gitlin (1974) have described specific IgG binding
by the human placenta which would, of course, be a necessary preliminary to
transplacental transfer. Second, specific proteins may be taken into cells for
processing therein. These may be situations in which coated vesicles lose their
coating shortly after endocytosis. The applicability of this phenomenon to
the recognition of antibodies by the immunological system is obvious, though
no morphological evidence of disturbance at the level of coated vesicles has
been described. The uptake of low-density lipoproteins by fibroblasts
(Goldstein & Brown, 1976) is followed by loss of vesicular coating, fusion with
lysosomes, and degradation of the protein and cholesteryl esters with the
release of free cholesterol, which then becomes available to the general metabolic pool. Third, specific proteins may be removed from an extracellular
situation where they are harmful. Ockleford (1977) has proposed a novel

hypothesis dealing with the non-rejection of the placenta which uses this argument. He has suggested that maternal IgG directed against foetal tissue is recognised and rapidly endocytosed by the syncytiotrophoblast so that the local concentration of antibody in contact with the foetal tissue, where it might mediate rejection, is kept at a low level. This might tie up with the same worker's observation that the coated vesicles lose their coating soon after endocytosis. One would postulate that these vesicles are merely concerned with the removal and degradation of IgG and that only a small minority of coated vesicles retain their coating to transfer antibodies to the foetus.

The role of non-specific bulk fluid uptake in endocytic vesicles, together with contained soluble protein and also non-specific protein adsorption by the glycoprotein lining of coated and non-coated endocytic vesicles, is of importance (see Chapter 7). Cells often require protein for nutritional and other purposes and in certain situations may obtain a proportion of this through coated vesicles. In this connection mention should be made of early embryonic (histiotrophic) nutrition. Before the establishment of the foetal circulation the nidus must obtain raw materials for its growth by the diffusion of small molecules from the extraembryonic membranes in contact with maternal tissues to the growing embryonic site. Although some of this material is undoubtedly provided by diffusion of solutes from the maternal circulation it seems likely that the bulk is provided by the breakdown of maternal macromolecules, provided by the endometrium, in the lysosomal system of the extraembryonic membranes themselves (trophoblast in the human situation). This material seems to be provided — at least in part — by endocytosis through coated vesicles (see Beck, 1976, for review). The ability of proteins to adsorb to cell surfaces may vary greatly with the substrate (Lloyd, Williams, Moore & Beck, 1975). It may also be that proteins stimulate pinocytosis by coated vesicles at sites such as the proximal convoluted kidney tubule in human glomerulonephritis (Ryabov, Plotkin, Freici & Nevorotin, 1978). Here it is likely that the mechanism is non-specific with respect to the type of protein taken up and has the effect of preventing excessive protein loss thus minimising the effects of the disease.

Apart from the endocytosis of specific proteins by coated vesicles it seems at least theoretically possible that the specific receptors which appear to be associated with the polygonal coat of these structures might also confer specificity upon the contents of Golgi-associated coated vesicles, which latter may have originated in the Golgi apparatus or in GERL (see p. 000). This could conceivably have some bearing upon the pathology of inherited storage disease (for review see Neufeld, Lim & Shapiro, 1975). It seems likely that the pathological manifestations of storage diseases are due to the congenital absence of specific acid hydrolases and it has been assumed that the fault lies in an inability of the affected cells to synthesise these enzymes. As far as this author knows there have never been any investigations directed to the ability of the Golgi apparatus to 'package' specific enzymes into primary lysosomes which have (at least by some authors) been equated with a certain type of coated vesicle (Friend & Farquhar, 1967). Theoretically, at least, it is possible that the deficiency of specific hydrolases in primary lysosomes may result from an inability of the Golgi apparatus to sequester these enzymes after their production, though of course no evidence presently exists to support such a hypothesis.

The possible role of coated vesicles in membrane turnover has been referred to earlier (p. 309). The complexity of this phenomenon has been discussed and a distinction has been made between direct reabsorption of the membrane of secretory vesicles and those cases where endocytosis is not confined to the sites of exocytosis (Kramer & Geuze, 1974). The situation is further complicated by the metabolic turnover of membranes in various cells (reviewed by Holtzman et al., 1977).

In cases where coated vesicles seem to be concerned with membrane resorption — possibly at synapses and in some gland cells — these membranes may be re-used directly or degraded in secondary lysosomes for re-use in the metabolic pool. In either case, but particularly in the former, disturbance of the recycling process could be associated with disturbances of secretion. The effect of (for example) black widow spider venom at neuromuscular junctions (Clark et al., 1972) deserves study from this aspect.

The postulated role of 'coated' plasma membrane and coated vesicles in Ca^{2+} sequestration has also been alluded to. If indeed the phenomenon is of major physiological importance it is difficult at this stage to envisage a clinically important condition associated with its disturbance per se, but future developments may well alter this point of view.

References

Abrahams, S. & Holtzman, E. (1973). Secretion and endocytosis in insulin-stimulated rat adrenal medulla cells. Journal of Cell Biology, 56, 540–58.

Amsterdam, A., Ohad, I., Schramm, M., Salomon, Y. & Selinger, Z. (1971). Concomitant synthesis of membrane protein and exportable protein of the secretory granule in rat parotid gland. Journal of Cell Biology, 50, 187–200.

Anderson, R.G.W., Brown, M.S. & Goldstein, J.L. (1977). Role of the coated endocytic vesicle in the uptake of receptor-bound low density lipoprotein in human fibroblasts. Cell, 10, 351–64.

Anderson, R.G.W., Goldstein, J.L. & Brown, M.S. (1977). A mutation that impairs the ability of lipoprotein receptors to localise in coated pits on the cell surface of human fibroblasts. *Nature, London*, 270, 695–8.

Bangham, D.R. (1960). The transmission of homologous serum proteins to the foetus and to the amniotic fluid in the rhesus monkey. *Journal of Physiology, London*, 153, 265–89.

Bangham, D.R., Hobbs, K.R. & Terry, R.J. (1958). Selective placental transfer of serum proteins in the rhesus. *Lancet*, ii, 351–4.

Beck, F. (1976). Comparative placental morphology and function. *Environmental Health Perspectives*, 18, 5–12.

Benedeczy, I. & Smith, A.D. (1972). Ultrastructural studies of the adrenal medulla of golden hamster: origin and fate of secretory granules. *Zeitschrift für Zellforschung und Mikroskopische Anatomie*, 124, 367–86.

Berridge, M.J. & Gupta, B.L. (1967). Fine structural changes in relation to ion and water transport in the rectal papillae of the blowfly, *Calliphora*. *Journal of Cell Science*, 2, 89–112.

Blitz, A.L., Fine, R.E. & Toselli, P.A. (1977). Evidence that coated vesicles from brain are calcium sequestering organelles resembling sarcoplasmic reticulum. *Journal of Cell Biology*, 75, 135–47.

Brambell, F.W.R. (1970). *The Transmission of Passive Immunity from Mother to Young. Frontiers of Biology 18*. Amsterdam: North-Holland.

Carpenter, G. & Cohen, S. (1976). ^{125}I-labelled human epidermal growth factor: binding, internalisation and degradation in human fibroblasts. *Journal of Cell Biology*, 71, 159–71.

Clark, A.W., Hurlbut, W.P. & Mauro, A. (1972). Changes in the fine structure of the neuromuscular junction of the frog caused by black widow spider venom. *Journal of Cell Biology*, 52, 1–14.

Clarke, C.A. (ed.) (1975). *Rhesus Haemolytic Disease*. Lancaster: Medical and Technical Publishing Co.

Culbertson, J.T. (1939). The immunisation of rats of different age groups against *Trypanosoma lewisi* by the administration of specific antiserum *per os*. *Journal of Parasitology*, 25, 181–2.

de Petris, S. & Raff, M.C. (1972). Distribution of immunoglobulin on the surface of mouse lymphoid cells as determined by immunoferritin electron microscopy: antibody induced, temperature dependent redistribution and its implications for membrane structure. *European Journal of Immunology*, 2, 523–5.

Douglas, W.W., Nagasawa, J. & Schulz, R.A. (1971). Coated microvesicles in neurosecretory terminals of posterior pituitary glands shed their coats to become smooth 'synaptic' vesicles *Nature, London*, 232, 340–1.

Ehrlich, P. (1892). Uber Immunität durch Vererbung und Säugung. *Zeitschrift für Hygiene und Infecktionskrankheiten*, 12, 183–203.

Franke, W.W., Luder, M.R., Kartenbeck, J., Zerban, H. & Keenan, T.W. (1976).

Involvement of vesicle coat material in casein secretion and surface regeneration. *Journal of Cell Biology*, **69**, 173–95.

Friend, D.S. & Farquhar, M.G. (1967). Functions of coated vesicles during protein absorption in the rat vas deferens. *Journal of Cell Biology*, **35**, 357–76.

Ghadially, F.N. (1975). *Ultrastructural Pathology of the Cell*, p. 493. London: Butterworth.

Gitlin, J.D. & Gitlin, D. (1974). Protein binding by specific receptors on human placenta, murine placenta and suckling murine intestine in relation to protein transport across these tissues. *Journal of Clinical Investigation*, **54**, 1155–66.

Goldstein, J.L. & Brown, M.S. (1976). Binding and degradation of low density lipoproteins by cultured human fibroblasts: comparison of cells from a normal subject and from a patient with homozygous familial hypercholesterolaemia. *Journal of Biological Chemistry*, **249**, 5153–62

Gray, E.G. (1975). Synaptic fine structure and nuclear, cytoplasmic and extracellular networks. *Journal of Neurocytology*, **4**, 315–39.

Grynszpan-Winograd, O. (1971). Morphological aspects of exocytosis in the renal medulla. *Philosophical Transactions of the Royal Society of London, Series B*, **261**, 291–8.

Halliday, R. (1955). The absorption of immune sera by the gut of the young rat. *Proceedings of the Royal Society of London, Series B*, **143**, 408–13.

Heuser, J.E. & Reese, T.S. (1973). Evidence for recycling of synaptic vesicle membrane during transmitter release at the frog neuromuscular junction. *Journal of Cell Biology*, **57**, 315–44.

Holtzman, E., Novikoff, A.B. & Villaverde, H. (1967). Lysosomes and GERL in normal and chromatolytic neurones of the rat ganglion nodosum. *Journal of Cell Biology*, **33**, 419–36.

Holtzman, E., Schacher, S., Evans, J. & Teichberg, S. (1977). Origin and fate of the membranes of secretion granules and synaptic vesicles: membrane circulation in serous gland cells and retinal photoreceptors. In *The Synthesis, Assembly and Turnover of Cell Surface Components*, ed. G. Poste & G.L. Nicholson, pp. 165–246. Amsterdam: Elsevier/North-Holland.

Kalina, M. & Rabinowitch, R. (1975). Exocytosis coupled to endocytosis of ferritin in parotid acinar cells from isoprenalin-stimulated rats. *Cell and Tissue Research*, **163**, 373–82.

Kallio, D.M., Garant, P.R. & Minkin, C. (1971). Evidence of coated membranes in the ruffled border of the osteoclast. *Journal of Ultrastructure Research*, **37**, 169–77.

Kendrick, N.C., Blaustein, M.P., Friend, R.C. & Ratzlaff, R.W. (1977). A.T.P.-dependent calcium storage in presynaptic nerve terminals. *Nature, London*, **265**, 246–8.

Kramer, M.F. & Geuze, J.J. (1974). Redundant cell-membrane regulation in the exocrine pancreas cells after pilocarpine stimulation of the secretion. In *Advances in Cytopharmacology*, ed. B. Caccarelli,

F. Clementi & J. Moldesi, vol. 2, pp. 87—98. New York: Raven Press.

Llinas, R., Blinks, J. & Nicholson, G. (1972). Calcium transport in presynaptic terminal of squid giant synapse: detection with Aequorin. *Science*, 172, 1127—9.

Lloyd, J.B., Williams, K.E., Moore, A.T. & Beck, F. (1975). Selective uptake and intracellular digestion of protein by rat yolk sac. In *Maternofoetal Transmission of Immunoglobulins*, ed. W.A. Hemmings, *Clinical and Experimental Immunoreproduction 2*, pp. 169—78. Cambridge University Press.

Loutit, J.F. & Sansom, J.M. (1976). Osteopetrosis of microphthalmic mice. A defect of the hematopoietic stem cell? *Calcified Tissue Research*, 20, 25.

Mollenhauer, H.H., Hass, B.S. & Morre, D.J. (1976). Membrane transformation in Golgi apparatus of rat spermatids. *Journal de Microscopie et de Biologie Cellulaire*, 27, 33—6.

Moxon, L.A., Wild, A.E. & Slade, B.S. (1976). Localisation of proteins in coated micropinocytic vesicles during transport across rat yolk sac endoderm. *Cell and Tissue Research*, 171, 175—93.

Neufeld, E.F., Lim, T.W. & Shapiro, L.J. (1975). Lysosomal storage diseases. *Annual Review of Biochemistry*, 44, 357—76.

Nickel, E. & Potter, L.T. (1970). Synaptic vesicles in freeze etched electric tissue of *Torpedo*. *Brain Research*, 23, 95—100.

Novikoff, P.N., Novikoff, A.B., Quintana, N. & Hann, J.J. (1971). Golgi apparatus, GERL and lysosomes of neurones in rat dorsal root ganglia studied by thick section and thin section cytochemistry. *Journal of Cell Biology*, 50, 859—86.

Ockleford, C.D. (1977). Antibody clearance by micropinocytosis: a possible role in fetal immunoprotection, *Lancet*, i, 310.

Ockleford, C.D. & Whyte, A. (1977). Differentiated regions of human placental cell surface associated with exchange of materials between maternal and foetal blood: coated vesicles. *Journal of Cell Science*, 25, 293—312.

Ockleford, C.D., Whyte, A. & Bowyer, D.E. (1977). Variation in volume of coated vesicles isolated from human placenta. *Cell Biology International Reports*, 1, 137—46.

Patton, S. (1974). Reversible suppression of lactation by colchicine. *FEBS Letters*, 48, 85—7.

Rodewald, R.B. (1970). Selective antibody transport in the proximal small intestine of the neonate rat. *Journal of Cell Biology*, 45, 635—70.

Rodewald, R.B. (1973). Intestinal transport of antibodies in the newborn rat. *Journal of Cell Biology*, 58, 189—211.

Roth, T.F. & Porter, K.R. (1964). Yolk protein uptake in the oocyte of the mosquito *Aedes aegypti*. *Journal of Cell Biology*, 20, 313—32.

Ryabov, S.I., Plotkin, V.Ya., Freici, I.I. & Nevorotin, A.J. (1978). Possible role of the proximal convoluted tubules of human kidney in chronic glomerulonephritis. *Nephron*, 21, 42—7.

Terris, S. & Steiner, D.F. (1975). Binding and degradation of ^{125}I-insulin by rat hepatocytes. *Journal of Biological Chemistry*, **250**, 8389–98.

Wild, A.E. (1976). Mechanism of protein transport across the rabbit yolk sac endoderm. In *Maternofoetal Transmission of Immunoglobulins*, ed. W.A. Hemmings, *Clinical and Experimental Immunoreproduction 2*, pp. 155–67. Cambridge University Press.

Williams, R.M. & Beck, F. (1969). A histological study of gut maturation. *Journal of Anatomy*, **104**, 487–502.

APPENDIX 1. NOMENCLATURE

The literature on coated vesicles has been dogged with nomenclature problems and Bowers (1964) lists a plethora of synonyms for the organelle. Since coated vesicles were first described, terms such as complex vesicles, fuzzy vesicles, spiny vesicles, bristle-coated vesicles, dense-rimmed vesicles, alveolate vesicles and acanthosomes have all been employed. Because the majority of the data contained in this volume refers to the one organelle we have used the single term 'coated vesicle' throughout. We propose that for *this* organelle the use of all the other terms be dropped.

Against such a background it was not surprising to observe recently the beginnings of a rift developing over choice of a name for the major protein composing the polygonal structure on these vesicles' outer (cytoplasmic) surface (Matus, 1976; Pearse & Bretscher, 1976). The name proferred by Matus for this protein is dependent on Gray's (1972) interpretation of the vesicle coat as a fixation artifact. Gray had observed that the coated vesicles in his preparations of sectioned material for transmission electron microscopy were surrounded by an electron-lucent area. He concluded that material which had occupied this area *in vivo* was contracted during preparation for electron microscopy onto the surface of smooth vesicles giving rise to the polygonally patterned surface structure. Biochemical evidence and fine-structural information based on improved fixation methods make the artifact hypothesis hardly tenable.

Pearse's (1975) work which showed that the polygonal structure was composed for the most part of a single protein (mol. wt 180 000), has since been repeated and confirmed by several independent groups working on coated vesicles isolated from several different tissues.

It is an acknowledged scientific tradition that discoverers have the right to name the objects of their discovery. Usually if the name is suitable it is adhered to. As the result of a suggestion made by Graeme Mitchison, Barbara Pearse chose the name clathrin for the protein she had isolated. The name clathrin has the merits of brevity and euphony. Apparently it is derived from the Greek word κλετθρον meaning a bar. Since that time the workers in the field have been content to use the term clathrin and we have supported its use in this volume.

As a result of the appearance of coated vesicles in transmission electron micrographs of median ultrathin sections, the term bristle coat was at one time widely applied to the structure on the vesicles' cytoplasmic surface. We now know that this image is a projection of a slice through the polygonal lattice of clathrin molecules. We have not continued the use of the term bristle coat on the grounds that it is misleading. The structures on the cytoplasmic surface of these vesicles – it is generally agreed – are not radially projecting bristles.

Micropinocytic vesicles of the type dealt with in this volume have been variously called coated because of: (*a*) the glycomolecular coating on the luminal surface of the vesicle membrane; and (*b*) the clathrin-composed polygonal lattice coating on the cytoplasmic surface of the vesicle membranes. We are aware of this problem and always try to specify the coat to which we refer.

Finally, as the result of an agreement between discussants at the April 1976 joint meeting of the British Society for Cell Biology and the Lysosome Club held at Nottingham University on the subject of endocytosis we have consistently used the word micropinocytic instead of the longer form micropinocytotic.

References

Bowers, B. (1964). Coated vesicles in the pericardial cells of the aphid *Myzus persicae* Sulz. *Protoplasma*, **59**, 351–67.

Gray, E.G. (1972). Are the coats of coated vesicles artifacts? *Journal of Neurocytology*, **1**, 363–87.

Matus, A.J. (1976). Coated vesicles and clathrin (reply). *Nature*, **263**, 95.

Pearse, B.M.F. (1975). Coated vesicles from pig brain: purification and biochemical characterisation. *Journal of Molecular Biology*, **97**, 93–8.

Pearse, B.M.F. & Bretscher, M.S. (1976). Coated vesicles and clathrin. *Nature*, **263**, 95.

APPENDIX 2. REFERENCES ADDED AT PROOF

Anderson, R.G.W., Vasile, E., Mello, R.J., Brown, M.S. & Goldstein, J.L. (1978). Immunocytochemical visualization of coated pits and vesicles in human fibroblasts: relation to low density lipoprotein receptor distribution. *Cell*, 15, 919–33.

Attie, A.D., Weinstein, D.B., Freeze, R.C., Pittman, R.C. & Steinberg, D. (1979). Unaltered catabolism of desialyated low-density lipoprotein in the pig and in cultured rat hepatocytes. *Biochemical Journal*, 180, 647–54.

Baker, J.B., Simmer, R.L., Glenn, K.C. & Cunningham, D.D. (1979). Thrombin and epidermal growth factor become linked to cell surface receptors during mitogenic stimulation. *Nature, London*, 278, 743–5.

Bartholeyns, J., Quintart, J. & Baudhuin, P. (1979). Inhibition of the discharge of endocytosed protein from phagosomes into lyosomes in hepatoma cells exposed to dimerized ribonuclease A. *Biochemical Journal*, 178, 433–42.

Beeck, H., Ullrich, K. & Von Figura, K. (1979). Effect of lectins on endocytosis and secretion of lysosomal enzymes by cultural fibroblasts. *Biochimica et Biophysica Acta*, 583, 179–88.

Berg, T., Tolleshaug, H., Drevon, C.A. & Norum, K.R. (1978). Uptake and degradation of proteins in isolated rat liver cells. In *Protein Turnover and Lysosomal Function*, ed. H.L. Segal & D.J. Doyle, pp. 417–30. New York & London: Academic Press.

Blest, A.D. & Maples, J. (1979). Exocytotic shedding and glial uptake of photoreceptor membrane by a salticid spider. *Proceedings of the Royal Society of London, Series B*, 204, 105–12.

Brown, M.S. & Goldstein, J.L. (1979). Receptor mediated endocytosis: insights from the lipoprotein receptor system. *Proceedings of the National Academy of Sciences, USA*, 76, 3330–7.

De Bruyn, P.P.H., Michelson, S. & Becker, R.P. (1978). Nonrandom distribution of sialic acid over the cell surface of bristle-coated endocytic vesicles of the sinusoidal endothelium cells. *Journal of Cell Biology*, 78, 379–89.

Carpenter, G. & Cohen, S. (1978). ^{125}I-labelled human epidermal growth factor: binding, internalisation and degradation in human fibroblasts. *Journal of Cell Biology*, 71, 159–71.

Carpentier, J.L., Gorden, P., Goldstein, J.L., Anderson, R.G.W., Brown, M.S. & Orci, L. (1979). Binding and internalization of ^{125}I-LDL in normal and mutant human fibroblasts. A quantitative autoradiographic study. *Experimental Cell Research*, 121, 135–42.

Cath, K.J., Harwood, J.P., Aguilera, G. & Dugau, M.L. (1979). Hormonal regulation of peptide receptors and target cell responses. *Nature, London*, 280, 109–16.

Charlwood, P.A., Regoeczi, E. & Hatton, M.W.C. (1979). Hepatic uptake and degradation of trace doses of asialofetuin and asialorosomucoid in the intact rat. *Biochimica et Biophysica Acta*, 585, 61–71.

D'Arcy Hart, P. & Young, M.R. (1979). The effect of inhibitors and enhancers of phagosome–lysosome fusion in cultured macrophages on the phagosome membranes of ingested yeasts. *Experimental Cell Research*, 118, 365–76.

Dean, R.T. (1978). Selectivity in endocytosis of serum and cytosol proteins by macrophages in culture. *Biochemical and Biophysical Research Communications*, 85, 815–19.

Dean, R.T. (1979). Effects of cytochalasin B on the pinocytosis and degradation of proteins by macrophages. *Biochemical Society Transactions*, 7, 362–4.

Dean, R.T. (1979). Macrophage protein turnover. Evidence for lysosomal participation in basal proteolysis. *Biochemical Journal*, 180, 339–45.

Dos Reis, G.A., Persechini, P.M., Ribeiro, J.M.C. & Oliveira-Castro, G.M. (1979). Electrophysiology of phagocytic membranes. II. Membrane potential and induction of slow hyperpolarizations in activated macrophages. *Biochimica et Biophysica Acta*, 552, 331–40.

Filopovic, I., Schwarzmann, G., Mraz, W., Wiegandt, H. & Buddecke, E. (1979). Sialic-acid content of low-density lipoproteins controls their binding and uptake by cultured cells. *European Journal of Biochemistry*, 93, 51–5.

Goldstein, J.L., Anderson, R.G.W. & Brown, M.S. (1979). Coated pits, coated vesicles, and receptor-mediated endocytosis. *Nature, London*, 279, 679–85.

Goldstein, J.L., Buja, M., Anderson, R.G.W. & Brown, M.S. (1978). Receptor-mediated uptake of macromolecules and their delivery to lysosomes in human fibroblasts. In *Protein Turnover and Lysosome Function*. ed. H.L. Segal & D.J. Doyle, pp. 455–77. New York & London: Academic Press.

Goldstein, J.L., Ho, Y.K., Basu, S.K. & Brown, M.S. (1979). Binding site on macrophages that mediates uptake and degradation of acetylated low density lipoprotein, producing massive cholesterol deposition. *Proceedings of the National Academy of Sciences, USA*, 76, 333–7.

Gorden, P., Carpentier, J., Cohen, S. & Orci, L. (1978). Epidermal growth factor: morphological demonstration of binding, internalization, and lysosomal association in human fibroblasts. *Proceedings of the National Academy of Sciences, USA*, 75, 5052–9.

Gregoriadis, G. (1978). Liposomes in the therapy of lysosomal storage disease. *Nature, London,* 275, 695.

Haigler, H.T., McKanna, J.A. & Cohen, S. (1979). Direct visualization of the binding and internalization of a ferritin conjugate of epidermal growth factor in human carcinoma cells A-431. *Journal of Cell Biology,* 81, 382-95.

Hatae, T. & Benedetti, E.L. (1978). Structural perturbations in the plasma membrane during concanavalin A endocytosis. *Cell Biology International Reports,* 2, 129-37.

Herzog, V. & Meiller, F. (1979). Membrane retrieval in epithelial cells of isolated thyroid follicles. *European Journal of Cell Biology,* 19, 203-15.

Hess, J.R. (1978). Frequency of surface microprojections and coated vesicles with increased malignancy in human astrocytic neoplasms. *Acta Neuropathologica, Berlin,* 44, 151-3.

Hoekstra, D. & Scherphof, G. (1979). Effect of fetal calf serum and serum protein fractions on the uptake of liposomal phosphatidylcholine by rat hepatocytes in primary monolayer culture. *Biochimica et Biophysica Acta,* 68, 109-21.

Juliano, R.L., Moore, M.R., Callahan, J.W. & Lowden, J.A. (1978). Concanavalin A promotes uptake of lysosomal hydrolases by human fibroblasts. *Biochimica et Biophysica Acta,* 513, 285-91.

Juliano, R.L., Moore, M.R., Callahan, J.W. & Lowden, J.A. (1979). Lectin-mediated uptake of lysosomal hydrolases by genetically deficient human fibroblasts. *Experimental Cell Research,* 120, 63-72.

Kadota, K. & Kadota, T. (1978). Low incubation temperature favors detection of depolarization-induced coated vesicles in motor axon endings in frog nerve-muscle preparations. *Tohoku Journal of Experimental Medicine,* 126, 399-400.

Kadota, K. & Kadota, T. (1978). Detection of depolarization-induced coated vesicles within presynaptic terminals in cat sympathetic ganglia maintained under a low temperature. *Brain Research,* 151, 201-5.

Kartenbeck, J. (1978). Preparation of membrane-depleted polygonal coat structures from isolated coated vesicles. *Cell Biology International Reports,* 2, 457-64.

Keen, J.H., Willingham, M.C. & Pastan, I.H. (1979). Clathrin-coated vesicles: isolation, dissociation and factor-dependent reassociation of clathrin baskets. *Cell,* 16, 303-12.

Linsley, P.S., Blifeld, C., Wrann, M. & Fox, C.F. (1979). Direct linkage of epidermal growth factor to its receptor. *Nature, London,* 278, 745-8.

Maxfield, F.R., Willingham, M.R., Davies, P.J.A. & Pastan, I. (1979). Amines inhibit the clustering of α_2-macroglobulin and EGF on the fibroblast cell surface. *Nature, London,* 277, 661-3.

Morris, G.P. & Harding, R.K. (1979). Phagocytosis of cells in the gastric surface epithelium of the rat. *Cell and Tissue Research,* 196, 449-54.

Nagasawa, J. (1977). Exocytosis: the common release mechanism of secretory granules in glandular cells, neurosecretory cells, neurons and para-neurons. *Archivum Histologicum Japonicum,* **40**, suppl, 31–47.

Oguchi, M., Komura, J. & Ofuji, S. (1978). Ultrastructural studies of epidermis in acute radiation dermatitis. Basal lamina thickening and coated vesicles. *Archives of Dermatological Research,* **262**, 73–81.

Papahadjopoulos, D., Mayhew, E., Taber, R. & Wilson, T. (1978). The use of lipid vesicles for introducing macromolecules into cells. In *Protein Turnover and Lysosome Function,* ed. H.L. Segal & D.J. Doyle, pp. 543–60. New York & London: Academic Press.

Pearse, B.M.F. (1978). On the structural and functional components of coated vesicles. *Journal of Molecular Biology,* **126**, 803–12.

Piasek, A. & Thyberg, J. (1979). Effects of colchicine on endocytosis and cellular inactivation of horseradish peroxidase in cultured chon-drocytes. *Journal of Cell Biology,* **81**, 426–37.

Prinz, R., Schwermann, J., Buddecke, E. & Von Figura, K. (1978). Endo-cytosis of sulphated proteoglycans by cultured skin fibroblasts. *Bio-chemical Journal,* **176**, 671–6.

Ryser, U. (1979). Cotton fibre differentiation: occurrence and distribution of coated and smooth vesicles during primary and secondary wall formation. *Protoplasma,* **98**, 223–41.

Sando, G.N., Titus-Dillon, P., Hall, C.W. & Neufeld, E.F. (1979). Inhibition of receptor-mediated uptake of a lysosomal enzyme into fibroblasts by chloroquine, procaine and ammonia. *Experimental Cell Research,* **119**, 359–64.

Schneider, Y.J., Tulkens, P., De Duve, C. & Trouet, A. (1979). Fate of plasma membrane during endocytosis. I. Uptake and processing of anti-plasma membrane and control immunoglobulins by cultured fibro-blasts. *Journal of Cell Biology,* **82**, 449–65.

Schneider, Y.J., Tulkens, P., De Duve, C. & Trouet, A. (1979). Fate of plasma membrane during endocytosis. II. Evidence for recycling (shuttle) of plasma membrane constituents. *Journal of Cell Biology,* **82**, 466–74.

Schook, W., Puszkin, S., Bloom, W., Ores, C. & Kochwa, S. (1979). Mechano-chemical properties of brain clathrin: interactions with actin and α-actinin and polymerization into basketlike structures or filaments. *Proceedings of the National Academy of Sciences, USA,* **76**, 116–20.

Schechter, Y., Hernaez, L. & Cuatrecasas, P. (1978). Epidermal growth factor: biological activity requires persistent occupation of high-affinity cell surface receptors. *Proceedings of the National Academy of Sciences, USA,* **75**, 5788–91.

Singer, S.J., Ash, J.F., Lilly, Y.W.B., Heggeness, M.H. & Louvard, D. (1978). Transmembrane interactions and the mechanisms of transport of proteins across membranes. *Journal of Supramolecular Structure,* **9**, 373.

Sinke, J., Bouma, J.M.W., Kooistra, T. & Gruber, M. (1979). Endocytosis and breakdown of ^{125}I-labelled lactate dehydrogenase isoenzyme M4 by rat liver and spleen *in vivo. Biochemical Journal,* **180**, 1–9.

Warchol, J.B., Herbert, D.C. & Rennels, E.G. (1978). Coated vesicles in the rat anterior pituitary gland. *Acta Histochemica, Jena,* **61**, 296–304.

Wild, A.E. (1979). Fc receptors and selective immunoglobulin transport across rabbit yolk sac endoderm. In *Protein Transmission Through Living Membranes,* ed. W.A. Hemmings, pp. 27–36. Elsevier/North Holland: Amsterdam.

Wild, A.E. & Richardson, L.J. (1979). Direct evidence for pH-dependent Fc receptors on proximal enterocytes of suckling rat gut. *Experientia,* **35**, 838–40.

Willingham, M.C., Maxfield, R.F. & Pastan, I.H. (1979). α_2Macroglobulin binding to the plasma membrane of cultured fibroblasts. Diffuse binding followed by clustering in coated regions. *Journal of Cell Biology,* **82**, 614–25.

Woodward, M.P. & Roth, T.F. (1978). Coated vesicles: characterization, selective dissociation and reassembly. *Proceedings of the National Academy of Sciences, USA,* **75**, 4394–8.

AUTHOR INDEX

Only the first-named author in any reference is indexed.
Page numbers of references in the reference lists are in *italic* type.

SUBJECT INDEX

NB. References to tables, figures and plates are in *italic* type.

absorptive epithelial cells, *see* columnar absorptive cells

Acanthamoeba castellanii, and quantitative study of pinocytosis in, 187, *202*

Acetabularia mediterranea, cyst formation of, and presence of coated vesicles in, 247, *252-3*

acetylcholine, 107, 108; and net movement of calcium ions, 117; and relationship of release to synaptic vesicles, 106, 108; storage of, by synaptic vesicles, 106

acid phosphatase activity, in vitellogenic oocytes, 7, 154-5

acid-phosphatase-positive coated vesicles, and intracellular transport of lysosomal enzymes, 39-40, 41, 43, 53-4

acinar cells, coated vesicles in, *29*

actin filaments, and direction of coated vesicle motility, 271

actomyosins, and a mechanism for membrane mobility, 261

adenosine triphosphate (ATP) binding, as a mechanism for membrane motility, 263

adrenal medulla, amino acid composition of clathrin isolated from, 285, *286*

adsorptive pinocytosis, 185-6, 191-2, 207-8; *see also* pinocytosis

Aedes aegypti: pinocytosis in oogenesis of, 199-200; protein incorporation in oocytes, 143

Agriolimax reticulatus, study of pinocytosis in foot epithelium of, 199

albumin, ^{131}I-labelled, and quantitative study of pinocytosis, 187, 192, *203*

albumin, serum: intestinal transmission of, 81; placental transmission of, 76, 77; yolk sac transmission of, 70-1, 72, 74, 75-6

algae, coated vesicles in, 55-7; *see also individual species*

α-pinosomes, 137

alveolate surface architecture of coated vesicles, in lower plants, 56-7

ameloblasts, coated vesicles in, *29*; and mucopolysaccharide transfer, 42

amino acid composition, of clathrin, 285, *286*

Ammi visnaga, 63

Amoeba proteus, and quantitative study of pinocytosis in, 186, 187, *202*

amphibians: phosphoprotein phosphatase activity in yolk platelets, 154; structural features of oocytes, 148; vitellogenic uptake in oocytes, 140, 152, 158-9, *see also Xenopus laevis*; *see also individual species*

angiosperms, coated vesicles in, 58; P-protein formation, 62-3

animal cell types, functional implications of coated vesicles in, 25-44, 244; functional pathways, 31-7; morphological aspects, 25-6; occurrence of, 26-30; structural and metabolic effects on function of, 37-42; *see also individual cell types*

Annelida, multivesicular bodies in oocytes of, 148

Apis mellifica, yolk protein precursors in, 143

Aptenia spp., 59

Arthropoda, 27; *see also individual species*

asialoglycoprotein recognition system, of liver cells, 206

ATP, *see* adenosine triphosphate

ATPase, calcium-activated, 116, 117, 118, 286-7, 311

autoradiography: and detection of protein incorporation into oocytes, 141-2, 147;